Worthington Hooker

Science for the School and family. Part I. Natural Philosophy

Second Edition

Worthington Hooker

Science for the School and family. Part I. Natural Philosophy
Second Edition

ISBN/EAN: 9783337025014

Printed in Europe, USA, Canada, Australia, Japan

Cover: Foto ©berggeist007 / pixelio.de

More available books at **www.hansebooks.com**

SCIENCE

FOR THE

SCHOOL AND FAMILY.

PART I.
NATURAL PHILOSOPHY.

BY

WORTHINGTON HOOKER, M.D.,

PROFESSOR OF THE THEORY AND PRACTICE OF MEDICINE IN YALE COLLEGE, AUTHOR OF "CHILD'S BOOK OF NATURE," "FIRST BOOK IN CHEMISTRY," "NATURAL HISTORY," "CHEMISTRY," ETC.

Illustrated by Numerous Engravings.

SECOND EDITION,
REVISED AND ENLARGED.

NEW YORK:
HARPER & BROTHERS, PUBLISHERS,
FRANKLIN SQUARE.
1881.

By Dr. WORTHINGTON HOOKER.

THE CHILD'S BOOK OF NATURE. For the Use of Families and Schools; intended to aid Mothers and Teachers in training Children in the Observation of Nature. In Three Parts. Illustrated by Engravings. The Three Parts complete in one volume. Small 4to, Cloth, $1 12; Separately, Cloth, Part I., 45 cents; Parts II. and IIL, 48 cents each.

    Part I. PLANTS.

    Part II. ANIMALS.

    Part III. AIR, WATER, HEAT, LIGHT, &c.

FIRST BOOK IN CHEMISTRY. For the Use of Schools and Families. Illustrated by Engravings. Revised Edition. Square 4to, Cloth, 48 cents.

NATURAL HISTORY. For the Use of Schools and Families. Illustrated by nearly 300 Engravings. 12mo, Cloth, $1 00.

SCIENCE FOR THE SCHOOL AND FAMILY.

    Part I. NATURAL PHILOSOPHY. Illustrated by numerous Engravings. Second Edition, Revised and Enlarged. 12mo, Cloth, $1 00.

    Part II. CHEMISTRY. Illustrated by numerous Engravings. Second Edition, Revised and Enlarged. 12mo, Cloth, $1 00.

    Part III. MINERALOGY AND GEOLOGY. Illustrated by numerous Engravings. 12mo, Cloth, $1 00.

Published by HARPER & BROTHERS, Franklin Square, N. Y.

☞ *Any of the above works sent by mail, postage prepaid, to any part of the United States upon receipt of the price.*

PREFACE.

DANIEL WEBSTER, in his Autobiography, speaks thus of his entering upon the study of law: "I was put to study in the old way—that is, the hardest books first—and lost much time. I read Coke on Littleton through without understanding a quarter of it. Happening to take up Espinasse's Law of Nisi Prius, I found I could understand it; and arguing that the object of reading was to understand what was written, I laid down the venerable Coke *et alios similes reverendos*, and kept company for a time with Mr. Espinasse and others, the most plain, easy, and intelligible writers. A boy of twenty, with no previous knowledge on such subjects, cannot understand Coke. It is folly to set him on such an author. There are propositions in Coke so abstract, and distinctions so nice, and doctrines embracing so many conditions and qualifications, that it requires an effort not only of a mature mind, but of a mind both strong and mature, to understand him. Why disgust and discourage a boy by telling him that he must break into his profession through such a wall as this? I really often despaired. I thought that I never could make myself a lawyer, and was almost going back to the business of school-keeping. Mr. Espinasse, however, helped me out of this in the way that I have mentioned, and I have always felt greatly obliged to him."

Here is most graphically depicted a defect which is now, as it was then, very prominent in all departments of education. It is even more so in early education than in that of the college and the professional school. Even in tender childhood pupils are put to studying books of which, as was true of Webster with his Coke on Littleton, they do not understand "a quarter part." If the rule is not "the hardest books first," there are many things in the books that it is not only hard but impossible for them to understand. And the hardest things are often put first. For example, in a very popular primary geography which lies before me the pupil is introduced to the world and its grand divisions at the outset, while he is taught about his own state and country only at the conclusion of the book. And this unnatural mode is the one very commonly pursued. Similar criticism can be passed upon most of the books used in teaching young children. Some of them are wholly useless. This is true of the grammars for primary schools. The formal statements, called the rules of grammar, are beyond the understanding of very young scholars, and therefore are useless burdens upon their memories. They are as useless to them as the three fourths of Coke which Webster could not understand was to him.

If we follow education upward from the primary school we find the same defect throughout the whole course. In the books which are used in teaching natural science it is especially prominent. Even in the elementary books, or compendiums, so called, formal propositions and technical terms render the study uninviting, and to a great extent unintelligible. The pupil is apt to be disgusted and discouraged, as Webster was with Coke on Littleton, and for the same reason.

Another defect intimately connected with that of which I have spoken is the very sparing and late introduction of the physical sciences. They are generally postponed to the latter part of the course of education, and then but little time is devoted to them. Generally, when a pupil designs to go through college, the study of these sciences is wholly neglected in his preparation, because a knowledge of them is not required for admission. Then in the college they are not attended to till the latter part of the course, and in the short time allotted to them there is so much to be learned that the teaching of them is a failure. Especially is this true of Chemistry and Geology.

This defect is a *radical* one. A thorough change should be effected in this respect in the whole course of education. The natural sciences should be made prominent from the beginning to the end, not only because they are of practical value, but also because they are as useful in their way for mental discipline as the study of mathematics and of language. They can be taught to some little extent to the youngest pupils. There are facts about air, water, and the various objects that they see around them, which they can understand if they be presented in the right manner. And the busy inquiries which they make after the reasons of the facts, and their appreciation of them if stated simply and without technical terms, show the appropriateness of such teaching. Children are really very good philosophers in their way. They have great activity not only of their perceptive but of their reasoning faculties also, to which due range should be given in their education.

Beginning thus, not a year should pass during the whole course when the pupil shall not be engaged in studying some one of the physical sciences to some extent. This continued attention to such studies in a reasonable amount, so far from interfering with the due prosecution of the other studies deemed so essential, *will so promote the pupil's advance in them as to more than make up for the time that is taken from them.* It will do this not only by the genial influence which such studies exert upon the mind, but by the contributions which they make to the knowledge of language and mathematics; for language is largely built up from natural objects and from the acquisitions of science, and there is an abundance of interesting applications of portions of the mathematics in the facts which the physical sciences develop to us.

I have said that the teaching of the natural sciences in our colleges is generally a failure, and it always will be so as long as the present plan is continued. In order to have it successful there must be *the same gradation in teaching them that we have in teaching language and the mathematics.* The college student needs to be prepared for the lectures which he hears on natural philosophy, chemistry, etc., and for his study of those branches, by previous familiarity with the simpler portions of them acquired in the school-room.

There is another very important reason for the early introduction of the physical sciences into education. By far the larger portion of pupils in our schools stop short of the college, or even the academy and high-school. That they should go forth into the world with no knowledge of the principles that lie at the basis of the arts in which so many of them are to engage is a shame and a wrong, if the communication of such knowledge be indeed practicable, as it undoubtedly is. Even those who are not to engage in

these arts will be greatly benefited by this knowledge, because in addition
to its constant practical applications in the management of life, it will con-
tribute to their mental power, and, what is no small consideration, to their
enjoyment; and it is, in fact, requisite to constitute them well-informed
persons.

If the views which I have presented be correct, there should be a series
of books on the natural sciences carefully adapted to the different periods
of the course of study. Those intended for the young beginner should be
exceedingly simple, and should not attempt to present anything like a full
view of the subjects treated. They should deal largely with familiar facts
or phenomena. The terminology of science and formal statements of principles, such as we often see in so-called compendiums, should have no place
in them, but should be gradually introduced as the series advances, and
should be made complete only in the concluding books.

It has been the object of the author to supply a part of such a series.
The first book in the series is the "Child's Book of Common Things,"
intended to teach the observation of familiar facts, or, in other words, the
beginnings of philosophy, to children as soon as they have got well started
in reading. Next comes the "Child's Book of Nature," which in its three
parts (Part I., Plants; Part II., Animals; Part III., Air, Water, Light,
Heat, etc.) extends considerably the knowledge of the philosophy of things
which the child has obtained from the first book in the series. Then follows
the "First Book in Chemistry." On a level with this is my "First Book
in Physiology." The next step in the gradation brings us to three books
under one title, "Science for the School and the Family;" Part I., Natural Philosophy; Part II., Chemistry; Part III., Mineralogy and Geology.
On a level with these is another book, "Natural History," and on another
still is to be written an "Introduction to Botany."

The three books, of which the present is one, are intended for the older
scholars in what are commonly called grammar-schools. At the same time
they are suited to scholars who are advanced to a higher grade who have
not gone through the previous books of the series. The preparation of
books especially adapted to high-schools and colleges I have left to others,
except in one branch of science, Physiology, on which I some years ago
published a work entitled "Human Physiology."

All of these books are from the press of Harper & Brothers except the two works on Physiology, published by Sheldon & Co., New York, and the "Child's Book of Common Things," published by Peck, White, & Peck, New Haven.

The general plan and style of these books are very different from what we see in most of the books for schools on the same subjects. The order of the subjects and the mode of developing them differ from the stereotype plan which has so generally been adopted. One prominent feature is the free use of illustrations from *familiar* phenomena. This leads the pupil to reason or philosophize about common things, thus giving an eminently practical character to his knowledge. At the same time it makes the books suitable for use in the family as well as the school, between which there should be more common ground than the present mode of education allows.

The style which I have chosen for all the books I have written for use in teaching is what may be called the *lecture style.* There are three other kinds of style which are more commonly used in school-books. The most common is what I term the *formal statement style.* In this principles and rules are stated, and then illustrations are given. This makes a formal and uninviting book. The bare skeleton of the science is generally for the most part presented, and the young pupil is apt to learn the statements by rote without understanding them. It is a style fitted only for books intended for advanced scholars. Another style is the *catechetical.* This is an un-natural mode of communicating knowledge; and, besides, it encourages learning by rote as the formal statement style does. In the third style, the *dramatic,* conversations are held between the teacher and some learners. The chief objection to this is that it undertakes to put in permanent shape what should be extemporized in the recitation.· What is needed in the book is simply clear and concise statement in an interesting style, and the living teacher and his scholars can best furnish the conversational element as the recitation goes on.

In the lecture style there may be and should be as much precision of statement as in the formal statement style, while it is more interesting, because it is the natural mode of communicating knowledge. In this style the facts are ordinarily so stated as to develop principles; while in the other the order is reversed, the principles being first stated and the facts given afterwards. One of the most successful books ever used in our colleges— "Paley's Natural Theology"—is in the lecture style, and it is a matter of surprise that this fact has had so little influence with those who have prepared books for instruction.

Whatever may. be true of advanced scholars, in teaching the *young* student in science bare, dry statement should be avoided, and the subjects should be presented in all their attractive features. I would not be understood as advocating the dressing up of science in adventitious charms. This is not necessary. Science possesses in itself an abundance of charms, which need only to be properly developed to attract the young mind; and the lecture style furnishes the best vehicle for such a development.

One grand essential for giving interest to any study is the presentation of the various points in the *natural order* in which they should enter the mind. *They should be so presented that each portion of a book shall make the following portions more interesting and more easily understood.* This principle, which is so commonly transgressed, I have endeavored to observe strictly in the preparation of these volumes.

Questions are inserted for those teachers who desire to use them. There is also an Index.

W. HOOKER.

January, 1868.

PREFACE TO THE SECOND EDITION.

In revising this work, its essential features, as fully explained in the Author's Preface, have been carefully preserved; at the same time, many portions have been entirely rewritten and much new matter has been added. The chapter on Galvanism, omitted in the Second Edition of Part II., Chemistry, has been revised and inserted in the present volume. Many new wood-cuts have been introduced, taken chiefly from German sources.

In the First Edition of this work the author makes frequent reference to Dr. Neil Arnott's *Elements of Physics;* the editor acknowledges his indebtedness to the Seventh Edition of the same for many of the illustrations of physical phenomena, and for suggesting the arrangement of matter in certain parts of this revision.

H. CARRINGTON BOLTON, Ph.D.

TRINITY COLLEGE,
  May, 1878.

· A 2

CONTENTS.

NATURAL PHILOSOPHY.

CHAPTER I.

MATTER.

1. Introduction.—Since you are about to begin the study of the series of books embraced under the general title "Science for the School and Family," and of which this work on Natural Philosophy forms the first volume, it is necessary that you should clearly understand what is meant by Science. This word literally means *knowledge;* but in the sense in which it is commonly employed, science denotes a systematic and orderly arrangement of knowledge. Superficial and incomplete information on any given topic, interwoven with fictions of the imagination, does not constitute science; on the contrary, science implies a profound, penetrating, and comprehensive knowledge based on general truths and fundamental principles. Since the human mind is capable of apprehending the phenomena of the whole universe of nature and of thought, and can subject them to the action of both the reason and the imagination, it is evident that science in its fullest signification is well-nigh boundless in its range and infinite in the variety of its material. To facilitate the study of so vast a subject as the sum of human knowledge, we naturally resort to systematic divisions and methods of classification. To enter upon an examination of the various systems of classifying the arts and sciences proposed by different authorities is quite foreign to our object, and we shall adopt without argumentation the views of the late Dugald Stewart, formerly Professor of Moral Philosophy in the University of Edinburgh. After carefully criticising the schemes of his predecessors, Professor Stewart concludes that the two most general heads on which to found an encyclopedical classification of science are MIND and MATTER. " No branch of human knowledge—no work of human skill—can be mentioned which does not obviously fall under the former head or the latter."

The sciences of mind and of matter are susceptible of subdivisions; the former embraces Pure Mathematics, Metaphysics, Mental and Moral Philosophy, Political Economy, Sociology, and other subjects upon which we do not dwell; the latter deals with the less abstract topics included in the sciences of Natural History, Astronomy, Chemistry, and Natural Philosophy or Physics.

Natural History includes within its extended limits the sciences of minerals, or Mineralogy, and of rocks, or Geology; of plants, or Botany; of animals, or Zoology; and of man, or Anthropology. Under this head, too, may be placed Medical Science, Physiology, etc. Astronomy, as you know, teaches about the heavenly bodies, their motions, magnitudes, and periods of revolution ; Chemistry deals with the internal composition of substances and their mutual reactions; and Natural Philosophy, or Physics, may be defined as that branch of science which treats of the properties and laws of matter. The term Natural Philosophy itself embraces a whole series of sciences,·as will appear from an examination of the following chapters.

2. **Effects of Matter on the Senses.**—The substance of

which material objects are made is called matter. Matter is often defined as anything perceptible by the senses—a statement which demands closer consideration. Some forms of matter can be perceived by all the senses; others can be perceived by only a part of them; some by only one. Air you cannot see, nor smell, nor taste; but you can feel it, and hear and see the effects of its motion. Sometimes matter affects only the sense of smell, or that with the sense of taste. Sea-air smells of salt; but the salt in the air is so finely divided we cannot see it. And yet it is the salt, entering the nostrils and coming in contact with the sensitive fibres of the nerves of smell, that produces the effect. Thus when we smell a flower, matter comes from it in particles so minute that no microscope can detect them, but they produce sensation when they strike upon the nerve.

Different kinds of matter constitute the substances of which bodies are made. These bodies are subject to change of state, form, and mutual relation; and a study of these phenomena is the province of Natural Philosophy.

3. **Forms of Matter.**—Matter appears in three forms: solid, liquid, and gaseous or aeriform—that is, like air. Sometimes matter is spoken of as having only two forms—solid and fluid. In this case fluids are divided into two classes, the elastic and non-elastic. The air and the various gases and vapors are the elastic fluids; while those which are called liquids are the non-elastic fluids. A football bounds because the air in it is an elastic fluid. If it were filled with a non-elastic fluid, as water, it would not bound. When water takes the form of steam it becomes elastic. Though it was formerly very common to use the expression elastic fluids, the division of matter into three forms is the one now usually recognized, liquids having been found to be feebly elastic.

Solids.—In solid matter the particles cannot be moved

about among each other; but each particle generally retains the same position in relation to those particles which are around it—in other words, it does not change its neighborhood. This is more true of some solids than of others. It is absolutely true of such hard solids as granite and the diamond. In these the particles always maintain the same relative position. But it is not so with gold or lead. By hammering these you can change greatly the relative position of their particles. India-rubber is a solid, but the relative position of its particles can be much altered in various ways.

Liquids.—It is characteristic of a liquid that its particles change their relative position from the slightest causes. It is in strong contrast with solids in this respect. When you move any portion of a solid body you move all the other portions of it, and generally in the same direction. But a body of liquid cannot be moved altogether as one body except by confining it, as, for example, in the case of a water-pipe or a syringe. And then, the moment that the water can escape, the particles use their liberty of altering their relative position. Since wind and other agents act continually upon water, no particle stays for any length of time in the neighborhood of the same particles. "Unstable as water" is, then, an exceedingly significant expression. Water is never at rest. A particle of it may at one time be floating on the surface of the ocean, and at another be in depths beyond the reach of man. It flies on the wings of the wind, falls in the rain, runs in the stream, is exhaled from a leaf, trembles in the dew-drop, flows in the blood of ·an animal or in the sap of a plant, and is always hurrying along in its ever-changing course.

Gases.—The particles of gaseous or aeriform substances move among each other even more freely than those of a liquid. Air, therefore, is more unstable and restless than water. Even when the air seems to be perfectly still its particles are moving about among each other. You can see this to be true if you darken a room, leaving a single shutter a little open. Where the light enters you will see motes flying about in every direction, which would not be the case if the air were really at rest. The particles of air have a greater range of travel than those of water; for the sea of atmosphere which envelopes the earth rises to the height of about fifty miles. How far water rises by evaporation we know not; but it is not at all probable that it rises to the uppermost regions of the atmosphere.

§ 4. **Filling of Spaces by Liquids and Gases.**—It is the freeness with which the particles of liquids and gases move among each other that enables them to insinuate themselves into spaces everywhere. They are ever ready to enter into any substances which have interstices or pores of such size as will admit them. Mingled with the grains of the soil are not only water, but air and gases. These are present also in all living substances, both vegetable and animal. Water forms the chief part of sap and of blood, and water is always accompanied by air and other gases. Part of the air we inhale enters the blood in the lungs, and courses with it through the system. The fishes could not live in water if no air were mingled with it. This can be proved by experiment. If you put a fish into a close vessel it will soon die, because it uses up all the air held in solution by the water. In an open vessel the fish is kept alive by the constant accessions of fresh air to the water. Advantage is taken of this in the preservation of fish in large aquaria, where air is constantly pumped into the water contained in the tanks.

Solution.—When solid substances dissolve in water or other liquids, the particles of the solids penetrate between the little particles of the liquid, and it is owing to the freedom with which the particles of water move about among each other that they are able to take in among them the minute particles of the solid. § 12.

5. **Relation of Heat to the Forms of Matter.**—Some kinds of matter are seen in all the three forms. Whether these shall assume one form or another depends on the amount of heat present. Thus when water is solid, ice, it is be-

cause a part of its heat is gone. Apply heat, and it becomes a liquid, water. Increase the heat to the boiling point, and it becomes steam, or an aeriform substance. Alcohol has only two forms—liquid and aeriform. It has never been frozen. Iron is usually solid; but in the foundry, by the application of great heat, it is liquefied. Mercury is liquid in all ordinary temperatures; but it often becomes solid in the extreme cold of arctic winters. A mercurial thermometer is of course useless under such circumstances, and the alcoholic thermometer is relied upon to denote the degree of cold. The difference between mercury, water, and iron in regard to the liquid state is this: Comparatively little heat is required to make mercury liquid, while more is required for this condition in water, and much more in the case of iron.

6. **The Nature of Matter Unknown.**—Let us now inquire, what do we know of the nature of matter? Can we say that we know anything of it? We may observe its phenomena and learn its properties; but, with our most searching analyses, we can no more determine the nature of matter than we can that of spirit. Newton supposed "that God in the beginning formed matter in solid, massy, hard, impenetrable particles." This he believed to be true of liquids, and even of gases, as well as of solids. In the gas these hard particles are much farther apart than in the solid. The supposition is a very probable one; but if it be true, it does not teach us what matter is, for it leaves us in the dark as to the nature of the particles. Newton further supposed that these particles have always remained unaltered amid all the changes that are taking place; these changes being occasioned by "the various separations and new associations and motions of these permanent particles." When, for example, anything is burned up, as it is expressed, not one of the particles is destroyed; they merely

assume new forms. Though most of the substance has flown off in the form of gas, the ultimate particles composing the gas are the same as when they made a part of the solid substance; and they may soon again become a part of some new solids. Such changes in the forms of matter are everywhere going on; and when you study Chemistry, Part II. of this Series, you will become familiar with them.

7. **The Constitution of Matter.**—The nature of the internal structure of matter cannot be experimentally determined, and we are again obliged to resort to hypotheses or suppositions. Two hypotheses of the internal constitution of matter have been proposed; according to the first matter is homogeneous throughout its mass, and presents no interior void, or, in other words, is continuous. This is the supposition of Descartes, an eminent French philosopher of the seventeenth century, but possesses so small a degree of probability that the scientific world has abandoned it for the second hypothesis. According to this, bodies consist of an agglomeration of an immense number of excessively small particles called *molecules;* these small particles do not touch each other, but are held in their places by reciprocal attraction; they are supposed to be continually in motion, the amplitude of their oscillations varying with the form of the body, solid, liquid, or gaseous. This hypothesis originated with ancient Greek philosophers about twenty-two centuries ago, but has been modified considerably in modern times. There are many reasons for accepting this view, some of which we will state briefly. In the first place, matter is divisible, as will be explained more fully in Chapter II., and it is difficult to comprehend its divisibility if it contains no void spaces. The solubility of solids is explained in § 4 by a reference to this hypothesis. Secondly, the expansion of bodies by heat, and their contraction on cooling, is readily explained by the molecular

hypothesis, for we may conceive that the spaces which separate the molecules become larger or smaller in consequence of the separation or approach of the latter. The fact that bodies assume three states, solid, liquid, and gaseous, is explained in a somewhat similar manner. Thirdly, when two substances are brought together it often happens that they intimately interpenetrate, and, each losing its characteristic property, they acquire new properties common to both. Chemical science affords us numerous examples of such combinations. *Sodium*, for instance, is a white lustrous metallic substance, as soft as wax, fusible at a low temperature, lighter than water, tarnishing very readily, and decomposing water at ordinary temperatures; *chlorine*, on the other hand, is a yellowish-green gas, heavier than air, of disagreeable odor, fatal to animals when breathed by them, possessing strong bleaching power, and soluble in water; and these two strangely dissimilar substances combine to form the white crystalline solid, common salt, so indispensable to man and animals.

Such facts as these are comprehensible if we adopt the view that matter is composed of exceedingly small particles, but it is impossible to explain them on the hypothesis of Descartes. Taking the example given, sodium and chlorine are each regarded as made up of minute particles having the properties named, and when combination takes place between individual particles they are associated in new forms and recognized as common salt.

Fourthly, the various phenomena of light and of heat, and certain chemical laws which will be explained in Part II., combine with the foregoing proofs to form an argument in favor of the molecular hypothesis not easily outweighed.

8. **Molecules.**—These particles of matter are so minute that they have never been seen by man; the smallest par-

ticle visible with the most perfect microscope is probably greater than $\frac{1}{4000}$ of a millimetre* in diameter, and it has been calculated that to see these molecules we should require a lens magnifying from 500 to 2000 times greater than any we now possess. We are about as far from seeing the largest molecules as we should be from reading with the unaided eye the letters on a page of this book at the distance of one third of a mile.

Eminent philosophers have estimated the size of molecules, and have obtained remarkable coincidence of results from independent and widely different data; from these calculations it may be concluded with a high degree of probability that in ordinary liquids or solids the diameter of the molecule is less than the ten-millionth and greater than the two-hundred-millionth of a millimètre.

Molecules are believed to be continually in motion, and with very great velocity, estimated to average seventeen miles a minute. This motion is in all directions, and of an oscillatory character, the particles flying to and fro through excessively small paths, the diameter of which varies with the nature of the substance.

Not only do molecules vary in size and velocity of motion, but they also differ among themselves in weight; their weight depends on that of the *atoms* (see next paragraph) composing them, and can be determined in accordance with laws explained in Part II., Chemistry.

Atoms.—Molecules are believed to be composed of still smaller particles of matter called atoms. The number of atoms forming a molecule varies greatly; in certain cases a molecule contains but one atom, in others several hundred. If the atoms composing the molecule are of one and the same substance, the molecule is said to be simple or elementary; if atoms of diverse kinds of matter unite to

---

* See Appendix, Metric System of Weights and Measures.

form a molecule, it is said to be compound. - Thus the molecules of copper are made up of atoms of copper solely, and copper is consequently regarded as an elementary body; on the other hand, molecules of sugar, of saltpetre, etc., are compound, the first named being made up of twelve atoms of carbon, twenty-two of hydrogen, and eleven of oxygen, and the second containing one atom of potassium, one atom of nitrogen, and three of oxygen.

Atoms of different kinds of matter vary in weight, those of gold and lead, for example, being much heavier than those of hydrogen.

By taking advantage of the extraordinary tinctorial power of certain aniline dyes, experiments have been made which show that an atom of hydrogen undoubtedly weighs less than 0.000,000,000,054 gramme; according to another authority the weight of a hydrogen atom cannot be less than 0.000,000,000,000,000,000,000,0075 of a gramme. It is impossible to conceive of such minute quantities, nor have these figures any practical value; we give them chiefly as a subject of curiosity.

The smallest particle of matter which can exist in a *free* state is the molecule; atoms exist in combination only, and are sometimes defined as the smallest particle of matter which can enter into the composition of a molecule.

The physical properties of bodies, hardness, transparency, elasticity, etc., depend mainly on their molecular relations; the chemical properties on their atomic relations. The study of the atomic composition of bodies, and of the laws governing their combination, is the province of chemistry, and will be fully explained in Part II.

9. **Matter acted upon by Forces.**—The constant changes of form and mutual relations of bodies are caused by external agents called forces. Gravitation, heat, light, and electricity are some of these forces. These physical forces were formerly regarded as exceedingly attenuated forms of matter, and, since they have no weight, were called im-

ponderable agents. At present, however, the opinion prevails that the phenomena caused by these forces result from the motions of the inappreciably minute particles of matter. It is believed also that there is in reality but one force in nature, and that heat, light, and electricity are different manifestations of this force. Just as a certain amount of matter was created, and continues to exist without diminution in quantity, in like manner a definite amount of force was created, and this, too, is indestructible, though manifesting itself in various ways to our senses. It will be shown farther on that the forces of heat, light, and electricity are mutually convertible and equivalent. They are all referred back to a common origin—motion of the molecules of matter.

The agency of the physical forces is of great importance and is very active, producing constant changes throughout nature; they are also obviously and immediately essential to life.

QUESTIONS.*

[The numbers refer to the sections.]

1. What is meant by Science? What is said about the classification of knowledge? Name some of the chief subdivisions.—2. What is said of the effects of matter on the senses?—3. What are the forms of matter? Illustrate the difference between elastic and non-elastic fluids. What is said of the union of the particles of a solid? Give the difference noted

------

* Teachers differ much in their plans of conducting recitations. Some are very minute in their questions; while others go to the other extreme, and merely name the topics, the pupils being expected to give in full what is said upon them. Neither of these plans should be adopted exclusively, but the mode of recitation should be much varied from time to time. This variety is somewhat aimed at in the questions which we have prepared, though in no case are the questions as minute as they should occasionally be made by the teacher. It would be well to have the pupils draw many of the figures upon the blackboard, and then recite from them. By drawing the simplest figures first sufficient skill may be acquired to enable the pupil to draw those which are quite difficult.

between different solids. How does a liquid differ from a solid? Give in detail what is said of water. What is said of the particles of gaseous substances? What of the atmosphere? What of the vapor in it?—4. What is said of the entering of liquids and gases into interstices? What of the mingling of gases with liquids? Give the illustration in regard to fishes. What is said of the solution of solids in liquids? What of the evaporation of water in the air?—5. Illustrate the influence of heat on the forms of matter. What is said of the thermometer? What of mercury, water, and iron in relation to the liquid state?—6. What is said of our knowledge of matter? What was the supposition of Newton about the composition of matter? What is said of the changes of matter?—7. What two suppositions have been made as to the internal structure of matter? With whom did the more probable one originate? Name the principal reasons for adopting the molecular hypothesis? Describe in full the chemical reasons. —8. What is said of molecules? How near are we to seeing them? How large are they believed to be? What of their motions? Of what are molecules composed? When are they elementary? Give examples. When are they said to be compound? Give examples. What is said of the weight of atoms of hydrogen? What properties of matter depend mainly on the molecular relations of bodies?—9. What forces act upon matter? What is said about the unity of force?

CHAPTER II.

PROPERTIES OF MATTER.

10. **Properties of Matter.**—All matter has properties or qualities; these differ in the different kinds of matter as well as in substances of the same class. Some of these properties are common to all kinds and forms of matter, and are called *universal properties;* others are peculiar to certain substances and are called *specific properties.* The properties of matter which we shall describe in this and the following chapters are:

Universal Properties.
- Extension and Figure.
- Impenetrability.
- Indestructibility.
- Inertia.
- Divisibility.
- Porosity.
- Compressibility and Expansibility.
- Elasticity.

Specific Properties...
- Hardness.
- Flexibility and Brittleness.
- Tenacity.
- Malleability.
- Ductility.

Besides the specific properties named, there are others which do not here require detailed explanations, such as solidity and fluidity, transparency and opacity, color, etc.

The universal properties are sometimes called *essential properties;* that is, properties of which no kind or form of matter can be destitute. The distinction between the two groups named will be explained in the next section.

11. **Extension and Figure.**—*Extension* is that property of matter by virtue of which a body occupies a limited portion of space and *figure*, the property by which it has some definite shape. You cannot conceive of any portion of matter, however small it may be, that does not occupy some space, and that has not shape or figure. This is, in fact, involved in the very idea of matter. A particle may be so small as to appear only as a point to the naked eye, but viewed through the microscope its shape becomes obvious. Even an atom must have length, breadth, and thickness, notwithstanding our inability to measure it or to see its shape with the most powerful microscope. The air is sometimes spoken of in common language as being shapeless. This is partly because it is invisible, and partly because no portion or body of air assumes any definite

B

shape. But air is continually forced into definite shapes by confinement in rooms, boxes, etc.; and then its extension in different directions can be measured as accurately as the extension of a solid. And, besides, the particles of which air is composed are undoubtedly solid, and we cannot conceive of their existence without attaching to them the idea of figure or extension.

Extension is an essential or universal property of matter, since no portion of matter, hard or soft, can be destitute of extension; on the other hand, hardness is not an essential property of matter, for some kinds of matter possess no hardness.

12. **Impenetrability.**—In common language one substance is said to penetrate another when forced into it by pressure or by blows. Thus a needle penetrates cloth, a nail penetrates wood, etc. But this .expression is not strictly correct; the needle does not enter the substance of the cloth, but goes between the fibres of it, pushing them to one side. And the nail goes between the fibres of the wood, and not into them; it does not occupy the same room as the fibres at the same time.

In like manner no atom of matter can penetrate or enter into any other matter, it can only push it out of the way, and then occupy its place. This property of matter whereby no two portions can simultaneously occupy the same place is called *impenetrability.* Impenetrability is usually accounted one of the universal properties of matter, but it is really a necessity of the existence of matter and of its extension, as explained in § 11.

Illustrations.—The property of impenetrability may be illustrated in many ways. If you hold a tumbler with its open end downward and press it into water, it will not fill with water, for the air in the tumbler prevents the water from rising; it cannot occupy the same space with the air. It fills indeed a portion of the tumbler, because the air is compressible to a certain extent (§ 17). If you introduce a glass funnel, *a*, Fig. 1, into a jar of water, *b*, the water will not rise to fill it so long as you close the opening, *c*, with your finger. But if you remove your finger, the water will rise to the level of the water outside of the funnel, pushing out the air before it. The following neat experiment illustrates the same point. Float a lighted candle, *a*, Fig. 2, on a large flat

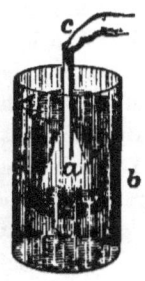

Fig. 1.

cork (weighted with lead so as to support the candle) in a jar of water, and place over it an open jar or receiver, *b*, provided with a stopcock, *c*. If you close the stopcock and press the receiver into the water, the candle will sink with it as represented in the figure, the

Fig. 2.

air preventing the water from entering the jar. As soon as you open the stopcock, however, the water will rush in, and the candle will appear to rise out of the water.

Other Illustrations.—The diving-bell, used for exploring the bottoms of rivers, lakes, etc., affords a good illustration of this property of matter. It consists of a metallic vessel, A, Fig. 3, shaped somewhat like a bell, and made sufficiently heavy to sink in water. It is lowered into deep water by means of a chain and cable, *r m q*. The water does not enter the bell any farther than the compressibility of the air allows. In order that the men in the bell may

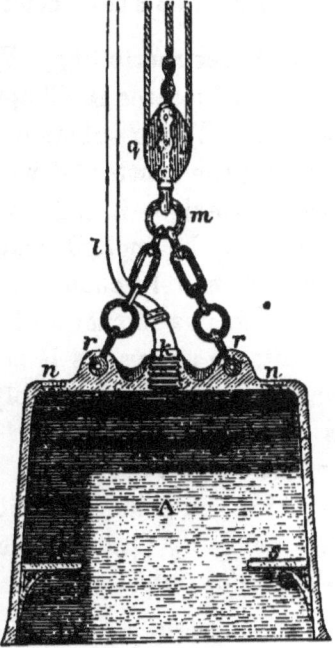

Fig. 3.

remain under water for some time, fresh air is supplied by the tube l, being forced in by a pump, and the vitiated air is allowed to escape through valves provided for that purpose. The seats, s, s, are for the accommodation of the men who descend in the bell to work at the bottom of the sea or river. By this means treasures are often recovered which would otherwise be lost. You will observe some resemblance between the diving-bell and the arrangement shown in Fig. 2, the receiver representing the bell, and the lighted taper the men within.

When a bullet is dropped into a glass of water it pushes the particles to one side, and occupies the room thus gained. If several bullets are thrown in, there is an evident rise of the water, and you may add enough to make it overflow. The same thing is true of the finest needle dropped into water; it does not penetrate the water, but, like the bullet, displaces the particles.

When any substance, sugar or saltpetre for example, is dissolved in water, its particles do not penetrate those of the water, but they enter the spaces between the particles. In more concise language, solution results from inter-molecular penetration. In like manner when particles of odorous substances are diffused in the air, they are in fact between its particles.

13. **Indestructibility.**—We have alluded to the indestructibility of matter in Chapter I., but it now requires amplification. When substances waste away, or are burned up rapidly, the matter of which they are composed does not cease to exist—it merely changes its form or condition. Gold may be melted, and even converted into vapor, but the form merely changes, and its substance is not lost. A candle grows shorter and lighter as it burns, but by means of suitable apparatus it is possible to collect all the gases, smoke, etc., which invisibly rise, and it is found that there is no loss in weight. In short, we cannot create matter, and we cannot destroy it; it is imperishable. The universe contains the same amount of matter as when first called into being by Omnipotence, and not the smallest particle will be put out of existence to the end of time.

We occasionally hear of new elements, new kinds of matter, discovered by chemists; but this signifies that the body discovered had previously escaped observation, either on account of its great rarity, or by reason of the difficulty of distinguishing it from its associated substances having similar properties. New elements are discovered, not created, by chemists. The indestructibility of matter is forcibly brought to the mind of the chemist, who witnesses in his dealings with matter such marvellous transformations, disappearances, and reappearances. The changes which the various kinds of matter undergo belong to the province of Chemistry, and will be fully explained in Part II.

14. Inertia.—Matter has no power to put itself in motion; when it is moved, it is acted upon by some force outside of the matter, or communicated to it in some way. When your arm moves, it is not the matter in your arm that causes its motion; it is the result of a force within you, exerted in obedience to the will. A book lying on a table has no power within itself of moving to another place; and if you reach out your arm, pick it up, and throw it across the room, you communicate force to it from without, and thus set it in motion. When air moves, it is set in motion by some force acting upon it, as when you blow it from your lungs or move it with a fan. When the wind blows, the air is set in motion by heat and the attraction of the earth, as will be explained in another part of this book.

Again, matter when set in motion has no power to stop itself. This inability of matter to move itself or to stop its motion is called *inertia*. If matter could stop itself, it would not be called inert. Owing to this inertness, matter once set in motion would keep moving forever were it not . stopped by some force; matter has no more tendency to stop moving when once put in motion than it has to begin motion when it is at rest. All motion would be perpetual if there were not forces opposing it. If there were only one body in the universe, and that were set in motion, it would move forever through empty space in a straight line; for there would be no matter anywhere to resist its motion or to attract it away from its onward course.

When a stone falls to the ground, it stops simply because the earth arrests it. If the earth were not in the way, the stone would move straight on until it reached the centre of the earth (§ 26). A stone thrown up in the air would keep on, and soon be out of sight, and never return to the earth, if it were not made to come down by forces acting upon it. One of these forces is the resistance of the air, which, from the moment the stone starts, is destroying its motion. Another force as constantly operating to retard the stone is the attraction, or drawing force, exerted by the earth upon it. This powerful though unseen force will be treated of fully in Chapter IV.

Advantage is taken of inertia in matters of every-day life. The massive fly-wheel of a stationary engine, once started on its course, continues to revolve by virtue of its inertia, and to give a steady regular motion to the machinery with which it is connected. The fly-wheel is made very heavy, because the heavier it is the greater resistance it offers to the friction which continually tends to stop its motion. When the locomotive of an incoming train is disconnected and shoots swiftly ahead, the train, by virtue of its inertia, or inability to stop itself, follows after until its motion is spent or an application of the brakes brings it to rest. It is owing to the inertia of matter that leaping from a rapidly moving train is dangerous; a person on the cars partakes of their motion, and, on jumping off, his feet come suddenly to rest while his body continues to move forward,

and he is thrown headlong to the ground.

The inertia of matter may be illustrated by a pleasing experiment. Balance a card on the neck of a bottle, and place a small coin on the card directly over the opening; by giving the card a quick, sharp blow with the finger, in a horizontal direction, the card will fly away and the coin will fall into the bottle. The coin does not move with the

Fig. 4.

card, because sufficient time does not elapse for the coin to partake of its motion.

The law of inertia was first recognized by Galileo, an eminent Italian philosopher, about the close of the sixteenth century. A correct comprehension of this important law was necessary before true explanations could be given of the laws governing the falling of bodies, of the vibrations of the pendulum, and of the motions of the planets in their orbits.

15. **Divisibility of Matter.**—Any portion of matter, whether solid or gaseous, may be divided into parts. Even if it be so small that it can be seen only with a powerful microscope, it could be still further subdivided, provided a sufficiently delicate instrument were available. A particle of matter the five-thousandth of a millimetre in diameter is no longer visible under a powerful microscope, and yet nothing but man's natural incapabilities prevents the division being carried yet finer. Whether or not there be a limit to the divisibility of matter is a question which has been discussed by philosophers in all ages. The theory prevailing at present is, that matter is not infinitely divisible, but is made up of definite ultimate parts called atoms, as explained in Chapter I.

Of the actual divisibility of matter we have numerous examples, in which the division is carried far beyond that which can be effected by any cutting instrument.

A gold-beater can hammer a grain (65 milligrammes) of gold into a leaf covering a space of fifty square inches (322.5 square centimetres). So thin is it that it would take 300,000 of such leaves, laid upon each other, to make the thickness of an inch. And yet so even and perfect is this thin layer of gold, that when laid upon any surface in gilding it has the appearance of solid gold. One fifty-millionth part of this grain of gold thus hammered out can be seen by the aid of a microscope which magnifies the diameter of an object ten times.

Recently the divisibility of gold has been carried much farther. Mr. Outerbridge, of Philadelphia, has obtained (by electric deposition) films of gold so thin that one grain of the metal would cover nearly four square feet (0.37 square metre). This is ten thousand times thinner than ordinary writing-paper, and 2,798,000 such films would measure only one inch. Gold-leaf of this tenuity is transparent and transmits a green light.

Further Illustrations. — A soap-bubble is a beautiful example of the minute division of matter. That thin wall which encloses the air is composed of particles of the soap and of the water mingled together. It is supposed to be less than one millionth of an inch in thickness.

The thread of the silk-worm is so minute that the finest sewing-silk is formed of many of these threads twisted together. But the spider spins much more finely than this. The thread by which he lets himself down from any height is made up of about 6000 threads or filaments, each coming from a separate hole in his spinning-machine. A quarter of an ounce (7.77 grammes) of the thread of a spider's web would extend 400 miles (643 kilometres).

Platinum, which is usually regarded as the heaviest known metal, can be drawn out into wire still finer than the web of the spider; 3000 feet (914.4 metres) of this wire weigh scarcely one grain (65 milligrammes), and a bundle of 140 of these wires would equal in thickness only a single silk-worm thread. See § 22.

Perhaps the most minute division of matter is exemplified in odors. A grain of musk will scent a room for years, and yet suffer no perceptible loss of weight. But during all this time the air is filled with fine particles coming from the musk.

The microscope reveals to us many wonderful examples of the minuteness of the particles of matter, both in the vegetable and the animal world.

If you press a common puff-ball a dust flies off like smoke. Examined with a microscope, each particle of this dust, which is the seed of the plant, is a perfectly round orange-colored ball. This ball is of course made up of very many particles, arranged in this regular form. Beautiful examples of various arrangements of the minute particles of matter are furnished by the pollen of different plants, as seen with the microscope.

Each particle of the dust which adheres to your fingers as you catch a moth is a scale with fine lines upon it regularly arranged. And if you look through the microscope at the wing of the moth, you will see, where the scales are rubbed off, the attachments by which they were held standing up from the surface of the wing, like nail-heads on a roof from which the shingles have been torn.

The organization of exceedingly small animals, as revealed by the microscope, furnishes us with wonderful examples of the minute division of matter.   A little of the dust of guano, examined through a powerful microscope, is seen to contain multitudes of shells of various shapes. These shells are the remains of animalcules that lived in the water, their destiny seeming to be in part to furnish food to other animals larger than themselves.   In the chalk formations of the earth are seen multitudes of such shells. They have been discovered even in the glazing of a visiting-card ; for they are so small that the fine grinding-up of the chalk does not wholly destroy them.   There are animals, both in the air and in the water, so small that it would take millions of them to equal in bulk a grain of sand, and a thousand of them could swim side by side through the eye of a common-sized needle.   Now all these animals are furnished with organs, constructed of particles of matter, which are arranged in them with as much order and symmetry as in the organs of our bodies.   How minute, then, must these particles be !

How do such facts extend our views of the power of the Deity !   The same power that moulded the earth, sun, moon, and the whole " host of heaven," gave form and life and motion to the millions which sport in every sunbeam ; the same eye that watches the immense heavenly bodies as they move on in their course looks upon one and all of these legions of animals in earth, air, and water, though unseen by human eyes, and provides that every particle shall take its right position, so that this part of creation may with all the rest be pronounced very good ;

B 2

and the same bountiful hand that dispenses the means of
life and enjoyment to the millions of the human race for-
gets not to minister to the brief life and enjoyment of each
one of these myriads of animalcules, though they seem to
be almost nothingness itself.

QUESTIONS.

10. What is said of variety in the properties of matter? Give the classification.—11. What is meant by extension? What by figure? Has
the air any extension or figure?—12. Illustrate the meaning of impenetrability. Describe the experiment with the funnel and a jar of water.
Also the experiment with the candle. State the arrangement of the diving-bell. Give the comparison between bullets and needles in relation to
penetration. What is said of solution?—13. Explain what is meant by
the indestructibility of matter. Can we create matter? What, then, is
meant by new elements?—14. What is inertia? Give illustrations of it.
Illustrate the fact that matter has no power to stop its own motion. What
stops a body set in motion? Illustrate by reference to a stone. What
advantage is taken of inertia. Describe an experiment illustrating the
inertia of matter. Who first recognized this law?—15. What is said of
the divisibility of matter? Give an example of the divisibility of matter
by reference to gold-leaf. What is said of the soap-bubble? What of
the thread of the silk-worm, and of the web of the spider? What of
platinum wires? What of odors? What is said of the dust of the
puff-ball? What of pollen? What of the dust rubbed from a moth's
wing? What of guano? What of the glazing of visiting-cards? What
of the minuteness of some animals? What is said of the Deity in relation to minute animals?

CHAPTER III.

PROPERTIES OF MATTER (CONTINUED).

16. Porosity.—The particles composing bodies of every description are surrounded by empty spaces; those substances which are called porous have quite large spaces in them. But even in those which are not commonly considered porous the particles are by no means close together. A celebrated experiment, tried at Florence in 1661, showed that the particles of even so dense a substance as gold are separated by spaces sufficiently large to let water pass through them. A hollow golden globe containing water was subjected to great pressure, and its surface was bedewed with the water that came out through the pores of the gold.

There are two kinds of pores—*sensible* pores, which can be distinguished by the naked eye or by the aid of the microscope, and *physical* pores, or interstices among the molecules alluded to in Chapter I. Sensible pores are visible in wood, sponge, pumice-stone, etc.; physical pores are invisible, but their existence is shown by the fact that substances can be compressed into a smaller bulk than they usually occupy. Solids can be thus compressed; some more than others. But the most compressible substances are the gases and vapors. The amount of space between their particles must be very large to allow of so great compression.

We can form some idea of the great amount of space in a gaseous or aeriform substance by observing the difference between water in its liquid

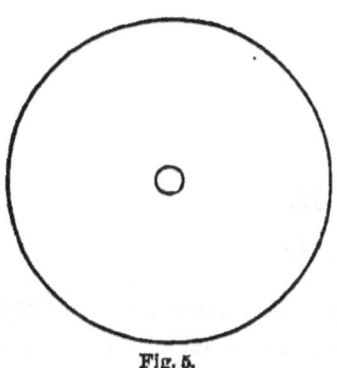

and in its aeriform state. A cubic centimetre of water, converted into steam, occupies 1696 times more room than before. The difference in proportion is exhibited in Fig. 5, the inner circle representing a volume of water, and the outer that of the steam into which it is converted. The water is not at all altered in its nature by being changed into steam. The particles are simply forced farther apart by the heat, and as soon as the heat is withdrawn they come together again to form water, or, in other words, the steam is condensed into water. It is plain, therefore, that the space between the particles is 1696 times as great in steam as it is in the water from which the steam is made.

Fig. 5.

When any substance, as sugar or salt, is dissolved in water, its particles are diffused through the intermolecular spaces. In like manner, when water evaporates, the particles of water are diffused through the spaces between the particles of the air. Animal and vegetable bodies are the most porous, being constituted of an immense number of interlacing channels through which during life nourishing fluids circulate. This is evident on examination of bone and of wood which abound in cells and partitions.

Density and Rarity.—The density of a substance depends upon the quantity of matter it contains in a given space. The more dense, therefore, a substance is, the greater its weight. A piece of lead is forty times heavier than a piece of cork of the same size. Mercury is nearly fourteen times heavier than an equal bulk of water. You see, then, that density must depend on the nearness of the molecules to each other. In so dense a substance as gold the molecules are all very close together; in wood there are spaces, some of which are so large that you can see them; and in air, steam, and the gases there is a great deal of space among the particles, so that we speak of their *rarity* instead of their density.

17. **Compressibility and Expansibility.**—Owing to porosity, matter may be compressed and expanded. Pressure applied to porous substances brings their particles nearer together, making them fill up in part their pores. You have a very familiar example of this in sponge. The greater the porosity of wood, the greater its compressibility. But even such dense substances as the metals can be compressed in some degree; that is, the interstices between their particles can be made smaller. Medals and coins have their figures and letters stamped upon them by pressure, just as impressions are made upon melted sealing-wax. The heavy and quick pressure required to do this actually compresses the whole piece of the hard metal, putting all the particles nearer together, so that it occupies less space than it did before it was stamped.

It might be supposed from the freeness with which the particles of liquids move among each other, and from the spaces which exist among them, that these substances could be easily compressed. But it is not so. The heaviest pressure is required to compress them even in a slight degree. Water can be compressed so very little that practically it is regarded as incompressible.

Although the interstices between the particles of-liquids cannot be varied by mechanical pressure, they can be by variations of temperature. Liquids are dilated or expanded by heat; that is, their particles are put farther apart. They are contracted or compressed by cold; that is, their particles are brought nearer together by the abstraction of heat. The most familiar example is the thermometer. The mercury rises in the tube when the heat increases the interstices between its particles, and it falls when the loss of heat allows the particles to come near together.

Aeriform bodies are more compressible than any other substances, showing that in their ordinary condition there is a great deal of space among their particles. While they are thus unlike liquids in compressibility, they are affected by heat in the same way.

18. **Elasticity.**—Closely allied to the compressibility of matter is its elasticity. We see this property strikingly exemplified in India-rubber. It occasions the rebounding of a ball of this substance when thrown against any immovable body—the floor, for example. When the ball

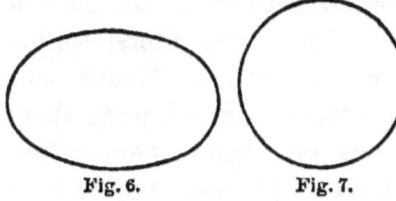

Fig. 6.   Fig. 7.

meets the resistance of the floor it is flattened, as represented in Fig. 6. Then, as it assumes the round shape, shown in Fig. 7, it pushes downward upon the floor. It is this sudden pushing downward that makes it rebound. It is as if there were a compressed spring between the ball and floor. It may be likened also to jumping. When a person jumps he bends his limbs at the thigh and knee-joints, and then, in straightening himself up, gives a sudden push, like that given by the ball as it assumes its round shape, and so is thrown forward or upward, according to the direction of the pushing force. The same flattening occurs in an ivory ball, though to a far less degree. You can prove this by experiment. Let a marble slab be moistened, and drop the ball upon it. Quite a spot will be made dry by the blow of the ball, showing that it touched more of the marble than it does when merely placed upon it.

When a stick or rod is bent, as soon as the bending force is withdrawn the stick becomes straight again from its elasticity. It is this elastic force of the bow, straightening it, that speeds the arrow. Observe in this case that while the particles on the concave side of the bent bow are brought nearer together, or compressed, those on the convex side are moved apart. This moving apart of the particles is often shown in India-rubber. You will see how very far apart particles in near neighborhood may be carried if you stick two pins close together in a strip of India-rubber, and observe their movements when you stretch it.

Some substances have so very little elasticity that practically they are considered as having none. Lead is one of these. A rod of lead when bent remains so, and a leaden ball does not rebound. While aeriform substances are the most compressible of all, they are also the most elastic. Compressed air returns to its usual condition the moment it is relieved from the pressure, and with a force proportioned to the amount of the pressure. So it is with steam and the gases. The varied results of this property of aeriform substances will claim our attention more particularly in some other parts of this book.

Glass is nearly perfectly elastic—that is, it will retain no permanent bend; but the limits of its elasticity are very small: it will not bend far without breaking. Hard bodies in general have a much smaller elastic limit than soft ones. This is evident on comparing the elasticity of steel, ivory, stone, glass, etc., with that of silk, catgut, India-rubber, and the like.

Elasticity may be defined as that property of matter by which its particles, when brought nearer together or carried farther apart by any force, return to their usual condition when the force is withdrawn. Closely connected with elasticity is the property of flexibility, which will be explained in the next section.

19. **Flexibility and Brittleness.**—If you bend a flexible body—a piece of wood, for example—it is obvious that the particles on the upper or convex side must be forced a little  farther apart, while those on the under or concave side

Fig. 8.

tle farther apart, while those on the under or concave side are brought a little nearer together (Fig. 8). But the wood does not break, because the particles that are thus moved a little apart still retain their hold upon each other. This is the explanation of what is called flexibility. On the other hand, the particles in a rod of glass cannot be put farther apart in this way. They are not actually in contact any more than are the particles of the wood, but they are in a *fixed* relative position; that is, a position which cannot be disturbed without a *permanent* separation of particles. If you attempt to bend the rod there is no slight separation of many particles, as in the bent wood, but a full and permanent separation in some one part of the rod. We call the property on which this result depends brittleness. Brittle substances are generally hard. Glass, while the most brittle of all substances, is hard enough to scratch iron.

Steel.—There are two kinds of steel, flexible and brittle. The steel of most cutting instruments is brittle. The steel of a sword-blade is quite flexible, and that of a watch-spring is so much so that we can wind it up in a coil. This difference is owing to a difference in the mode of cooling the steel. If it be cooled suddenly, it is brittle; if slowly, it is flexible. The process by which it is cooled slowly is called *annealing.* The explanation of all this is quite simple. The steel being expanded by heat—that is, its particles being put farther apart than they usually are—when they are suddenly brought together again they have not time to arrange their relative position properly. Brittleness is, therefore, the result. But, on the other hand, when the cooling is effected gradually, time is given for the arrangement.

Steel suddenly hardened is too brittle for common use. A process called tempering is therefore resorted to for diminishing the brittleness. The steel is reheated after the hardening, and is then allowed to cool slowly. The degree in which the brittleness is lessened depends on the degree of heat to which the steel is subjected. It can be entirely removed by a red heat, for then the particles have a full opportunity to readjust themselves; and the more the heat comes short of this point the less thorough will be the adjustment, because the less perfectly are the particles released from their suddenly taken position. In lessening the brittleness we diminish hardness also, and therefore the tempering is varied in different cases according to the degree of hardness desired.

Annealing of Glass.—Glass for economical uses is always annealed. If this were not done our glass vessels and window-panes would be exceedingly brittle, and be constantly breaking. Articles made of glass are annealed by being passed very slowly through a long oven which is very hot at one end, the heat gradually lessening towards the other end.

We have a striking example of brittleness induced by sudden cooling in what are called "Prince Rupert's Drops." These are made by dropping melted glass into cold water, and they usually have a shape resembling that of Fig. 9. If you break off ever so small a bit of the point of one of these drops, the whole will at once shiver to pieces. That is, the sudden arrangement of the particles is so slight and unnatural that the disturbance of the arrangement in a small part suffices to destroy the arrangement of the whole, very much as a row of bricks is thrown down by the fall of the first in the row. Faraday says that these drops were not, as is commonly supposed, invented by Prince Rupert, but were first brought to England by him in

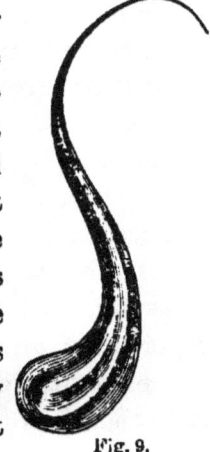

Fig. 9.

1660. They excited much curiosity at that time, and were considered "a kind of miracle in nature." But you see that this, like many other wonders, is capable of an easy explanation.

The so-called "tempered glass," invented by a Frenchman named La Bastie, affords another example. La Bastie's process consists in heating the glass to a certain temperature, and passing it through oil or fatty materials; glass articles thus treated are rendered tough enough to stand rough usage, such as dropping on a wooden floor from a height of ten feet, and even hammering to a certain extent. And yet the glass is in a peculiar condition, and, when broken, is shivered into thousands of pieces, much as is the case with Prince Rupert's Drops.

20. **Hardness.**—This property seems to depend upon some

peculiar arrangement of the particles of matter. We should suppose that the densest substances would be the hardest. But it is not so. Iron is the hardest of the metals, but its particles are not so close together as those of gold, which is quite a soft metal. And gold is about four times as heavy as the diamond, which is so hard as to cut glass easily. Common flint is hard enough to scratch glass, but will not cut it so well as the diamond.

Advantage is taken of the different degrees of hardness possessed by minerals in determining their species. In the following table a number of minerals, whose degrees of hardness is very uniform, are arranged so as to form a convenient scale, by reference to which the hardness of any substance can be determined. It is only necessary to secure specimens of the minerals named, and to ascertain which of these ten the body under trial will scratch. Since, however, it is not always possible to obtain a complete set of these minerals, we have added remarks showing approximately their hardness:

SCALE OF HARDNESS.

1. Talc; easily scratched by the finger-nail.
2. Gypsum; not easily scratched by the nail; does not scratch a copper coin.
3. Pure limestone (calcite); is scratched by a copper coin.
4. Fluor-spar; not scratched by a copper coin.
5. Apatite; scratches glass with difficulty, but is easily scratched by a knife.
6. Feldspar; scratches glass easily; is scarcely scratched by a knife.
7. Quartz; not scratched by a knife.
8. Topaz; harder than quartz.
9. Corundum; harder still.
10. Diamond; is scratched by no other substance.

The property of hardness depends on some circumstances not perfectly understood, for a substance may be hard or soft according to the manner in which it is treated. That this is the case with steel has been mentioned (§ 19).

21. Tenacity.—The power possessed by substances which causes them to resist being pulled asunder, termed tenacity, depends on the degree of attraction between the particles. By attraction we mean a disposition in particles to come together, this disposition being manifested in opposition to any force tending to draw them apart. Tenacity does not exist at all in gaseous substances. The particles of air and of steam, for example, show no disposition to cling together—that is, have no tenacity. This property is weak in liquids; it is only strong enough in water to enable its particles to hang together in the shape of a drop. It is strong in solids, enabling their particles not only to hold together in large quantities, but also to hold up heavy weights suspended to them. It is strongest of all in steel.

Various metals and other substances have been tested to ascertain their comparative tenacity. It was done in this way: Rods were made of the metals, etc., all of the same size, having, in fact, a cross-section of one square inch. Weights were suspended to them, and additions were made to the weights little by little till the rods broke. The table below was made by placing against each substance the greatest weight that its rod would sustain:

TABLE SHOWING COMPARATIVE TENACITY OF MATERIALS.

Cast steel	45 to 60 tons.
Best wrought iron	25 to 30 "
Cast iron	6 to 13 "
Copper	9 to 26 "
Platinum	8 "
Ash-wood	8 "
Silver	5 "
Beech-wood	5 "
Gold	4½ "
Zinc	2 tons.
Tin	about 1½ to 2 "
Lead	1 ton.

Some animal substances have great tenacity, as the thread of the silk-worm, hair, wool, and the ligaments and tendons of our bodies and of other animals.

"The gradual discovery," says Dr. Arnott, "of substances possessed of strong tenacity, and which man could yet easily mould and apply to his purposes, has been of great importance to his progress in the arts of life. The place of the hempen cordage of European navies is still held in China by twisted canes and strips of bamboo; and even the hempen cable of Europe, so great an improvement on former usage, is now rapidly giving way to the more complete and commodious security of the iron chain—of which the material to our remote ancestors existed only as useless stone or earth. And what a magnificent spectacle it is, at the present day, to behold chains of tenacious iron stretched high across a channel of the ocean, as at the Menai Strait between Anglesea and England, and supporting an admirable bridge-road of safety, along which crowded processions may pour, regardless of the deep below or of the storm; and beneath which ships, with sails full-spread, pursue their course unmolesting and unmolested."

↘ 22. **Malleability and Ductility.**—Those metals which can be hammered into thin plates are called malleable. Gold furnishes us with the best illustration of this property. We have already mentioned (§ 15), that a single grain of gold can be hammered out into a sheet the one 300,000th part of an inch in thickness. An alloy of 20 parts of gold and 22 of silver is equally malleable. Silver, copper, and tin are quite malleable; but the thinnest leaves of tin are the one 1600th part of an inch in thickness. Most of the other metals are very little so, and some of them are not at all, breaking at the first blow. A substance is said to be *ductile* when it can be drawn out into wire. The principal metals that have this quality are platinum, silver, iron, copper, and gold, and in the order named. The celebrated English chemist Dr. Wollaston obtained a platinum wire only the one 30,000th part of an inch in diameter by the following ingenious process. A small platinum wire was soldered within a cylinder of silver, and the compound wire was drawn out in the usual way as fine as possible. The silver was then dissolved off by immersing the wire in nitric acid, and the platinum core remained about half the thickness of the thread of a spider's web. Melted glass is very ductile. It can be drawn out into a very fine thread; and when this thread is cut and arranged in bunches, it resembles beautiful white hair. In hammering metals into plates, or drawing them into wire, there is a considerable change of relative position in the particles, similar to that which we have in fluids, though nothing like so free. In this change of position, those particles that remain in close neighborhood have a remarkable tenacity or attraction, preventing their separation. In welding two pieces of iron, which is done by the blacksmith by hammering them together when red-hot, there must be enough movement among the particles to permit those of one piece to mingle somewhat with those of the other.

23. Usefulness of Variety in Properties of Matter.—The various properties of matter brought to view in this and the preceding chapters are providential adaptations to the necessities of man. Each substance has those properties which best fit it for his use. Iron, for example, designed by the Creator to be both the strongest and most extensively useful servant of man among the metals, is therefore provided in great abundance, and has those strong, decided, and various qualities which fit it for the services it is to perform. Gold and silver, on the other hand, designed for services less extensive, and in a great measure ornamental, are provided in very much less quantity, and have properties admirably adapting them to the services for which they are so manifestly intended. The same can be substantially said of all other substances, and especially of those very abundant ones air and water. And it may be remarked also that the ingenuity of man is continually discovering new modes of bringing the various properties of matter into his service. We will give but a single illustration—the tempering of steel. "This discovery," says Dr. Arnott, "is perhaps second in importance to few discoveries which man has made; for it has given him all the edge-tools and cutting instruments by which

he now moulds every other substance to his wishes, and to which he owes
all his modern mechanical improvements. A savage would work for
twelve months with fire and sharpened flints to fell a great tree or carve
a rough canoe, where a modern carpenter, with his tools of hard steel,
could accomplish the same object better in a few days."

QUESTIONS.

16. What is meant by the porosity of matter? Show that gold is
porous. Name the kinds of pores, and explain by illustrations. What
proof is there that all substances have spaces in them? What is said of
the amount of space in gases and vapors? Give the statement in regard
to steam. What is said of solutions of solids in fluids? What of evaporation? What of animal and vegetable bodies? Upon what do density
and rarity depend?—17. What is said of compressibility? Illustrate.
What of the incompressibility of liquids? How is the position of the
particles of liquids affected by a change of temperature? Which are the
most compressible substances?—18. Explain elasticity by reference to a
rubber-ball. Illustrate by reference to a bent stick. What is said of
the degrees of elasticity in different substances? Define elasticity.—19. Illustrate what is meant by flexibility. What of brittleness? Give examples of flexible and brittle steel. Explain the actual difference between them. Explain the tempering of steel. What is said of the annealing of glass? What of Prince Rupert's Drops? What of "tempered
glass?"—20. Upon what does the hardness of bodies depend? What use
is made of the different degrees of hardness in minerals? Name the
typical minerals.—21. Define tenacity. What is said of the comparative
tenacity of substances? Which metal is the strongest? Which the
weakest? What is said of the value of tenacious bodies?—22. What
is the difference between malleability and ductility? Give examples.—
23. What is said of the usefulness of the variety of properties in matter?
What of the importance of steel?

CHAPTER IV.

ATTRACTIONS OF MATTER.

24. Matter attracts Matter.—We have already stated that matter is acted upon by forces, and we will now explain this more fully. The minute particles of matter of which bodies are composed do not touch each other, but even in the densest substances are surrounded by void spaces, and these particles are held in their place by attraction between them, each particle of matter attracting every other particle. This property invariably accompanies matter of every form and under all circumstances. And since tangible masses are made up of small particles, what is true of the latter is equally true of the former. Every body in the universe attracts with greater or less force every other body, however near or distant they may .be from each other. Sun, earth, moon, and stars attract each other; and this power binds them together as they roll through space. This force is generally called the attraction of *gravitation,* a name given to it because we have such common examples of its influence in the fall of bodies to the earth; they are said to *gravitate* towards the earth.

Whether the mysterious force which binds the minute particles of matter together to constitute masses is the same as that which controls the motions of celestial bodies is as yet unproved. That an attraction actually exists between small masses when they are brought exceedingly close to each other is easily shown. Thus if two cork balls coated with varnish be placed on the surface of water near to each other, their attraction will soon bring them together. Thin globes of glass will exhibit the same attraction. So, also, floating pieces of wood are apt to be found together; and when a ship is wrecked, the parts of the wreck collect in tangled masses here and there on the surface of the sea as soon as it becomes calm. The gravitation between particles and masses of matter may possibly be identical, and our appreciation of it depends upon their relative size and distance. For convenience of distinction, different names have been given to attraction according to the distances at which it acts.

Gravitation is the attraction existing between matter at great or appreciable distances, as between the heavenly bodies, or between the earth and a stone thrown into the air.

Cohesion is the attraction between molecules of the same kind of matter binding them together to form masses.

Adhesion is the attraction between molecules of dissimilar matter, as exhibited in cements. A peculiar kind of adhesion is known as *capillary attraction.*

Chemical attraction is the force which binds together the atoms of a molecule (§ 8). Its study is the province of chemistry and will be fully treated in Part II.

Explanations and illustrations of the phenomena resulting from these attractions will occupy the remainder of this chapter and the succeeding one.

25. **Gravitation.**—A stone falls to the ground for precisely the same reason that the two cork balls approach each other when floated on water (§ 24). It falls owing to the attraction which the earth and the stone have for each other; in other words, the attraction is mutual. If you hold a stone in your hand and thus prevent its falling, you simply resist a power which is pulling it down. If it were possible to suspend the mutual attraction of the earth and the stone, you could release your hold of the stone, and it would remain suspended in the air until the attraction had

been restored. The attraction of the earth and the stone is really mutual; but the earth is so immense in comparison with the stone that its motion towards the stone is exceedingly small, and may practically be considered as naught.

This may be clearly illustrated by a comparison of the force of attraction with the force of muscular action. Suppose a man in a boat pulls on a rope which is made fast to a ship lying loose at the wharf, and in this way draws his boat towards it. He does not consider that he moves the ship at all; but in reality he does, for if, instead of one, a hundred or more men in boats pull upon the ship, they will make the motion apparent. In the case of the single boat, the motion of the ship is as real as when a hundred boats are pulling it, but it is only the one-hundredth part as great. Now let the ship represent the earth, and the little boat some object, as a stone, attracted by it. The earth and the stone move towards each other, just as the ship and the boat do. And if, as we multiplied the number of boats, we should multiply the bulk of the stone till it is of an immense size, it would by its attraction have a perceptible influence upon the earth.

Observe in regard to the illustration that it makes no difference whether the man pull in the boat or in the ship. In either case he exerts an equal force on the ship and the boat, making them to approach each other. So it is with the attraction between the earth and the stone. It is a force exerted equally upon both. Its effect on the earth is not manifest, because it is so much larger than the stone; just as the effect of the man's exertion is not manifest upon the ship, because it is so much larger than the boat.

Proportion of the Mutual Motions of Attraction.—Let us pursue the illustration a little farther. If a man stand in a boat, and pull a rope made fast to another boat of the same size and weight, both boats, in coming together, will move over the same space. Just so it is with the attraction between two bodies having the same quantities of matter or equal masses —they attract each other equally, and therefore meet each other half-way. Suppose, however, that one boat is ten times as great and as heavy as the other. The small boat would move ten times as much as the large one when the man brings them together by pulling the rope. In like manner, if a body one tenth as large as the earth should approach it, they would attract each other, but in coming together the body would move ten times as far as the earth. In the case of falling bodies, even though they may be of great size, the earth moves so slightly to meet them that its motion is wholly imperceptible. It has been calculated that if a ball of earth the

C

tenth of a mile in diameter were placed at the distance of a tenth part of a mile from the earth, and let fall, the motion of the earth would be only the one eighty-thousand-millionth ($\frac{1}{80\ 000\ 000\ 000}$) part of an inch.

26. Attraction Towards the Earth's Centre.—All bodies are attracted towards the centre of the earth. This is

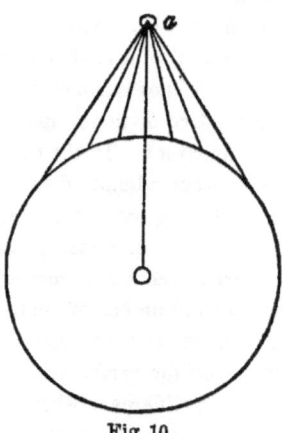

because the earth is spherical. Let the circle, Fig. 10, represent the earth, and a a body attracted by it. The lines drawn from the body to the earth represent the attractive force exerted by the earth upon the body. It is obvious from these that there is as much attraction on the one side of the line drawn from the body to the earth's centre as there is on the other. The attractive force, then, of the earth as a whole is exerted upon the body in the direction

Fig. 10.

of this middle line. It tends to draw it, therefore, towards the centre. Consequently, a plumb-line points towards the centre of the earth, and it is evident that two weights suspended by two strings do not hang perfectly parallel to each other. The difference is so slight in an ordinary pair of scales that it cannot be perceived. But if it were possible to suspend in the heavens a beam so long as to stretch over a large extent of the earth's circumference, as represented in Fig. 11, the scales attached to it would be very far from hanging parallel to each other. Substances suspended in different parts of the globe are hanging in different directions, and those which are hung

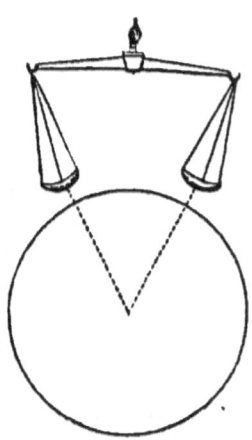

Fig. 11.

. up by our fellow-men on the opposite side of the earth are hanging directly towards us.

Up and Down.—All falling bodies fall towards the centre of the earth, and the remarks made in relation to suspended weights are similarly applicable. Up and down are merely relative terms—*up* being from the centre of the earth, and *down* towards it. As the earth moves round on its axis, the same line of direction which we call upward at one time is downward at another. This may be illustrated by Fig. 12. Let the circle represent the circumference of the earth. In the daily revolution we pass over this whole circle. If we are at D at noon, we are at E at six o'clock, and at F at midnight. If, therefore, the ball A be dropped from some height at noon, the line in which it falls will be at right angles

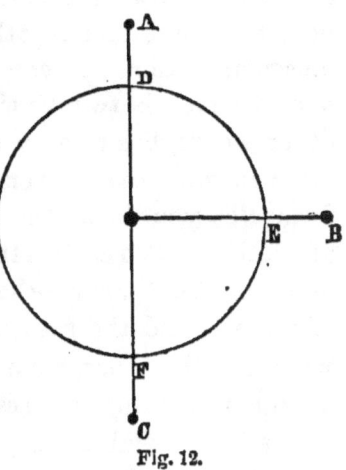

Fig. 12.

to a line in which it will fall if dropped from the same height at six o'clock; for this height will have moved in this same time from A to B. If it be dropped from the same height at midnight, its line of direction will be directly opposite to the first; for the place of the experiment will have moved in that time to C.

It is not always true that falling bodies tend exactly towards the centre of the earth. The centre does not attract them, but it is the substance of the whole earth; and since this is irregular in its density and form, the attraction will be irregular also. Thus it is found by accurate experiments that a plumb-line suspended in the neighborhood of a mountain is attracted by it, and will not hang exactly parallel with another suspended at some distance from the mountain. The difference is not, however, enough to have any practical bearing.

27. Weight.—That which we call weight is not a property · of matter, but merely the resisted attraction of the earth. If two bodies fall to the earth, and one of them contain ten times as many particles of matter as the other, ten times as much force of gravity is required, and is actually exerted, to bring it to the ground. This will appear plain to you if you bear in mind that a body falls because it is drawn down by the force of attraction, and then compare this force to any other force, as, for example, that of muscular action. If you draw towards you two weights, one of which is twenty times as heavy as the other, or, in other words, has twenty times as great a quantity of matter, you must exert twenty times as much strength on the former as you do on the latter. So it is with the force of attraction. The earth attracts a body having twenty times more matter than another with twenty times the amount of force. And the first body will have twenty times the weight of the other, for it will make twenty times the pressure upon anything that resists the force with which the earth draws it. Weight, then, is *the amount of resistance to the attraction existing between the earth and the body weighed.* If you place a substance in one side of a pair of scales, it goes down because of the attraction between it and the earth. By placing weights in the other side until the scales are balanced, you find how much is needed to counteract the resistance caused by the attraction of the substance and the earth for each other; or, in other words, you find out how much it weighs. In doing this you use certain standard weights; that is, certain quantities of matter which have been agreed upon by mankind, and are called by certain names, as ounces, pounds, grammes, kilo-grammes, etc. When a spring-balance is used, the spring has been tested by these standard weights, and its scale marked accordingly.

Weight not Fixed, but Variable. — Weight does not depend alone upon the density of the body weighed, but also upon the density of the earth. For the attraction causing the resistance which we call weight is a *mutual* attraction, and is in proportion to the quantities of matter of both the body and the earth. If, therefore, the density of the earth were increased twice, three times, or four times, the weights of all bodies would be increased in the same proportion; that is, the force with which the earth would attract them would be twice, three times, or four times as great as now. This would not be perceived by any effect on balances, for the weights and the articles weighed would be alike increased in weight. But it would be perceived in instruments that indicate the weights of bodies by their influence on a spring. These would disagree with scales and steelyards just in proportion to the increase of the earth's density. It would be perceived also in the application of muscular and other forces in raising and sustaining weights; every stone would require twice, three times, or four times the muscular effort to raise it.

28. **Weight Varies with Distance.** — The nearer two bodies are to each other, the greater the mutual attraction. The nearer a body is to the earth, the greater the attraction that draws it towards the earth—in other words, the greater is its weight. The force of gravity, or weight, is greatest, therefore, just at the surface of the earth, and it diminishes as we go up from the earth. As we leave the surface of the earth, the force of gravity lessens in such a proportion that it is always *inversely* as the square of the distance from the centre of the earth. In other words, the force of gravity increases or decreases at the square of the rate that the distance decreases or increases. This requires still further explanation. If the distance from the centre of the earth to its surface, which is 4000 miles, be called 1, then 4000 miles from the earth would be called 2, or twice as far from the centre, and 8000 miles from the earth would be 3, 12,000 miles from the earth would be 4, and so on. The squares of these numbers are 1, 4, 9, 16, etc. Now, since weight decreases *inversely* as the square of the distance, any object weighing

one pound on the surface of the earth would weigh but $\frac{1}{4}$ pound at the distance of 4000 miles, and only $\frac{1}{9}$ pound at 8000 miles.

An object weighs less on the summit of a high mountain than in the valley below, because it is farther removed from the great bulk of the earth, and is therefore not so strongly attracted. The difference, however, is but small; a man weighing 250 pounds in the valley would weigh but half a pound less on the summit of a mountain four miles high.

We have spoken of weight only in relation to the earth, but weight is an attribute of bodies everywhere, for wherever matter is found there must be attraction.

The weight of the substances on the surface of the different heavenly bodies varies according to the quantity of matter in, or density of, those bodies. Since the moon is much smaller than the earth, a body which weighs a pound on the surface of the earth would weigh much less than a pound on the moon. And since the sun is much larger than the earth, the same body carried there would weigh much more than a pound.

If we knew the exact densities of the sun and the moon and the earth, as well as their size, we could estimate exactly the difference in the weights which any body would have in them; for the attraction which causes the resistance called weight is in direct proportion to the quantity of matter, and the quantity of matter depends on both density and size.

29. Cohesion.—That form of attraction which binds together the molecules of a body is called cohesion.

Cohesion is stronger in some solids than in others. The mason with his trowel easily divides a brick; but he cannot do this with a piece of granite, for its particles have a greater attraction for each other than those of the brick. A blow which would break a glass dish would not injure a copper one of the same thickness. A weight that would hang securely from an iron wire would break a lead wire of the same size; that is, it would tear the particles apart, because they are not strongly attracted to each other. Cohesion has different modes of action in different solids. It therefore fastens their particles together in different ways, and thus occasions the physical properties which are so useful to us—tenacity, elasticity, ductility, flexibility, etc.

Cohesion is exerted only between molecules of the same kind; when two masses are made to cohere they must be of similar material and must be pressed very closely together, because the attractive force is exerted only at inappreciable distances. For this same reason also it is only the surface particles which influence the cohesion.

Examples of cohesion of masses are numerous: two highly polished surfaces of glass may be made to stick together as if glued, and can only be separated by sliding one off the other.

Before rubber tubing was a commercial article, it was made by a simple process based upon its cohesive property. A piece of sheet rubber of suitable length and width is wrapped around a glass or wooden rod, and a strip cut off, where the edges lap, with a pair of long scissors; by pressing together the freshly cut surfaces, they cohere firmly, making a perfect tube.

The manufacture of various articles which are made by compressing powders until they form solids, as in the case of graphite for lead-pencils, sawdust for wooden ornaments, brick-dust for tiles, etc., are examples of practical applications of cohesion.

If you cut two bullets so as to give to each a very smooth flat surface, you can make them cohere quite strongly by pressing them together, especially if you give a little turning motion at the same time that you press, for this will bring the particles on the surfaces in close contact. If the balls of lead are quite large and furnished with handles, as

represented in Fig. 13, it will require considerable force to separate them when they have been thus pressed together.

Fig. 13.

30. **Cohesion in Liquids.**—In liquids the attraction between the particles is very feeble compared with that in solids. The strength of the attraction of particles of steel is

about three million times that of the particles of water. The estimate is made in this way: We find that a steel wire will sustain a weight equal to 39,000 feet (11.887 kilometres) of the wire. But a drop of water hanging to the end of a stick cannot be more than one sixth of an inch (42 millimetres) in length; that is, water will hold together by the attraction of its particles only to this extent, which is a little less than the three-millionth part of the length of steel wire which could hang without breaking.

The freedom with which the particles of a liquid move among one another is due to the comparative feebleness of cohesion, and to the fact that the molecules are more widely separated than in solids. This mobility of liquids varies considerably according to the intensity of the cohesive power; in *limpid* liquids, such as ether, alcohol, naphtha, etc., the force of cohesion is very feeble, while *viscid* liquids, such as oil, molasses, glycerin, etc., are sluggish in their motions, being hampered by greater cohesion of their particles. For this reason, too, drops of viscous liquids are much larger than those of mobile ones poured from the same bottle; sixty drops of water fill the same measure as one hundred of laudanum when poured from a lip of the same size. A knowledge of this and similar facts is of importance to physicians and druggists.

31. Globular Shape of Drops of Liquid.—Since the particles of a liquid move thus freely among each other, the

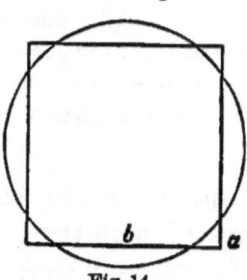
Fig. 14.

attraction of cohesion disposes them to assume a globular or spherical shape. The reason of this can be made plain by Figs. 14 and 15. The outside of a perfect sphere is all at the same distance from the centre; and the circumference of a circle is equidistant from the centre, as represented in Fig. 14.

But this is not true of all parts of the surface of a cube or of a square: *a*, for example, is farther from the centre than *b*. Now in a drop of liquid all the particles are attracted towards the centre, for in that line from each particle lies the largest number of particles to attract it. This can be made obvious by taking some point in the drop, as represented in Fig. 15, and drawing lines from it through the centre and in other directions. If *a* be the point in the drop, it is plain that the line from it through the centre is longer than *a b* or *a c*. Therefore a particle, *a*, will be attracted towards the centre rather than in the direction

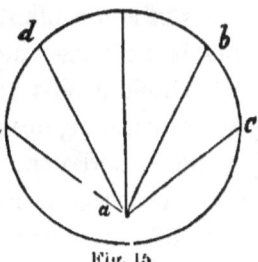

Fig. 15.

a b or *a c*, because there are more particles in the direction of the centre, and the more particles there are the stronger is the attraction. But this is not all. The particles in the line *a c*, tending to make *a* go towards *c*, are balanced by the particles in the line *a e*, tending to make it go towards *e*. The two lines of particles, therefore, together tend to make it go in a middle line between them; that is, towards the centre, just as two strings pulling equally, the one to *c* and the other to *e*, would make a body, *a*, move in a middle line between these two directions. The same can be shown of the two lines of particles *a b* and *a d*, and so of any other two alike in situation on each side of the line through the centre. The tendency of every particle is, then, to move towards the centre, and a globular form results.

32. **The Spherical Form in Different Liquids.**—The disposition to form a sphere is seen more distinctly in mercury than in any other liquid. If you drop a little of it upon a plate it separates into globules, which roll about like shot. Why does water behave differently? Why do the drops

C 2

of water hang upon the window-pane, showing only in an imperfect way their disposition to a globular arrangement? It is because the particles of water have a greater attraction for other substances, and less attraction for each other, than the particles of quicksilver. Water sometimes exhibits its disposition to form spheroidal drops on the leaves of some plants, and rolls about in balls like mercury. This is because there is something on the surface of the leaf which repels rather than attracts the water. If you put your finger, however, on one of these drops, it will alter its shape, and your finger will be moistened, because there is an attraction between the particles of your skin and those of the water. Take another illustration of this difference in attraction. If you drop a little oil upon the surface of water it will float about in round drops. This is because the water repels the oil. But when oil is spilled upon wood or cloth their particles have so strong an attraction that they unite, instead of gathering into little round masses as they do on the surface of water.

Manufacture of Shot.—We have a good example of the tendency of fluids to form spherical drops in the manufacture of shot. Melted lead is poured into a large vessel, in the top of the shot-tower, having holes in its bottom, from which the metal falls in drops. Each drop, as it whirls round and round in its fall, takes the globular form. By the time that it reaches the end of its journey, about two hundred feet, it becomes so far cooled as to be solid, and as it is received in a reservoir of water, its globular form is retained. Bullets cannot be made in this way, because a quantity of melted lead sufficient to make a bullet will not hold together in a globular form.

33. **Spherical Form of the Earth and the Heavenly Bodies.** —It is supposed that the sun, moon, earth, and all the heavenly bodies were once in a liquid state, and that they owe their spherical shape to this fact. As they whirled on in their course, the liquid mass gradually cooled, and at length they acquired their present state. How all the

mighty changes could be effected in our earth, converting it from a liquid into a body with a solid crust, having such a diversity of substances in it, and so variously arranged, with its depressions containing water, and the whole covered with its robe of air fifty miles in thickness, we cannot fully understand. And yet there are some portions of the process which chemistry and geology have revealed to us, giving us some glimpses of the wonders which, during the lapse of ages, God wrought in our earth in preparing it for the habitation of man.

34. **Crystallization.**—The attraction of cohesion is not in all cases uniformly strong in all directions around a molecule, and when the particles are free to move they often assume a more or less regular arrangement, becoming *crystalline*. This happens most frequently when a substance passes from a liquid state to a solid one, and when it is deposited from a solution.

The process of crystallization is readily studied by slowly cooling saturated solutions of certain chemical substances: alum, saltpeter, sulphate of copper, borax, and other substances. There is an immense variety of crystalline forms, the study of which is pursued in connection with the science of mineralogy; we can here merely indicate a few of the forms which substances assume. Common salt crystallizes in cubes, Fig. 16;

Fig. 16.

alum in octahedra, or eight-sided figures, Fig. 17. Crystalline forms

Fig. 17.

are also assumed by many minerals: the bright-red garnet crystallizes in the form shown in Fig. 18, and quartz takes the form of Fig. 19. All the

Fig. 18.

Fig. 19.

precious stones have a crystalline structure, and even the common rocks under your feet exhibit the same crystalline disposition in detail which you see in the mass.

Water, when it changes into a solid, shows the same disposition, of which the crystals of the snow and the frost-work on our windows are familiar examples. When snow forms, the water of the clouds is suddenly crystallized by the cold air, the particles taking their regular places more readily and certainly than if they were guided by intelligence, because in obedience to an unerring law established by the Creator. Examples of this sudden crystallization of water are common. The water in a pitcher may remain fluid, although it is cooled down to the freezing-point, and even below it, if it be kept perfectly still. But on agitating the pitcher the water at once becomes filled with a net-work of ice-crystals. The stillness of the water prevented its particles from taking the crystalline arrangement needed for the formation of ice; and the shaking of the particles assisted the motion necessary to the assumption of a crystalline form.

35. **Frost and Snow.**—The frost-work on our windows is a wonderful exhibition of the variety of forms that crystallization can produce. It sometimes presents figures like leaves and flowers, such as are chased on vessels of silver, but much more delicate and beautiful. So varied and fantastic are the forms in which these water-crystals are arranged, that it is very natural to ascribe them, as is done universally in the dialect of the nursery, to the ingenuity of a strange and fanciful spirit. Every snow-flake is a bundle of little crystals as regular and beautiful as the

crystals which you so much admire in a mineralogical cabinet. And there is great variety in the grouping of these crystals. Figs. 20 and 21 show some of these forms as they appear under the microscope.

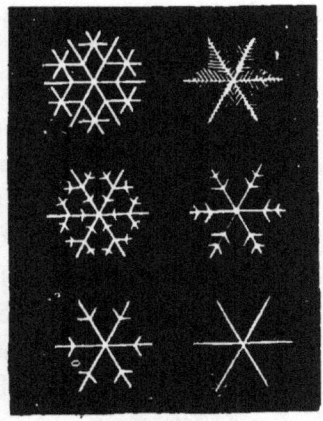

Fig. 20.

It is a very quick operation by which the particles of water in the clouds thus marshal themselves, as if by magic, in these regular forms. But a quicker operation is that by which hail is formed—so quick that the particles have not time to arrange themselves in crystalline forms, but are huddled together without order. The brilliant and glistening whiteness of the snow is owing to the reflection of light from its minute crystals. In the arctic regions the beauty of the snow is often much greater than with us. "The snow crystals of last night," says Captain M'Clintock in his "Discovery of the Fate of Sir John Franklin," "were extremely beautiful. The largest kind is an inch in length; its form exactly resembles the end of a pointed feather. Stellar crystals two tenths of an inch in diameter have also fallen; these have six points, and are the most exquisite things when seen under a microscope. In the sun, or even in moonlight, all these crystals glisten most brilliantly; and as our masts and rigging are abundantly covered with them, the *Fox* never was so gorgeously arrayed as she now appears."

Fig. 21.

Order in Nature.— We see in this general tendency to crystallization a striking

illustration of the fact that the Almighty is a God of order.
Disorderly arrangement is never seen except where there
is an obvious necessity for it. And even when there is apparent disorder, a little examination generally shows that
essentially there is order. The rocks that give so much
variety to scenery seem to be piled up in confusion, yet order has evidently reigned in their construction. Pick up a
common stone, and on breaking it you will see the crystalline arrangement in its interior. Nay, more, much of the
very soil is made up of separated and broken crystals.

Amorphous Bodies.—Substances which possess no regularity of structure are termed *amorphous*, that is, without
crystalline form. Glue, soap, clay, chalk, and many minerals are amorphous. Some substances may occur at one
time in a crystalline state and at another without any trace
of regularity of form. Carbonate of lime is one of these,
being crystalline in limestone, Iceland spar, and various
minerals, while in the form of chalk it is amorphous. Metals, too, may occur both amorphous and crystalline. The
attraction of cohesion, which produces crystalline forms,
leads to peculiarities of structure which receive special
names, as "hard," "brittle," etc., as already explained in
Chapter III. _____

QUESTIONS.

24. What is said of the attraction of matter? What is the force generally called, and why? Show that attraction exists between small masses.
Name and define the different kinds of attraction.—25. What is said of
gravitation? Illustrate the fact that attraction is mutual. Illustrate also
the proportions of the mutual attractions. What is said of the motion of
the earth?—26. Explain why bodies are attracted towards the earth's
centre. How does this affect plumb-lines suspended at some distance from
one another? Show that *up* and *down* are only relative terms. Why do
falling bodies deviate from a line drawn exactly to the earth's centre?—
27. What is weight? Give the comparison in regard to muscular force.
What is said of scales and weights? What of using springs in weighing?

—28. What would be the effect on weight if the density of the earth were increased? In what ways would this be perceived?—29. What is said of the variation of weight with distance? Explain the law. What is said of the difference of weight on mountains and in valleys? What is said of the weight of bodies on the moon?—30. What is said of cohesion? Give examples of cohesion. Describe the experiment with two bullets.—31. What is said of cohesion in liquids? What of the mobility of liquids? What causes some liquids to be limpid and some viscid? Explain by reference to Figs. 14 and 15 the globular form of drops.—32. Give the difference between mercury and water in regard to the spherical form. What is said of drops of water on leaves? What is said of oil in reference to attraction? Describe and explain the manufacture of shot.—33. What is said of the spherical form of the earth and the heavenly bodies?—34. What is said of crystallization? State the examples cited. What is said of the crystallization of water? Give and explain the example of sudden crystallization.—35. What is said of frost-work? What of snow? What is stated in regard to the snow-crystals of the arctic regions? What is said of order in nature? What of amorphous bodies?

CHAPTER V.

ATTRACTIONS OF MATTER (CONTINUED).

36. **Adhesion.**—Adhesion is the attraction between different kinds of matter, as between solids and liquids, or between glue and wood. When a glass article is broken you cannot unite the pieces, however accurately you may bring them together, or however firmly they may be pressed. This is because the power of cohesion acts strongly only when the molecules are brought very near together; and it is impossible to bring the particles on the two surfaces of a broken piece of glass as near together as they were before the fracture. If it were possible so to do, no crack would be visible. We are obliged therefore to resort to some kind of cement; this causes the two surfaces

to adhere because, while soft, it insinuates itself among the particles of glass, and on drying becomes a bond of union between the broken fragments. In adhesion as well as in cohesion only the surface layer of molecules exert any influence, consequently a mere film over a surface suffices to alter its adhesive power, as when a glass is greasy.

Examples of the adhesion of solids are familiar: silver and gold may be made to adhere to iron by a very great and sudden pressure. The iron must be made very smooth, and the silver or gold plate very thin. A powerful blow brings the particles of the thin plate into such nearness to those of the iron that union is affected, or, in other words, they attract each other sufficiently to be united. Similarly, a sheet of tin and one of lead can be made to adhere so as to form one sheet by the pressure of the rollers of a rolling-mill.

37. **Adhesion of Solids and Liquids.**—The attraction which solids and liquids have for each other furnishes us with many interesting phenomena. The adhesion of drops of water to glass and other solids is a familiar example of this attraction. If you dip your hand into water, it is wet on taking it out, because your skin has sufficient attraction for the water to retain some of it. A towel will retain more of it for two reasons: owing to the interstices between its fibres it presents much more surface to the water (see § 39, Capillary Attraction), and it has none of the oily substance which on your skin, though in small quantity, serves somewhat to repel the water.

If you dip your hand into mercury the latter will not adhere to it, and it would seem that the skin has an attraction for water and none for mercury. This, however, is only apparent, for a small globule of mercury will adhere to the finger, though if it be brought in contact with a larger amount of mercury the globule leaves the finger and loses itself in the liquid. This shows that liquids wet solids when the adhesion of the liquid to the solid is greater than the cohesion of the liquid.

The attraction of solids and fluids for each other is shown very prettily in the experiment represented in Fig. 22. A plate of glass is attached by strings to one end of a balance, and weights just sufficient to balance it are placed in

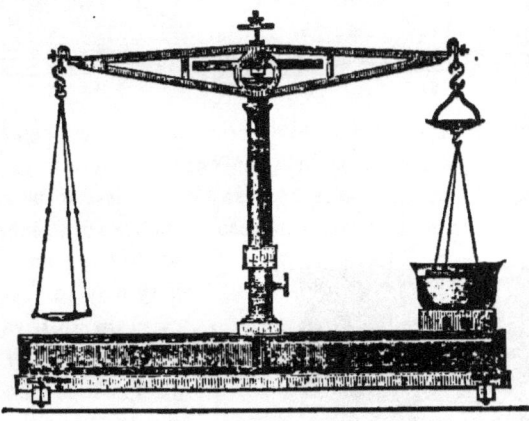

Fig. 22.

the opposite scale. When the glass is brought in contact with water, it will require additional weight in the scale to separate the glass from the water. This experiment, however, does not measure the adhesion of the glass and water accurately, because we cannot detach the plate of glass clean and dry; it merely measures the force necessary to overcome the cohesion of the liquid.

Further Illustrations.—When you see stems of plants rising above the surface of stagnant water you will observe that the water is considerably raised about them. This is from the attraction between them and the water. For the same reason water rises higher at the sides of a tumbler than in the middle. If you immerse a piece of glass in water, the water will rise at its sides as represented in Fig. 23. If you immerse two

Fig. 23.

pieces together, as in Fig. 24, the water will rise higher between them than on the outside, because the particles between are attracted by two surfaces, while those outside are attracted by only one. It is for the same reason

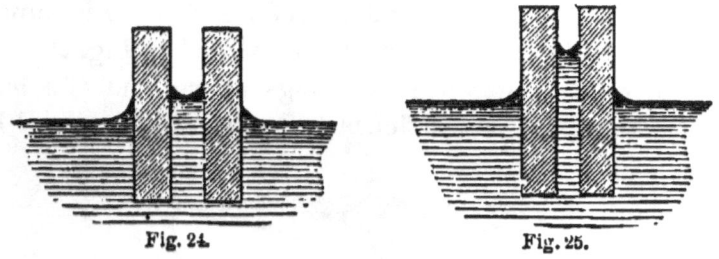

Fig. 24.   Fig. 25.

that two men can raise a weight higher than one of them can alone. And if the pieces of glass be brought quite near together, as in Fig. 25, the water will be raised higher still, because there is less to be raised by the two surfaces. Just as two men can raise a small weight higher than they can a large one. The same thing may be beautifully illustrated in this way: Let two pieces of glass, as represented in Fig. 26, be immersed in colored water, with two of their edges joined together, the opposite edges being separated. The height to which the fluid rises will make a curved line, it being lowest at the edges which are separated, and highest at the edges which are joined.

Fig. 26.

38. Rise of Liquids in Tubes.—For the same reason that water rises higher between plates of glass than outside, it will rise higher within a tube than on the outside. The diagram in Fig. 27 will make this clear. It represents a transverse section of a tube, enlarged so that the demonstration may be plain. Consider the case of one particle on the inside and another on the outside at equal distances from the glass. It is evident that the particle a is not so near to as many particles of the glass as is the particle b.

The lines drawn show this. The longest lines extending from the particles *a* and *b* to the glass are equal in length; that is, *a e* and *a f* are equal to *b g* and *b h*. It is clear, therefore, that all the glass between the lines at *c* and *d* is as near to the particle *b* as the glass between the lines at *e* and *f* is to the particle *a*. But this is not all. The particle *b* is

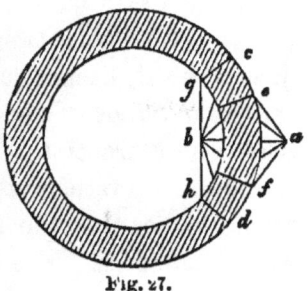

Fig. 27.

near enough to the whole inside of the tube to be attracted by it, while very little attraction is exerted upon *a* by any part of the glass beyond that which is included between *e* and *f*. The same difference can be shown with regard to all the particles on the inside of the tube compared with those outside. The former are nearer to more particles of the glass than the latter, and therefore are more strongly

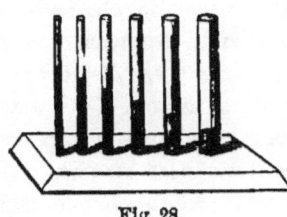

Fig. 28.

attracted. Again, the nearer the plates of glass, the higher the water rises between them; so the smaller the tube, the higher will the water rise in it. You can try the experiment by immersing in water glass tubes of different diameters, as represented in Fig. 28. It is obvious that the particle *b* (Fig. 27) would not be very strongly attracted by the part of the tube opposite if the tube were a large one; but it would be if the tube were very small, for then it would be quite near to that part. Since glass is not wet by mercury, a tube plunged into this liquid causes a depression without and within. Figs. 29 and 30 show the contrast between water and mercury.

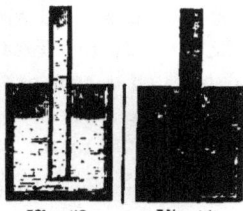

Fig. 29.　　Fig. 30.

39. Capillary Attraction.—The term *capillary* (derived from the Latin word *capilla*, hair) has been commonly applied to the attraction exhibited under the circumstances just noticed, because it is most obvious and was first observed in tubes of very fine bore. The same term is used when the attraction is seen in the rising or spreading of a liquid in interstices as well as in tubes. Thus capillary attraction causes the rising of oil or burning-fluid in the wicks of lamps. The liquid ascends in the interstices, or spaces, between the fibres, as it does in the spaces of tubes.

Other Examples.—If you let one end of a towel lie in a bowl of water, the other end lying over upon the table, the whole towel will become wet from the spreading of the water among the fibres in obedience to capillary attraction. If you suspend a piece of sponge so that it merely touch the surface of some water, or if you lay it in a plate with water in it, the whole sponge will become wet. So, too, if you dip the end of a lump of sugar in your tea, and hold it there a little time, the whole lump will be moistened. In very damp weather the wood-work in our houses swells from the spreading of water in the pores of the wood in obedience to capillary attraction. Especially is this the case in basement rooms, where the water can ascend from the ground in the pores of the walls, as well as from the damp air. In watering plants in pots, if the water be poured into the saucers, it will pass through the earth by capillary attraction. For the same reason plants and trees near streams grow luxuriantly, being abundantly supplied with water, which rises to their roots through the pores of the soil. The disposition of the wood to imbibe moisture in its pores has sometimes been made use of very effectually in quarrying out millstones. First a large block of stone is hewn into a cylindrical shape. Then grooves are cut into it all around where a separation is desired, and wooden wedges are driven tightly into them. These are then moistened with water, and eventually swell so much as to split the stone in the direction of the grooves. Blotting-paper furnishes an illustration of capillary attraction, the ink being taken up among the fibres of the paper. Ordinary writing-paper will not answer as a blotter, because the sizing fills up the interstices between the fibres.

As already stated (§ 38), whenever a body is wet by a liquid, a rise of its surface ensues; but when otherwise, a depression takes place. Thus a sewing-needle washed with alcohol is easily wet by water when placed on its surface, and sinks immediately; when, however, the same needle is somewhat greasy, so that it can make a depression, it will float. Some insects which skip about on the

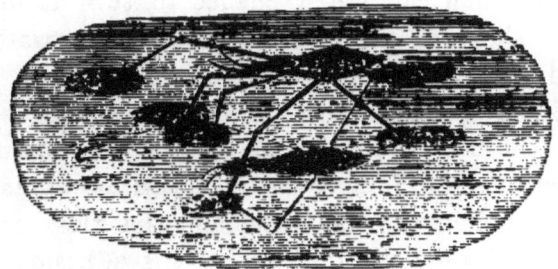

Fig. 31.

surface of the water are protected from being wet by it. The feathers of water-fowl are always slightly oily, and thus they remain quite dry even when swimming in the water.

40. **Opposition between the Modes of Attraction.** — Although adhesion and gravitation are essentially the same thing, we see them continually acting in opposition to each other. Abundant illustrations might be given, but we will cite only a few.

If you pour water out of a tumbler, there is a struggle between the attraction of adhesion and gravitation for the mastery—the attraction of adhesion tending to make the water adhere to the tumbler, and run down its side, as in Fig. 32, and gravitation tending to make it fall straight down. But when water is poured out of a pitcher, as in Fig. 33, the lip of the pitcher acts in favor of the at-

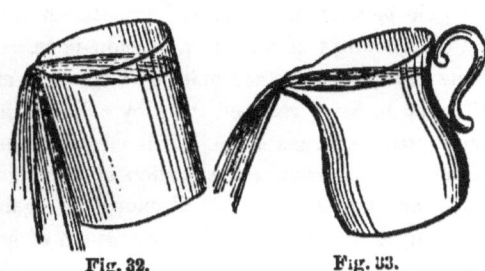

Fig. 32.   Fig. 33.

traction of gravity; for the water would have to turn a very sharp corner to run down the outside of the pitcher in obedience to adhesion. In pouring water from a tumbler, we can often, by a quick movement, throw the water, as we may say, into the hands of gravity before the attraction of adhesion can get a chance to turn it down the tumbler's side. If you can only make the water *begin* to run from the tumbler without going down its side there will be no difficulty; for there is an attraction of cohesion between the particles of the water, tending to make them keep together, which in this case acts against the adhesion between the water and the glass, and therefore acts in favor of gravitation. It is adhesion together with cohesion that forms the drop on the lip of a bottle as we drop medicine—cohesion between the particles of the liquid, and adhesion between the latter particles and those of the glass. It is gravitation, on the other hand, that makes the drop fall, it becoming so large that the force of gravity overcomes the adhesion between the drop and the bottle.

Size of Drops Influenced by Gravitation.—Were it not for the attraction of gravitation, there would be no limit to the size of drops of any liquid. When the drop reaches a certain size, it falls because it is so heavy; or, in other words, because with its slight adhesion the attraction of the earth brings it down. Now if this attraction could be suspended, and the attraction of adhesion left to act alone, particles of water might be added to the drop to any extent, and they would cling there. You can see the struggle between adhesion and gravitation very prettily illustrated if you watch the drops of rain on a window-pane. If two drops happen to be quite near together, they unite by attraction, and then, being too large to allow of its being retained there by adhesion in opposition to gravitation, the united drop runs down. If it meet with no other drop, it soon stops, because by adhesion some portion of it clings to the glass all along its track, and thus becomes small enough to again admit of suspension. It is owing to the influence of the attraction of gravitation that the drops of different liquids differ in size, the heavier yielding small, and the lighter large ones. You have another illustration of a similar character in the adhesion of chalk to a black-board or any surface. The chalk crayon itself cannot adhere, for the attraction of the earth does not permit it. But small quantities of it can adhere for the same reason that water adheres to surfaces in small quantities. Dust also clings to the vertical sides of furniture, though a lump of earth would not.

41. Size of Solid Bodies Limited by Gravitation.—We can illustrate the limitation of size in solid masses by Figs. 34 and 35. Suppose that a and b, Fig. 34, are two pieces of timber projecting from a post, b being twice as large as a. It is evident that b cannot support twice as much weight as a, for gravitation is dragging it downward from its connection with the upright post with twice the force that it does a. The case is still stronger

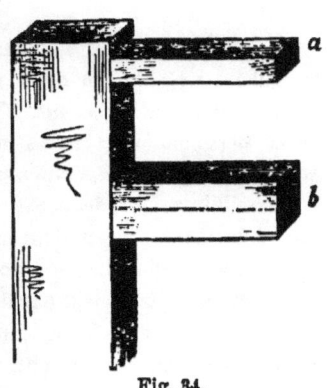

Fig. 34.

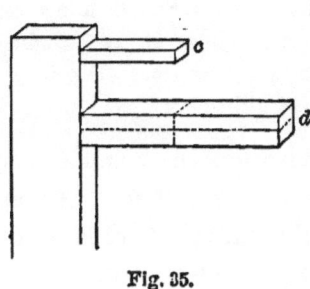

Fig. 35.

when, as represented in Fig. 35, the larger timber is twice as long as the smaller. Here d has four times the bulk of c. But it cannot support four times as much weight at its end, not only because its own weight presses it downward, but because half of its weight is at a greater distance from the place of attachment than the smaller beam is. Gravitation here operates in opposition to cohesion in such a way that the projecting timber, if carried to a certain size, will fall by its own weight, either breaking in two or tearing away from its attachment. This tendency is very commonly resisted in buildings and other structures by

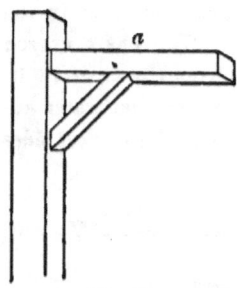

Fig. 36.

braces, as represented in Fig. 36. Here the weight of the horizontal timber at some distance on each side of a is made to press upon the upright post instead of directly downward.

The above Principles Transgressed by Man.— Man often transgresses these principles in his structures. For example, a building settles because the foundation is not strong enough to bear the superincumbent weight. In other words, the force of gravitation is not sufficiently taken into account. When a very tall building is erected, the lower portions ought to be made of very cohesive substances. Firm granite is therefore an appropriate material for the lower story of tall brick buildings. At least the walls of the lower stories of such buildings should be made thicker than usual, to resist properly the force of gravitation in the weight above. Stores intended to bear much weight on their floors are often built without due regard to the cohesive force required to sustain the weight. Long timbers are sometimes supported only at the ends, when their own weight, to say nothing of what may be brought to press upon them, requires that they should be supported at other points. While in modern buildings the timbers are often too small, in some old buildings the upper timbers are so heavy as to lessen rather than increase the strength of the structure. Especially is this true of the unsightly beams which in some very old houses we see extending along the ceilings. Many other examples could be given, but these will suffice.

42. Adhesion, Cohesion, and Gravitation the Same.—We again refer to the statement made in § 24, that cohesion, adhesion, and gravitation are only different modes of action of the same power, viz., the attraction which matter everywhere has for matter. At first thought it would appear that there is something peculiar in the attraction of particles when they are brought together so as to adhere. For if we take any substance—a piece of glass, for example—its particles seem to be held together by an attraction vastly stronger than that attraction which inclines different bodies to move towards each other. If you break the glass, however closely you may press the two pieces together, they will not unite again. It would seem, at first view, that there must be some peculiar arrangement of the particles which is destroyed by breaking the glass. But we can readily account for the facts in another way. The attraction between bodies of matter is greater the nearer we bring them together. The nearer, for example, the moon is to any portion of the earth, the greater the attraction which it exerts, as seen in the tides; and if it were much nearer to the earth than it is, our tides would prove very destructive. What is true of masses is also true of the particles of which they are composed. Though their attraction is comparatively feeble when at a distance from each other, it increases not in the arithmetical, but the geometrical ratio as they approach; so that when they are exceedingly near together the attraction is very powerful. It must be remembered in regard to the pieces of broken glass that you cannot bring the particles on their surfaces as near as they were before the glass was broken; and the attraction being inversely as the square of the distance, a little distance must make a great difference. The particles of some substances you can bring so near together as to cause adhesion, as in the case of the two bullets (§ 30). That their adhesion depends merely upon their particles being brought near to each other appears from the fact that the smoother you make the surfaces, the more strongly will they adhere. And the reason that liquids and semi-liquids adhere so readily to solid substances is that their particles, moving freely among each other, have thus the power of arranging themselves very near to the particles of the solid. Thus, when a drop of water hangs to glass, all the particles of water in that part of the drop next to the glass touch, or rather are exceedingly near to, the particles of the glass.

D

43. Chemical Attraction.—The kinds of attraction hitherto explained in this work belong to the study of Natural Philosophy, but another kind, known as *interatomic* or chemical attraction, is capable of producing the most wonderful effects. The former produce chiefly mechanical effects, while chemical attraction goes farther and affects the composition of substances. For example, the attraction between the two gases oxygen and hydrogen, which makes them combine to form water, belongs to Chemistry; while that which makes the particles of water cohere is in the province of Natural Philosophy. You will learn more about chemical attraction in Part II. of this series.

Variety in the Results of Attraction.—It is one and the same force, then, which binds the particles of a pebble together, and makes it fall to the ground—which " moulds the tear" and "bids it trickle from its source"—which gives the earth and all the heavenly bodies their globular shape, and makes them revolve in their orbits. How sublime the thought that one simple principle which gives form to a drop extends its influence through the immensity of space, and so marshals "the host of heaven" that, without the least interruption or discord, they all hold on their course from year to year and from age to age! Thus Omnipotence makes the simplest means produce the grandest and most multiform results.

QUESTIONS.

36. What is adhesion? Why can you not make the surfaces of broken glass adhere? Explain the cementing of glass. How may silver and gold be made to adhere to iron? What is said of the adhesion of tin and lead? —37. What is said of the adhesion of liquids to solids? What of the action of mercury? Describe an experiment showing the adhesion of solids and liquids. What is said of stems in stagnant water? Explain Figs. 23, 24, and 25. Explain Fig. 27.—38. Explain the rise of fluids in tubes by

Fig. 28. How does mercury act with respect to tubes plunged into it?—39. What is meant by *capillary* attraction? Give familiar examples of the rising of liquids in interstices. Describe and explain the process of getting out millstones. How does a blotter differ from writing-paper? Describe the experiment with a sewing-needle. What is said of certain water insects?—40. What is said of the various results of attraction? Explain fully why you can pour water from a pitcher easier than from a tumbler. Explain the operation of the quick movement by which you prevent water from running down the side of a tumbler in pouring it out. What is said of dropping from a vial? How is the size of drops limited? What is said of the movements of drops on window-panes? Why do the drops of different liquids vary in size? Give the illustration about chalk. Give that about dust.—41. Illustrate the limitation of size in solid masses. Show how these principles are transgressed by man.—42. Give the summary referring to the connection between the different ways in which attraction is exerted.—43. Wherein does chemical attraction differ from the other kinds? What is said of the variety in the results of attraction?

CHAPTER VI.

CENTRE OF GRAVITY.

44. Centre of Gravity Illustrated.—If you support a ruler on your finger as in Fig. 37, it balances when there is just as much weight on one side as on the other. Now just over your finger, in the middle of the ruler, there is a point called the centre of gravity; or, in other words, the centre of the weight of the ruler. This point is indicated in the figure. There is as much of the weight of the ruler on the one side of

Fig. 37.

this point as on the other, and also as much above it as below it. If your finger should be a little to the one side or
the other of this point, the ruler would not be balanced, and
would fall. When balanced, it does not fall, simply because
this central point is supported by being directly over the
end of the finger. The whole weight of the ruler, then,
may be considered as practically concentrated at that point,
for all the downward pressure of the ruler is there exerted. When the ruler is balanced on the finger as repre-

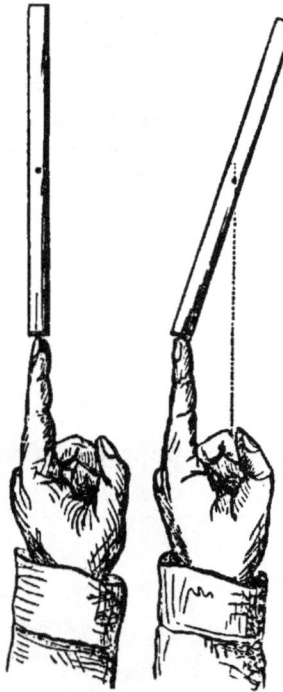

sented in Fig. 38, it will maintain
its position so long as its centre of
gravity is directly over the point
of the finger. If it be to the one
side or the other, as in Fig. 39,
it is not supported, and the ruler
therefore falls. You see, then, that
when a body is balanced, the centre of gravity lies directly *over* the
point of support. If, on the other
hand, a body is suspended, the centre of gravity is directly *under* the
point of support.

If a plumb-line from the centre
of gravity of any body could be
prolonged into the earth, it would
go directly to its centre. The body
may be considered as making all its
pressure from its centre of gravity

Fig. 38.   Fig. 39.   towards the centre of the earth, in

obedience to the attraction of gravitation. The best
definition, then, that we can give of the centre of gravity is, *that point in a body from which proceeds its
pressure as a whole towards the centre of the earth.* It
is that point, therefore, the support of which insures the

support of the whole body. And in speaking of the weight of a body, or its downward pressure, we may consider all the matter composing it as collected or concentrated in that point. The body, therefore, can be balanced in any position in which this point is supported, as shown in Figs. 37 and 38. When a body is suspended, it is at rest only when the centre of gravity is directly under the point of support. Thus, if you have a circular plate suspended at E, Fig. 40, it will not be at rest when moved to the one side or the other, as represented by the dotted lines, but only when the centre of gravity, c, is directly under the point E.

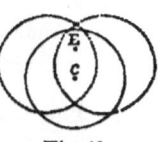

Fig. 40.

45. **How to Find the Centre of Gravity of a Body.**—If you take a piece of board, and suspend it at a, Fig. 41, and

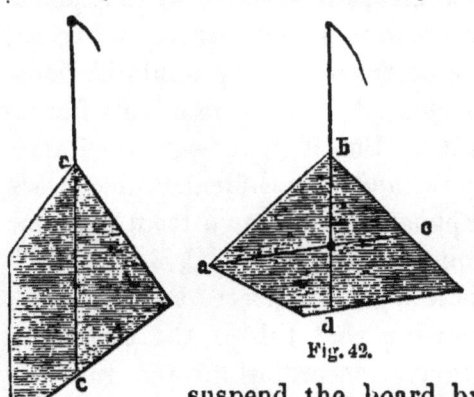

Fig. 42.

Fig. 41.

hang a plumb-line from the same point, the centre must be somewhere in that line. But exactly at what point it is you do not know. How will you ascertain this? Mark the line a c on the board, and suspend the board by another point, as in Fig. 42. Since the centre of gravity must be somewhere in the plumb-line as it now hangs, of course it is where the two lines ac and bd cross, and the board suspended by a cord attached at this point will remain evenly balanced.

Scales and Steelyards.—When two bodies are connected by a rod or bar, the centre of gravity of the whole is somewhere in the connecting rod. If the two bodies be equal in weight, as in Fig. 43, the centre of gravity is

exactly in the middle of the rod, as marked. But if the bodies are unequal, as in Fig. 44, the centre of gravity is nearer to the larger body than

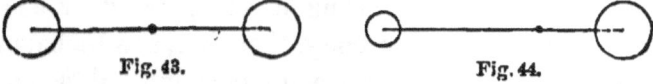

Fig. 43.   Fig. 44.

to the smaller. In weighing a body in one pan of a balance by means of
weights placed in the other, we have a case parallel to that of Fig. 43. The
centre of gravity of the body weighed, the weights and the pans, as a whole,
is midway between the scales, at the point of support. In the steelyard
the heavy body to be weighed is nearer the centre of gravity than the small
weight on the long arm, and so the case is similar to that of Fig. 44.

46. **The Centre of Gravity of a Body not Always in the
Body itself.**—The centre of gravity of a hollow ball of
uniform thickness is not in the substance of the ball, but
it is in the centre of the space within the ball, for the line
of the ball's downward pressure is situated at that point.
If the ball had a framework in it, as represented in Fig. 45,

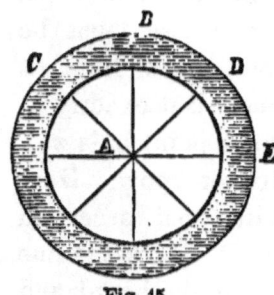

Fig. 45.

the centre of gravity would obviously be at A, the centre of this frame-
work. But if there were no frame-
work, and perpendicular lines were
supposed to be drawn from different
points of suspension, C, B, D, and E,
these would intersect at the point A,
showing that this is the centre of
gravity, according to the rule for
finding it given in § 45. In like manner, the centre of
gravity of an empty box, or of an empty ship, is an imaginary point in the space inside. In a hoop it is the centre
of the hoop's circle.

47. **The Centre of Gravity Seeks the Lowest Point.**—
The centre of gravity always assumes the lowest place
which the support of the body will allow. In a hanging
body, therefore, it is always directly under the point of
suspension. To reach one side or the other of this posi-

tion, it must rise. This the attraction of gravitation forbids, and if by any force it is made to rise, this attraction at once brings it back. This is manifest in the case of a suspended ball, Fig. 46. If the ball be moved to b, it will, on being let go, return to its first position, simply because its centre of gravity, in obedience to the earth's attraction, seeks the lowest place possible.

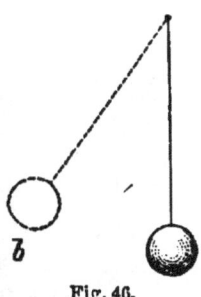

Fig. 46.

From inertia (§ 14), it moves beyond this point, and continues to vibrate back and forth for some time; but when its motion is stopped, it hangs perpendicularly; that is, in such a way that its centre of gravity shall have the lowest possible position. Many illustrations of this point might be mentioned.

When a rocking-horse is at rest, its centre of gravity is directly over the point at which it touches the floor, for in that position the centre of gravity is as low as possible. If it be rocked, the centre of gravity is moved to a higher point, and for this reason it rocks back again. The same is seen in the swing, the cradle, the rocking-chair, etc. Most interesting illustrations are found in the Loggan Stones, as they are called, several of which are seen on the rugged parts of the British coast. An immense rock, loosened by some of the forces of nature, rests with a slightly rounded base on another rock which is flat, and it is so nicely balanced that one person alone has sufficient strength to set it rocking. Similarly balanced "rocking-stones" are found near Salem, Massachusetts, in Great Barrington, Massachusetts, and in many other localities.

48. **Further Illustrations.**—An egg lies upon its side because the centre of gravity seeks the lowest position. When on its side, the centre of gravity is at its lowest point, as is manifest by a comparison of Fig. 47 with Fig. 48. Children have a toy, called a witch, which illustrates the same thing in another way. It is constructed of some light substance, as pith, with a shot or bullet fastened in one end. It always stands up on its loaded end, and can-

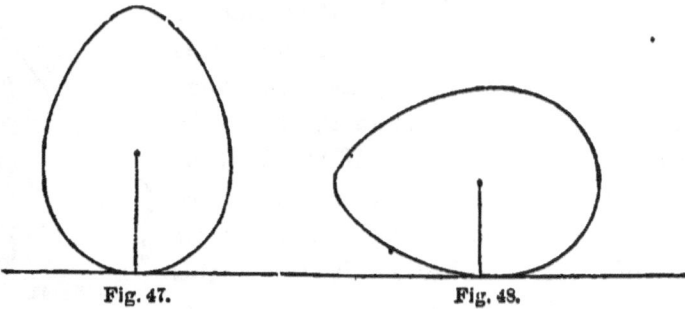

Fig. 47.   Fig. 48.

not be made to lie down on its side, because the centre
of gravity would not then be at the lowest point. The
figure of a fat old woman, Fig. 49, loaded with lead at
the bottom, is another form of this toy. If the figure be
thrust over to one side, as shown by the dotted lines,
the centre of
gravity is
raised, and
the upright
position is at
once resumed.

Fig. 49.

Fig. 50.

If the toy were not loaded, it would
lie in the position represented in Fig.
50, just as the egg lies on its side.

Certain curious cases which at first sight appear to be
exceptions to this law are really interesting proofs of it.
If a light wooden cylinder loaded with lead on one side
be placed upon an inclined plane with the lead in the position e (Fig. 51), the lead, in fall-
ing to c, will cause the cylinder to
roll up the incline. What is apparently a rolling up-hill is really
a falling of the centre of gravity.

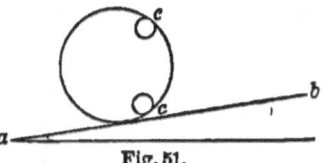

Fig. 51.

For the same reason a billiard-ball placed on the smaller
ends of two billiard-cues laid on a table with their points

in contact and their larger ends slightly separated, will roll towards the large ends, apparently rolling up-hill; actually, however, the centre of gravity falls as the ball rolls along the cues.

Fig. 52.

Curious Experiments.—You cannot hang a pail of water on a stick laid upon a table as represented in Fig. 52, for the centre of gravity is not supported. But if you place another stick as a brace, in the manner represented in Fig. 53, so as to push the pail under the table, it will hang securely, because the centre of gravity will be brought under the point of suspension. In an entirely similar manner a needle passed through a cork into which a fork is thrust (as shown in Fig. 54), may be suspended on the edge of a table. We have another illustra-

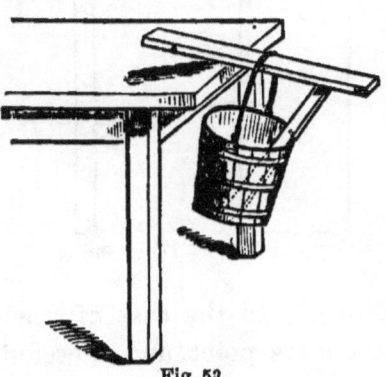

Fig. 53.

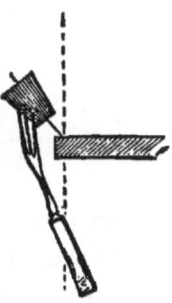

Fig. 54.

tion in the common toy represented in Fig. 55. The horse, made of very light material, stands securely, because the centre of gravity of the whole is in the heavy ball, which is under the point of suspension. If the horse be made to rock back and forth, the centre of gravity in the ball moves in a curved line, as in the case of a ball suspended by a string (Fig. 46). It is at its lowest place only when the horse is at rest. The hanging of a cane with a hook-shaped handle on the edge of a table is to be explained in the same way.

Fig. 55.

D 2

- 49. Stability of Bodies.—The firmness with which a body stands depends upon two circumstances—the height of its centre of gravity and the extent of its base. The lower the centre of gravity, and the broader the base, the firmer the body stands. A cube, represented in Fig. 56, is more stable —that is, less easily turned over—than a body shaped like that in Fig. 57, because it has a larger base. The contrast is still greater between Figs. 56 and 58. The reason of the stability of a body with a broad base is found in the fact that in turning it over the centre of gravity must be raised more than in turning over one of a narrower base. The curved lines indicate the paths of the centres of grav-

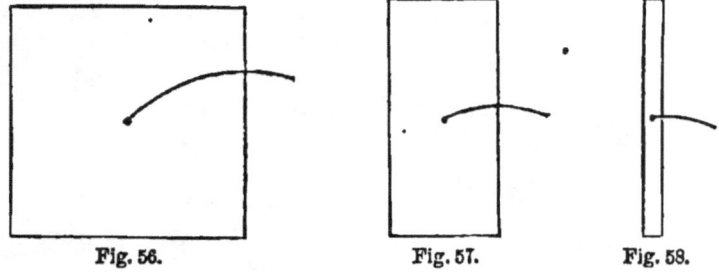

Fig. 56.   Fig. 57.   Fig. 58.

ity as the bodies are turned over. In the case of a perfectly round ball, the base is a mere point, and therefore the least touch turns it over. Its centre of gravity does not rise at all, but moves in a horizontal line, as shown in Fig. 59. The pyramid is the firmest possible structure, be-

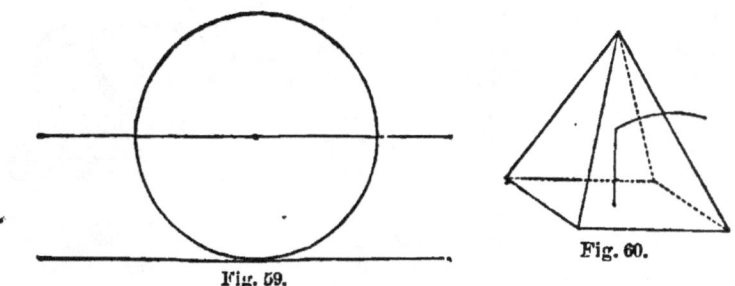

Fig. 59.   Fig. 60.

cause it possesses in the highest degree the two elements a broad base and a low position of the centre of gravity. On both these accounts the centre of gravity must ascend considerably when the body is turned over, as shown in Fig. 60.

50. Unstable Bodies.—When a body does not stand upright, its stability is diminished because only a portion of the base is concerned in its support. In Fig. 61 the base is broad, but the body is so far from being upright that the centre of gravity bears upon the very extremity of the base on one side, as indicated by the perpendicular line. A small force

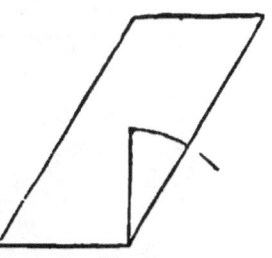

Fig. 61.

will turn it over, because the centre of gravity need not ascend the least when this is done. You see, then, that the less upright a body is, the less of the base is of service in its support. The famous tower of Pisa, Fig. 62, one hundred and thirty feet high, overhangs its base fifteen feet. Whether it was built intentionally in this way to excite wonder and surprise, or whether it settled on one side after its completion, has long been discussed; we are inclined to the former theory, for what would otherwise have been a very unsafe structure is rendered stable and safe by the arrangement of its materials. Its

Fig. 62.

lower portion is built of very dense rock, the middle of brick, and the upper of a very light porous stone. In this way the centre of gravity of the whole structure is made to have a very low position.

Familiar Illustrations. — You are now prepared to understand a fact which common experience teaches every one, that the taller a body, and the narrower its base, the more easily is it overturned. This is exemplified in the two loads, Fig. 63. The base is the space included by the wheels. The centre of gravity is so high in the tall load that a perpendicular line drawn from it falls outside of the base if the cart reaches a considerable lateral inclination of the road. But the smaller load, under the same circumstances, is perfectly secure. A high carriage is more easily overturned than a low one, for the same reason. A stage, if overloaded on its top, is very unsafe on a rough road. Stability is given to articles of furniture by making their bases broad and heavy, as shown in tables supported by a central pillar, candlesticks, lamps, etc. The tall chairs in which children sit at table would be very insecure if the legs were not widely separated at the bottom, thus widening the base of support.

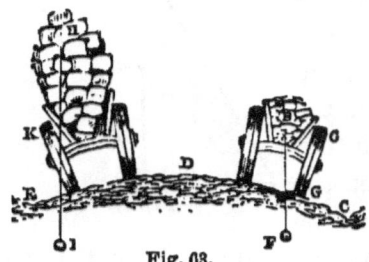

Fig. 63.

51. Support of the Centre of Gravity in Animals.—The base of support which quadrupeds have, viz., the space included between their four feet, is quite large; and this is one reason why they are able to walk while yet very young. A child does well who can walk at the end of ten or twelve months, for the supporting base is quite small compared with that of a quadruped. It requires skill, therefore, in the child to manage the centre of gravity in standing and walking, and this is gradually acquired. It is on account of the smallness of the base furnished by the feet that the statue of a man is always made with a large base or pedestal. Although we exert considerable skill in walking, it is by no means so great as that which the Chinese ladies require with their painfully small feet. Still more skill is exercised by one who has two wooden legs, or one who walks on stilts. The base made by the feet can be varied much by their position. If the toes be turned out and the heels brought near to each other, the base will not be so large as when the feet are straight forward and far apart, as is manifest in Figs. 64 and 65. It is for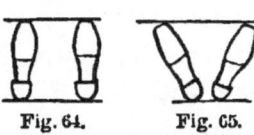

Fig. 64.   Fig. 65.

this reason that the child, in his first attempts at standing and walking, instinctively manages his feet as in Fig. 64.

52. **Motions of the Centre of Gravity in Walking.**—In walking, the centre of gravity is alternately brought over one foot and the other, and so moves in a waving line. This is very manifest as you see people before you going down the aisle out of a church. When two are walking together, if they keep step the two waving lines of their centres of gravity run parallel, as in Fig. 66, and they walk easily;

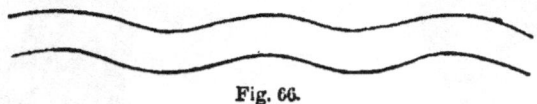

Fig. 66.

but if they do not keep step these lines run as in Fig. 67, and the movement is both awkward and embarrassing.

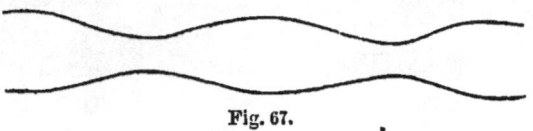

Fig. 67.

This line of movement of the centre of gravity is always slightly waving *upward* also, as seen in Fig. 68. In the

Fig. 68.

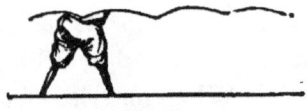

Fig. 69.

case of a man with wooden legs, the line would not be gently waving, but somewhat angular, as represented in Fig. 69.

53. **The Centre of Gravity and Attitudes.**—The object of various attitudes assumed under different circumstances is to keep the centre of gravity over the base of support. A man with a load on his back would not stand straight, but would assume the position of Fig. 70, so that the centre of gravity of his load may be directly over his feet. A man carrying a pail of water in his left hand leans to the right, and raises his right hand in order to bring the centre of gravity over his feet (Fig. 71). In ascending a hill a man appears to lean forward, and in descending to lean

Fig. 70.   Fig. 71.

backward; but in fact he is in both cases upright in reference to the plain on which the hill stands. A perpendicular line drawn from his centre of gravity strikes the ground midway between his feet, that is, in the middle of the base, and if prolonged would go straight to the centre of the earth. When one rises from a chair he draws his feet backward, and then bends his body forward to bring the centre of gravity over the feet. Unless this is done, it is impossible to rise, at least deliberately, as you will find by trying the experiment. A man standing with his heels close to a wall cannot stoop forward and pick up anything, for the wall prevents him from moving any part of his body backward, and therefore when he stoops forward, the centre of gravity being brought in advance of the base, he loses his balance and falls. A man who did not understand this undertook to stoop in this way to pick up a purse containing twenty guineas, which he was to win if he succeeded, the forfeiture in case of failure being ten guineas. Of course his lack of knowledge as to the principles of the centre of gravity made him lose his wager.

Great skill is exhibited by the rope-dancer in supporting the centre of gravity. He carries a long pole in his hands, loaded at each end, and when he inclines to one side he throws it a little towards the other side, that the reaction may restore his balance. Similar skill is seen in feats of balancing, as, for example, in balancing a long stick upright on the finger. In these cases the centre of gravity is very little of the time directly over the point of support. It is kept in constant motion nearly but not quite over this point—this unstable equilibrium, as it is called, being vastly less difficult to maintain than stable equilibrium; that is, keeping the balance in one unvarying position. It is the motion of the top that makes it stand upright upon its point—a very beautiful example of unstable equilibrium. The centre of gravity revolves around a perpendicular line, at exceedingly little distance from it at first, but greater and greater as its motion becomes less rapid, till at length the centre of gravity gets so far from this line that the top falls. For a similar reason an intoxicated man may not be able to keep himself up if he undertakes to stand still, and yet may do so if he keep moving.

54. Centre of Gravity in Floating Bodies.—The same principles which apply to the centre of gravity in bodies standing on a firm basis apply also to floating bodies. That the centre of gravity may be low in a loaded vessel the heavy part of the cargo is put underneath, and generally ballast of stone or iron is necessary for the same purpose. In large

flat-boats, the base of support being extensive, there is not the same need of taking care that the centre of gravity be low. If a ship be laden in part with an article which will dissolve in water, there is much danger, if the ship should leak, lest this portion of the cargo be dissolved and pumped out with the bilge-water; this would alter the trim of the vessel by removing the centre of gravity from over the middle line, and bringing it too far forward or carrying it too far back, making the ship wholly unmanageable. Four large English ships, in part loaded with saltpetre, were supposed to have been lost from this cause in 1809 off the Isle of France. The immense ice-islands, or icebergs, which float about in summer in the polar regions, by melting irregularly often change the place of their centre of gravity, and in turning over present one of the most sublime spectacles in nature. A mountain of ice, extending high in the air and deep in the sea, suddenly turns over, and produces a rolling of the ocean which is often felt at the distance of many leagues.

. QUESTIONS.

44. Show what we mean by the centre of gravity by Figs. 37, 38, and 39. Give the definition of centre of gravity, and explain it. What is shown by Fig. 40?—45. How can we find the centre of gravity of a body? What is said of scales and steelyards?—46. State what is represented by Fig. 45.—47. Illustrate the fact that the centre of gravity seeks always the lowest point. Give the illustrations of the rocking-horse, the swing, etc. What is said of the Laggan Stones?—48. Why does an egg lie on its side? Give the illustrations from toys. Explain how a ball may be made to roll up an incline. Describe the experiment with a pail. And that with a toy horse.—49. Upon what two things does the stability of a body depend? What is said of the stability of bodies whose shapes are represented in Figs. 56, 57, and 58? What of that of a round ball? Why is the pyramid the firmest of all structures?—50. What is the relation of upright position to stability? What is stated of the tower of Pisa? Give some familiar illustrations.—51. What is said of the support of the centre of

gravity in animals? What is said of the skill exercised in walking? What of the mode of walking in a child?—52. What of the motions of the centre of gravity in walking? What is said of the walking of a man with wooden legs?—53. Illustrate the management of the centre of gravity in different attitudes. Describe and explain the way in which one rises from a chair. State and explain the wager case. What is said of unstable equilibrium? Give the illustrations.—54. What of the centre of gravity in floating bodies? What is said of icebergs?

CHAPTER VII.

MOTIONS OF MATTER.

55. Matter, Motion, Force.—When a ball is rolled over the floor a superficial observer sees but little occasion for scientific discussion, but a philosophical mind finds therein a symbolic illustration of certain phenomena of nature ever present with us, and a comprehension of which is of the highest importance. Were the sentence, "A ball rolls," critically analyzed by a student of grammar, he would tell us of the three parts of speech represented, and of their relation to each other; in like manner the student of the laws of nature, analyzing the same sentence, would tell us that it embodies three facts, and that their mutual relations, intelligently studied, cover the whole groundwork of Natural Philosophy. We will endeavor to explain our meaning by dissecting the sentence.

"A ball rolls" leads us, in the first place, to consider the ball itself; the phrase being indefinite, we have no information as to the material of which the ball is made, whether of wood, iron, or rubber; whatever the substance may be, it is called *matter*, as explained in Chapter I. The ball, then, abstractly considered, is merely an indefinite quantity of matter having a spherical form, the latter idea being associated with the term ball itself.

In the second place, this sentence leads us to regard the ball as changing its place with reference to some other body not mentioned. The ball "rolls," *i. e.*, is in motion, or moves from one place to another in a particular manner known as rolling. *Motion*, then, in the abstract, is a *change of place*, and is so defined. Now we have already learned that matter is of itself inert (§ 14), and cannot put itself in motion, hence there must be some external cause for the rolling of the ball; and this leads us to the third point, viz., the idea of *force*. That which causes motion in matter is called force: the *force* of the explosion of gunpowder sets the bullet (*matter*) in *motion;* the *force* of a violent wind uprooting a tree also sets *matter* in *motion*. A clear understanding of these three phenomena, and of their mutual relations as governed by laws, is essential to the study of Natural Philosophy, and their discussion will occupy us throughout this Chapter.

Force.—We have already stated that force is that which tends to move matter. When a body begins to move, changes the style of motion it acquired, or ceases to move, it is the result of one or more forces acting upon it from without. Force is not an attribute of matter like divisibility or hardness (Chapter II.), but merely a tendency to put it in motion; we say a *tendency*, because force may exist where there is no actual motion. Thus a huge rock may rest quietly for years on the sloping hill-side, prevented from moving by a small quantity of earth in front of it, but let this obstacle be removed by shovelling, or by a sudden flood of water, and the rock will roll down the hill with immense force, crushing everything in its path. The magnet, about which you will learn more in Chapter XX., affords another illustration of the correct idea conveyed by the word force. A magnet has the power of attracting to itself pieces of iron; if the magnet lies on the table, and no iron objects are near it, the fact of its possessing this peculiar force is not apparent—the force is sleeping as we may say; but bring near to the magnet some iron nails or some steel filings, and this sleeping force is aroused, and manifests itself by drawing the iron articles towards the magnet.

56. **Motion Universal.**—The material universe is in ceaseless motion. The rising and setting of the sun, the changes of the seasons, the falling of the rain, the running of rivers into the ocean, the ascent of water into the air by evaporation, the wind moving in silence or rushing on in its might, are familiar examples of motion constant and everywhere present. But with all this motion, sometimes in conflict and often variable, order and regularity reign. The forces causing motion, though various in their operation, are kept by the Creator from producing confusion and disorganization by a few simple laws, which regulate the movements both of atoms and of worlds.

The principal of these causes of motion are the forces mentioned in Chapter IV.; we will briefly recapitulate them. Attraction is the most universal of the causes of motion in the universe. While it binds atom to atom, it also binds system to system throughout the immensity of space; and while it makes the stone fall to the ground, it moves the countless orbs forever onward in their courses. It is this which causes the tides to flow and the rivers to run down their slopes to the ocean, and thus by keeping up the never-ending motion of water all over the earth in seas, lakes, rivers, and the millions of little streamlets, diffuses life and beauty over the vegetable world, and gives to man the vast resources which we see developed in the numberless applications of water-power and navigation.

Heat is everywhere uniting its influence with the other forces to cause motion. It is heat that produces all the motions of the air, termed winds. It is heat that causes the rise of the water all over the earth in evaporation, so that it may be collected in clouds, again to descend to moisten the earth and keep the ever-flowing rivers full.　Heat applied to water gives to man one of his best means of producing motion in machinery.

Light and electricity are also manifestations of this universal force, as will be shown in Chapters XVII. and XVIII.; these are, to a certain extent, productive of motion.

The agencies which Chemistry reveals to us are ever at work causing motion among the particles of matter; and though they generally work in silence, they sometimes show themselves in tremendous explosions, and in convulsions of nature.

Busy life is everywhere producing motion, more especially in the animal world.　It gives to the myriads of animals, great and small, that swarm the earth not only the power of moving themselves, but also the power, to some extent, of moving the material world around them.

57. **Varieties of Motion.**—The different kinds of motion have received distinguishing names; the following list embraces the principal varieties, with examples taken from familiar sources:

Varieties of Motion.	Examples.
Slow	The sun's shadow.
Swift	Lightning.
Straight	A stone dropped into a well.
Curved	The path of a stone in the air.
Uniform	The hands of a clock.
Variable	Winds, animal motions, etc.
Accelerated	Gradually increasing motion.
Retarded	Gradually diminishing motion.

Whether motion is slow or swift is altogether a relative matter; a boy may run very swiftly, yet he moves slowly compared with a race-horse, and the horse in turn cannot compete with the locomotive, while the speed of the latter is as nothing compared with the inconceivably rapid motion of electricity.　The rate of motion is called *velocity*, and it is measured by the space traversed in a given time.　Velocities are compared by reference to the distance travelled in one second, taken as a standard of time, very swift velocities being expressed in miles per second, and slower ones in feet per second; this will be understood by examining the following table:

TABLE OF COMPARATIVE VELOCITIES.*	Miles in one second.
Light..	192,500
Electricity....................................not less than	200,000
Electric currents in telegraph wires.........................	12,000

	Feet in one second.
Relative motion of the sun in space..........................	205,920
Mean rate of the earth's centre in its path around the sun...	101,061
Sound traversing solid bodies................................	11,280
A 24-pound cannon-ball (maximum)........................	2,450
Rifle-ball (maximum)...	1,600
A point at the surface of the earth under the equator........	1,525
Volcanic stones projected from Etna..........................	1,250
A point at the earth's surface, latitude of London...........	950
The most violent hurricane	146
Flight of a swallow..	134
A hurricane	117
Locomotive running 65 miles per hour......................	95
An ordinary race-horse	42
Flight of a crow...	37
A brisk wind...	26
The fastest sailing vessel....................................	15
A carriage travelling six miles an hour..............nearly	9
A man walking...	6

Straight motion is one which does not change its direction at any point; *curved* motion, on the other hand, is continually changing its direction. These require no special explanations, but to the latter we shall refer again.

When a moving body passes over equal distances in each second of time, it is said to have a *uniform motion.* Our standard of uniform motion, with which we compare and

* Condensed from Arnott's "Elements of Physics," Seventh Edition.

measure all other motions, is that of the earth round its own axis. "Here we have a huge spinning-top, which, not for hours or days, but for unknown ages, has kept up its original speed practically undiminished. All our notions of *time* are based on the regularity with which the earth turns round."

Watches and clocks are contrivances for obtaining a uniform motion which can be compared with that of the earth, and for marking off smaller intervals than can be conveniently observed in the revolution of the earth. There are very few cases of uniform motion in the world, other forces than that which started the uniform motion constantly retarding or otherwise modifying it. Motion which is not uniform or regular is said to be *variable.* In determining the velocity of a body having a variable motion, we must observe the rate of motion at various equal intervals of time, and average them. The distance traversed by a sailing vessel or steamer is ascertained by frequently " throwing the log," by means of which the speed at definite times is obtained, and calculating the average velocity.

When a moving body passes over gradually diminishing distances in equal intervals of time, its motion is said to be *retarded.* Examples of uniformly retarded motion are familiar: a ball rolled along the ground moves more and more slowly under the influence of gravitation and the resistance offered by the air until it finally comes to rest. A train detached from a locomotive has its motion uniformly retarded owing to the same causes; should brakes be applied and then suddenly released, its motion would also be retarded, but not uniformly. Opposed to retarded motion is *accelerated* motion, in which the velocity of a body continually increases until external forces bring it to rest; as, for example, when a stone is dropped from a height, it falls 16 feet in the first second, 64 feet in the next second, 144 feet in the third second, and so on, the motion being *uniformly accelerated*, owing to the action of gravity.

58. **Motion and Rest.**—Though we use the term rest in opposition to motion, it is obvious from some of the illustrations given that rest is only a relative term, for not a particle of matter in the universe is at rest. When we are sitting still we call ourselves at rest, though we are moving every hour, by the revolution of the earth on its axis, 1000 miles eastward, and 68,000 miles in our annual journey round the sun. Why, then, are we so insensible to these rapid motions? It is partly because the motions are so uniform, but chiefly because all things around us, our houses, trees, and even the atmosphere, are moving along with us. If we were moving along alone, even at a slow rate, while all these objects were standing still, we should be conscious of our motion, as when we ride along in a carriage objects at the roadside do not appear to move along with us.

This can be made more clear and impressive by a familiar comparison. A man on board of a steamboat, by confining his attention to things within the boat, may, after a while, be almost unconscious of the boat's moving, if the water be smooth, though the boat may be going at the rate of fifteen miles an hour. If he be reading in the cabin, he will think as little of his motion as he would were he reading in his parlor at home. Should he be blindfolded, and turned around a few times, it would be impossible for him to tell the direction in which the boat is going. Now the case is similar with a man on the earth—he is unconscious of the motion of the earth for the same reason that the man in the boat is unconscious of the boat's motion. All objects around him are moving along with him, as the objects around the man in the cabin of the boat are moving along with him. We can carry the parallel farther. While the man sits in the cabin he knows not how fast the boat moves, nor even whether it moves at all. He must look out to decide this, and even then he may not be able to tell whether the boat moves, or whether he merely sees the water running by it. We are often actually deceived in this respect. A steamboat struggling against wind and wave may appear to those on board to be advancing when it is really stationary, or even when it is losing ground. So when we look at the sun, we know not whether it is the sun or the earth that is moving. Mere vision, without reasoning on the subject, leads one to think that it is the sun that moves.

59. **Absolute and Relative Motion.**—The motion of a body is said to be *absolute* when it is considered without relation to the position of any other body. Its motion is said to be *relative* when it is moving with respect to some other body. Absolute rest is unknown, for no spot in the universe is known to be without motion. But a body may be relatively at rest, that is, in a fixed relative position to other bodies. Every body is in a state of absolute motion, and yet it may be in a state of relative rest. All objects that appear to us to be at rest have a very rapid absolute motion. They appear to be at rest merely because they have the same rapidity and direction of absolute motion that we have ourselves. And all the motions which are apparent to the eye are only slight differences in the common absolute motions, of which, though they are so exceedingly rapid, we are entirely unconscious. Thus, if you stand still, and another at your side walk at the rate of three miles an hour eastward, you both have a common absolute motion of 1000 miles in every hour, and he merely adds three miles to his thousand—you move 1000 miles, and he 1003. So if you sit still in your parlor, and your friend travel eastward at the rate of 20 miles an hour, you move every hour 1000 miles, and he 1020. And if he travel westward at this rate he really travels more slowly than you do—he has an absolute motion eastward of 980 miles, while you move 1000. At the same time you are both whirling on in the annual journey around the sun at the rate of 68,000 miles an hour.

60. **Compound Motion.**—From what has been shown in the preceding section, it is evident that a body may partake of two motions at one and the same time, and these motions

may be in the same direction or in different directions. If a man travelling on a steamboat walk towards the bow, he will move forward with the boat at the same time; suppose the steamboat moves at seven miles an hour, and he walks at the rate of three miles an hour, during the time that he passes from the stern to the bow of the boat his total velocity will be ten miles per hour; if, on the other hand, he turn about and walk back to the stern at the same rate, his velocity will be only four miles an hour. That is, if we refer his motion to the banks of the river on which the steamboat moves; but if we refer his motion to the steamboat only, his velocity will be three miles an hour, no matter in what direction he walks. This is an example of simultaneous motions in the same direction; we will now give one of simultaneous motions in different directions.

If a man attempt to row a boat straight across a swiftly running river, he will reach a point not directly opposite to that from which he started, but below. Two forces act upon the boat: the current tending to carry it straight down the stream, and his rowing tending to carry it straight across. The boat will go in neither of these directions, but in a line between them. Let A B, Fig. 72, represent the bank of the river, from which he starts at A, with the bow of the boat pointing to C, on the opposite bank. Suppose, now,

Fig. 72.

that in the time that it takes him to row across the current would carry him down to B if he did not row at all. He will in this time, by the two forces together, reach the point D, opposite to B, his course being the line A D. If the wind blow upon a vessel in such a way as to carry it eastward, and a current be pushing it southward, the vessel will run in a middle line, viz., southeast. For the same reason, if a boy kick a foot-ball already in motion, it

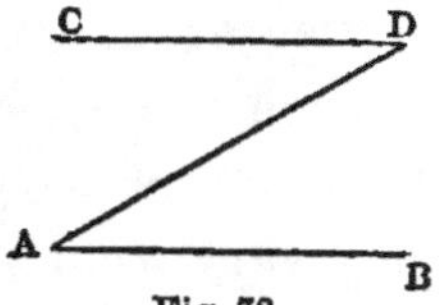

E

will not be carried in the direction in which he kicks it, but in a line between that direction and the direction in which its former motion was carrying it. In swimming, flying, rowing, etc., we have examples of compound motion, the middle line between the directions of the forces always being taken by the body moved.

If we take Fig. 72, illustrating the movement of the boat,

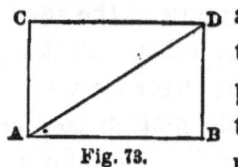
Fig. 73.

and draw two lines, one from A to C and the other from B to D, we shall have the parallelogram A C D B, Fig. 73, in which the line A C represents the force of the rowing, A B the force of the current, and A D the path of the boat. You see, then, that if we wish to find in what direction and how far in a given time a body acted upon by two forces will move, we are to draw two lines in the direction of these forces, and of a length proportionate to the distances to which they would move it in that time; then by drawing two lines parallel to these we shall have a parallelogram, and the diagonal of this will represent the distance and the course of the moving body. If a body be acted upon by two equal forces and at right angles to each other, the figure described will be a square, as you see in Fig. 74. If they vary from a right angle, the figure will vary in the same proportion from the square figure, as seen in Figs. 75

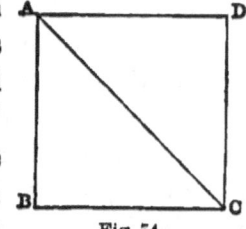
Fig. 74.

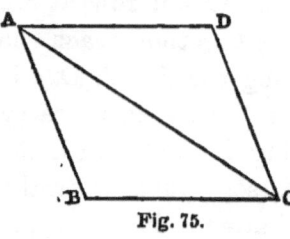
Fig. 75.

and 76. In the three figures, A B and A D represent the two forces and A C the resulting motion. You observe by these diagrams that the nearer the two forces are to the same direction, the farther will they move the body. This is shown by the different lengths of the diagonals in Fig. 74 and Fig. 76. The more nearly, therefore, the

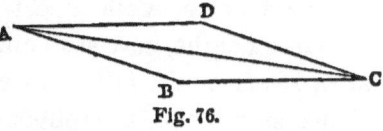

Fig. 76.

wind coincides with the current, the more rapidly will a vessel be carried along before the wind. When, on the other hand, the angle at which two forces act upon a body is much greater than a right angle, they will propel it but a small distance. Thus, if two forces act on a body at D in the directions D A and D C, Fig. 77, they will move it

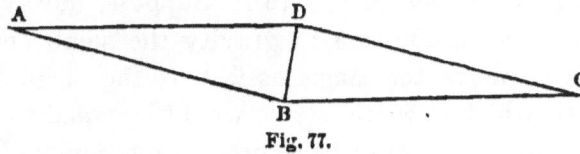

Fig. 77.

only the distance represented by the diagonal D B. This diagram represents the motion of a vessel sailing almost directly against a current by a wind the force of which is equal to that of the current, while Fig. 76 represents the motion of a vessel where wind and current, being of equal force, very nearly coincide. In the above diagrams we have supposed the forces to be equal; but the same truth can be shown in regard to unequal forces.

61. **Momentum.**—The momentum of a body is its quantity of motion. In estimating the momentum of any body two things must be considered—its velocity, and its quantity of matter or weight. A bullet fired from a gun has a vastly greater force, or power of overcoming obstacles, than one thrown by the hand, owing to its greater velocity. Now suppose the weight or quantity of matter to be increased ten times, and that it moves with the same velocity as before; it will have ten times as much force as before, and will overcome ten times as great an obstacle. For this reason, a small stone dropping upon a man's head may do

but little harm, while one ten times as large, falling from the same height, may stun and perhaps kill him. But if the large stone could fall with only one tenth of the velocity of the small one, the effect of both would be the same. The rule for calculating the momentum of a moving body is to multiply its weight by its velocity. Using the above illustration for an example, suppose the weight of the small stone be 1 ounce, and that of the large one 10 ounces. If they fall from a height of 16 feet, the force with which the large one will strike will be expressed by 160 (16 × 10), that of the small one by 16 (1 × 16). Suppose, however, that by some force in addition to gravity the small one could be made to move ten times as fast as the large one, the force with which it would strike would be equal to that of the large one, and would be expressed by the number 160.

We will illustrate this in another way. Let a and b, Fig.

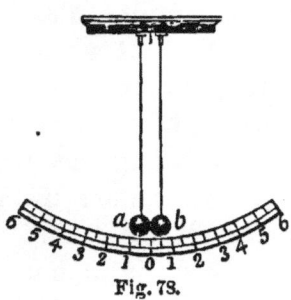

Fig. 78.

78, be two balls of clay of equal size hanging over a graduated arc. Now if b be let fall from the top of the arc 6, on striking against a it gives half of its motion to a, and they both move on together. But how far will they go? To 3 on the other side of the arc. Why? Let the quantity of matter in each ball be called 1, and the motion of b 6. The momentum will, therefore, be 6. Now the momentum of the two together will be the same after the blow as that of b was before it. But the quantity of matter is twice as great, and must be called 2. Therefore the motion must be represented as 3, to make the momentum 6 (2 × 3). But suppose that b is twice as large as a. Falling from 6, its momentum would be represented by 12 (2 × 6). After it has struck a, the momentum of the two together would be the same as that of b before the stroke;

but the quantity being 3, the motion would be represented by 4. They would, therefore, move to 4 on the arc.

Examples.—A few examples illustrating momentum, as dependent upon weight and velocity, will suffice. If a musket-ball of an ounce weight were so far spent as to move with a velocity of only a foot in a second, its force would be so small that if it hit any one it would do little harm. But a cannon-ball weighing a thousand ounces moving at this slow rate would have a very great force—equal, in fact, to the momentum of an ounce ball moving 1000 feet in a second.—If a plank push a man's foot against a wharf, he will scarcely feel it; but if the plank, instead of being alone, is one of a thousand planks fastened together in a raft, and the whole move with the same velocity, the force will be increased a thousand-fold, and the plank will crush the foot. And if the one plank, when alone, should move a thousand times as fast as the whole raft, the same result would follow.— So soft a substance as a candle can be fired through a board from the momentum given to it by an immense velocity.—Perhaps there is no better example of the great force given to a substance by an enormous velocity than we have in the wind. So light a thing is air that people think of it as almost nothing. But let it be set in rapid motion, and the velocity gives to it a force, a momentum, which will drive ships upon the shore, throw over buildings, and tear up trees by the roots. In this last example we see beautifully illustrated the meaning of the expression quantity of motion. In the moving air each particle does its share of the work in the destructive effects mentioned. Each particle, therefore, may be considered as a *reservoir* of motion, and the quantity of motion in any case depends upon the quantity which each particle has and the number of the particles.

62. Relation of Force to Velocity.—It would seem, at first thought, that the motion produced in any body must be in exact proportion to the force producing it; that is, that twice the force which produces a given velocity would double that velocity, and three times would treble it, etc. This is true where there is no resistance to motion, as in the case of the heavenly bodies moving in their orbits. But in all motions here upon the earth there is resistance; and the greater the velocity, the greater the resistance. If, therefore, you increase the velocity of any body, you not only have to communicate more motion to it, but you must overcome also the increased resistance. The rate of increase of force for increased velocities has been very accurately ascertained. A boat moving from B to A, Fig. 79, we will suppose, displaces a quantity of water represented

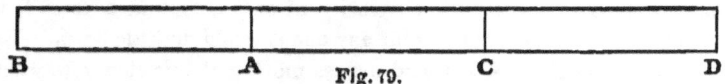

B   A   Fig. 79.   C   D

by the space between the two lines extending from B to A. Now if it move from B to C, it displaces twice the bulk of water B C; and as it is displaced in the same time that B A was, each particle is displaced with twice the velocity. Double the force is required to displace a double portion of water, and to do this with double the velocity the force must be doubled again. So if the boat be made to move three times as far in the same time—that is, from B to D— three times the quantity of water is displaced, and each of these three portions, B A, A C, and C D, is displaced with three times the velocity. The force required, then, to do this is nine times that required to carry the boat from B to A in the same time. It is plain, therefore, that with velocities represented by the numbers 1, 2, 3, 4, etc., the forces requisite to produce these velocities must be as the squares of these numbers; viz., 1, 4, 9, 16, etc. This law is a very important one, in a practical point of view. For example, it shows us how much larger a quantity of coal is required to produce in steamboats a high velocity than a moderate one. Its application, too, to the science of gunnery is important.

When the weight of a moving body is multiplied by its velocity, we obtain (§ 61) its momentum; when the weight is multiplied by the square of the velocity, we obtain the force with which a body strikes a resisting substance. This is directly deduced from the explanation just given.

63. **Accelerated Force.**—You have learned in § 57 some-

thing of the varieties of motion: these are obviously the result of the action of corresponding forces on matter. Thus we have uniform, accelerated, and retarded forces. If the momentum remain the same, independently of time, the force is uniform; if the momentum increase, the force is accelerated; and if diminished, the force is retarded. Were there no obstacles to motion, such as resistance of the air, etc., we might have uniform forces; but since we do not meet with absolutely isolated or free matter, all moving forces are more or less variable. Even in the production of very rapid motions the force is seldom instantaneously applied, but is rather gradual in its action; the motion is not the result of a single impulse, but a succession of impulses is required to accumulate sufficient momentum to overcome the resistance opposed. The action of gunpowder upon a bullet issuing from a gun is apparently an instantaneous and single impulse, but it is not really so. The great velocity given to the bullet is due to the continued impulse of the expansive force of gases produced from the powder, and it therefore depends much on the length of the barrel. If this be short, the force of the powder is not confined long enough to the bullet to give it a great velocity.

It is on the same principle of continued action that a man lifts his hammer high when he wishes to inflict a heavy blow. In this case both gravitation and the muscular power of the arm exert their force on the hammer through the whole space. A horse in kicking does the same thing, and by the great length of the leg the velocity given to the foot by this continued action of the muscles is very great. An arrow is not shot by a single momentary impulse of the bowstring, but the string, by following it through a considerable space, gives it a continued impulse.

One of the best examples of accelerated force is the attraction of gravity. You know that the greater the elevation from which a body falls, the greater is its velocity, and, therefore, the greater the force with which it strikes. Why is this? If it fell because of a single impulse drawing it towards the earth, this would not be the case; and if there were no air in the way, the velocity would be uniform. But the resistance of the air would retard the velocity; so that if a number of bodies should receive the same impulse at different elevations, the one farthest off would be most retarded, and, therefore, come down slower than all the rest. In this case, the higher the elevation from which a man should fall, the less would be the injury. But a body does not come to the ground by a single impulse, but by a succession of impulses, or, rather, a continued impulse. Every moment that the body is coming down it is drawn upon by the attraction of the earth, and this continued action causes an increase in the rapidity of motion.

Expressing this in somewhat different language, we may say that gravity is an accelerating force. Of this examples are innumerable: "A person may leap from a chair with impunity; if from a table, he receives a harder shock; if from a high window, a topmast of a ship, or the parapet of a high bridge, he will probably fracture bones; and if he fall from a balloon at a great height, his body will be literally dashed to pieces."

Water falling from a height acquires a power proportional to the elevation. The same is true of meteoric stones, which approach the earth with such immensely accelerated velocity that they become heated in their passage through the atmosphere, and bury themselves deep in the earth when they strike its surface.

64. **Gravity a Uniformly Accelerating Force.**—An accelerating force may be uniform or variable; gravity is not only an accelerating force, but it is uniform in its rate of increase.

A stone dropped from a height falls through a distance of 16 feet in one second, 64 feet in two seconds, 144 feet in three seconds, and so on. Now, after the stone has passed through the 16 feet—that is to say, at the end of the first second of time, its velocity is 32 feet per second; at the end of two seconds, 64 feet per second; at the end of three seconds, 96 feet per second, and so on. Thus the velocity at the end of 2, 3, 4 seconds is double, triple, quadruple, etc., that at the end of one second; that is, its rate of increase is uniform, viz., 32 feet per second.

This law holds good for all bodies, no matter whether they are heavy or light. At first sight it seems very paradoxical that a ball weighing one pound dropped from a height will reach the ground just as quickly as one weighing ten times or one thousand times as much. And yet such is the case; *gravity causes all unimpeded bodies to fall with equal rapidity, without reference to their weight.* Any one who throws a feather and a bullet into the air, however, observes that the bullet falls to the ground long before the feather, and will be disposed to dispute the statement just made. Such a one must remember that the resistance of the air must be taken into account, as we will now proceed to show.

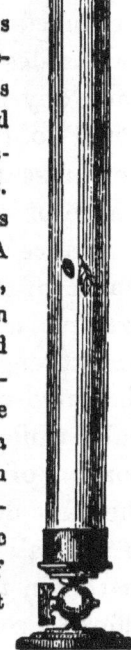

When a stone is thrown into the air, its upward motion is gradually destroyed by the attraction of the earth and the resistance of the air. Observe, now, why it descends. It is from the action of one of the causes which arrested its upward flight — the attraction of the earth. In its descent it is retarded by the resistance of the air, as it was in its ascent. This retardation is very obvious in the case of substances which present a large surface to the air, as a feather. A small piece of lead will outweigh many feathers, and, therefore, since its quantity of matter is so much greater in proportion to its surface than that of a feather, it will fall to the ground much more quickly. That this is owing wholly to the resistance of the air can be proved with the air-pump. Suppose that you have a tall receiver, Fig. 80, on the air-pump, and a piece of lead and a feather are placed at its upper part in such a way that they can be made to fall at the same instant. Exhaust the air, and then let them fall. They will go down side by side, as represented by the figure, and reach the bottom of the receiver at the same time, because there is no air to resist the progress of the feather.

The toy called the water-hammer illustrates the same thing. When water falls through the air, the resistance of the air tends

Fig. 80.

E 2

to separate its particles, as we see in the falling of water thrown up by a fountain. In the water-hammer, which is a closed tube containing a little water and no air, when the water is made to fall from one end to the other, as there is no air to divide it, it falls as one mass, and gives a sharp sound like the blow of a hammer. An instrument essentially like this can be made with a thin glass flask. Put a little water into it, and, after heating it to boiling over a spirit-lamp, cork the flask tightly, and then leave the water to cool. As all the space above the water was filled with steam when the flask was corked, it is a vacuum now that the steam is condensed.

65. Retarded Force.—We have seen (§ 63) that force is never instantaneously communicated, and that a succession of impulses are required to communicate motion. In like manner, no force can be instantaneously arrested, and a gradual resistance to motion is necessary to make it disappear. Examples showing the gradual nature of the retardation of force are numerous. It is by the gradual or continued resistance of the air that the motion of a cannon-ball is destroyed. Now if, instead of this gradual resistance, any hard substance, as a block of granite, were opposed to the progress of the ball, it would be at once broken asunder. We see, then, the reason that a hard substance of moderate thickness does not offer so effectual a resistance to a body moving very rapidly as some substance of a more yielding kind and of greater bulk. For example, a bale of cotton will arrest a ball which would pass through a plank, for the cotton, yielding easily, permits the force of the ball to be felt and resisted by a larger bulk, while the wood, not yielding, opposes but a small portion of its whole bulk to the force of the ball, and therefore does not arrest it; in other words, the momentum of the ball is communicated to a much larger quantity of matter in the cotton than in the wood. These principles afford a ready explanation of a feat which is sometimes performed. A man lies upon his back, and, having an anvil carefully placed upon his chest, allows some one to strike a heavy blow with a hammer upon the anvil, and no injury is received. Why? Because the momentum, or force, of the hammer is diffused throughout the bulk of the anvil, and then again throughout the bulk of the yielding chest. The man takes good care to have his lungs well filled with air at the moment of the blow, for this increases the bulk and elasticity of the chest, and thus promotes the diffusion of the momentum. If the blow of the hammer were received directly upon the chest, great injury would be done, for the force would then be spent upon one small spot alone.

The principles above elucidated are applied by men instinctively in their common labors and efforts. Watch a man catching bricks that are tossed to him. As he receives the bricks in his hands he lets his hands and the bricks move together a little way, so that he may gradually arrest the motion of the bricks. To do it suddenly would give him a painful lesson on momentum. So when a man jumps from a height he does not come to the ground in a straight position. This would cause a sudden and therefore a painful arrest of the motion of the whole body. To avoid this he comes to his feet with all the great joints of his body bent, so that the different portions approach the ground successively, his head having its motion arrested last.

QUESTIONS.

55. Explain the relations of matter, motion, and force as seen in the rolling of a ball. Define motion. Define force. Show that force may exist without motion. Give the illustration of the magnet. — 56. What is said of the universality of motion? What are the principal causes of motion? How does attraction act? Explain the influence of heat and of other forces.—57. Name the varieties of motion, and give examples of each. How is velocity measured? How compared? Name the rate of motion of light. Of electricity. Of sound. Of a violent hurricane. Of a man walking. Illustrate uniform motion. Illustrate variable motion. Give examples of uniformly retarded motion. Of accelerated motion. —58. Show that motion and rest are relative terms. Give the comparison of the steamboat.—59. What is the difference between absolute and relative motion? What is said of absolute rest?—60. What is said of com-

pound motion. Give the example of a man walking the deck of a moving steamboat. Illustrate simultaneous motions in different directions. Explain the principles illustrated by the parallelograms, Figs. 73, 74, and 75. —61. What is the momentum of a body? Give the rule for calculating momentum, and give an example. Give the illustration of two balls of clay. Also of the cannon-ball. And of the plank. What is said of the wind as a reservoir of motion ?—62. What is said of the relation of force to velocity ? — 63. What is said of accelerated force ? Show that rapid motions are usually caused by a succession of impulses. Give the illustrations of the hammer, of the horse, and of the arrow. Show that gravity is an accelerating force. Give illustrations.—64. Show that gravity is uniformly accelerating. State the law. Explain why bodies of different weights fall with equal rapidity. Describe the experiment in a vacuum. Give the illustration of the water-hammer.—65. What is said of retarded force ? Give examples. Explain the anvil trick. Mention illustrations taken from every-day life.

CHAPTER VIII.

MOTIONS OF MATTER (CONTINUED).

66. Course of Bodies Thrown into the Air.—When any body —a stone, for example—is thrown straight upward into the air, it does not, in reality, go up or come down vertically. If it did, it would come down at a great distance from us. Suppose it takes two seconds for it to go up and to reach the ground. If we stand at the equator, in that two seconds we move from the point where we threw up the stone nearly 3000 feet eastward; and, therefore, if the stone rose and fell vertically, it would fall 3000 feet westward of us. Why, instead of this, does it fall at our feet? Because when thrown into the air it not only has the upward motion given by the hand, but also the forward motion of the earth. It is a case similar to that of a man on board of a steamboat, who, though the vessel move fifteen miles an

hour, tosses up his ball or orange and catches it as well as
if he were on land. This he could not do if both he and
the orange did not have the same forward motion as the
boat. If a man fall from a mast-head, he reaches the deck
at the foot of the mast when the vessel is sailing rapidly,
just as if it were lying still at the wharf. If he did not
by inertia (§ 14) retain the forward motion which he had
in common with the vessel, he would fall at some distance
behind the mast.

The Earth and the Atmosphere.—The air being held to the earth by attraction, it has a motion in common with the earth. It revolves with
the earth just as the tire of a wheel revolves with the wheel. This being
so, our winds are nothing but slight variations of this constant rapid whirl
of the aerial coating of the earth. If the atmosphere were suddenly to
stop whirling round with the earth, we should move through it with a
velocity of 1500 feet a second; and the destructive effect upon us would
be the same as if the earth were standing still while the air moved over
its surface with this fearful velocity. A thoughtless man, not reflecting
that the atmosphere moved with the earth, proposed rising in a balloon, and
waiting till the country to which he wished to go should pass under him,
and then to descend to the earth.

67. **Path of Projectiles.**—If we consider the connection
between the motion of the earth and the course of a body
thrown into the air, and ascertain its actual path, we find
that it forms a peculiar curve.

Anything thrown into the air is called a *projectile ;* and
the path which it follows is
that of a *parabola.* Suppose
a stone be thrown by a man
standing at A, Fig. 81, in the
direction A C E G, it will deviate from a straight line
on account of the attraction
of gravity, and actually describe the parabolic curve

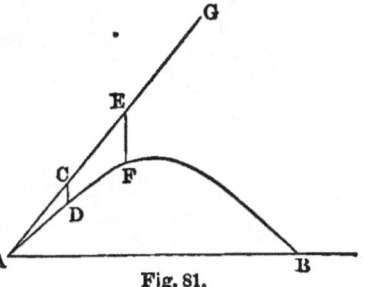

Fig. 81.

A D F B. If the stone, having reached the point C, has fallen towards the earth a certain distance, represented by C D, when it reaches the point E, twice as far from A, it will fall a distance not twice, but *four* times as great as C D, viz., E F, and so on, thus forming the curve.

Of course, it is understood that, besides the propelling force of the arm and the attraction of gravity, a third force acts upon the stone, viz., the resistance of the air; but this being in direct opposition to the first, it only retards the motion, and does not tend to turn it from its straight course.

If the stone be thrown horizontally, it also describes a parabola. If the propulsive force be very great, as in the case of a bullet discharged from a gun, the path will appear to be straight; but this is not so. The force of gravity pulls the bullet towards the ground from the instant that it leaves the gun. This deviation is very slight, however, and for short distances the bullet may be considered as moving in a straight line. When, however, a marksman shoots at long range, he must make allowance for this bending-down of the motion. Accordingly, for the sake of precision, a double sight is provided in modern guns, as shown at A and B, Fig. 82. This arrangement secures

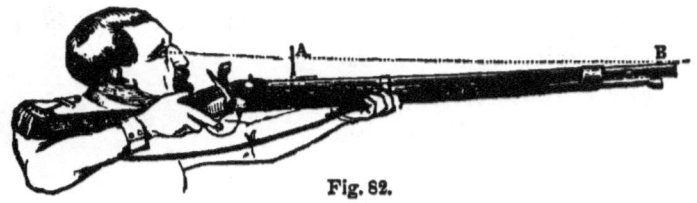

Fig. 82.

the pointing of the gun a little above the level of the object aimed at, that level being indicated by the dotted line.

Let us further study the path of a projectile impelled

horizontally. In the case of the musket-ball just mentioned, we have seen that two forces act upon it, viz., the projectile force given by the powder and the force of gravitation. The force of gravity being always the same, the shape of the curve which the projected body describes must depend on the force with which it is projected. This is very strikingly exemplified in the curves described by the different streams of water in Chapter XII. But whether the projectile force be great or small, the moving body thrown horizontally will, in every case, reach the ground in the same time. Thus, if two cannons stand side by side on a height, one of which will send a ball a mile and the other half a mile, the two balls, if fired together, will reach the ground at the same instant, though at first thought it would seem that the ball which travels twice as far as the other would take a longer time to accomplish it. This is because the *horizontal* force of the ball does not oppose in the least the *downward* force of gravity. If it were thrown upward instead of horizontally, the projectile force would be opposed to gravity, and in proportion as the direction came near to being vertical. As horizontal force does not interfere with the action of the force of gravity, it follows that, if a ball be dropped at the instant at which another is fired, both will reach the ground at the same instant. This can be made clear by Fig. 83. Suppose it takes three seconds for a ball to fall from the top of a tower to its foot. In the first second it falls to *a*. The ball projected horizontally from the cannon, being operated upon by the same force of gravity, will fall just as far, and will be on a level with it at *b*. Both balls fall farther and farther each second, both being accelerated in the same degree because it is done by the same force. The projected ball will reach *d* when the falling ball is at *c*, and the plane at *f* when the falling ball is at *e*, the foot of the tower.

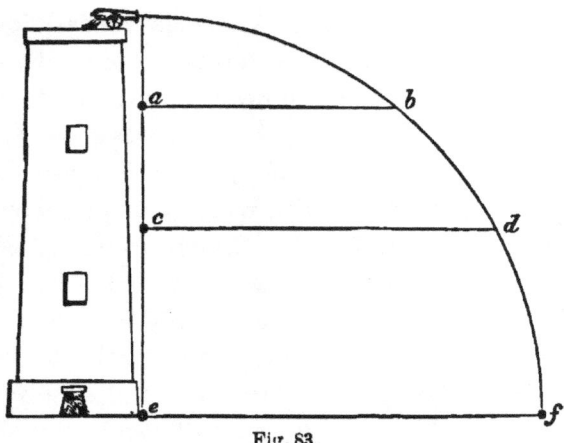

Fig. 83.

The same holds true in all cases. A bullet dropped from a level with the barrel of a gun, paradoxical as it may seem, will fall to the ground no sooner than one which is shot from the gun.

68. **All Falling Bodies really Projected.** — When a body falls from any height, it does not, as you have already seen in § 66, fall in a straight line, as it appears to do. It falls in a curved line, for, like all projectiles, it is acted upon by a horizontal force as well as the force of gravity. But what is this horizontal force? It is the motion which the body has in common with the earth in its rotation on its axis. In this rotation the height from which the body falls goes to the eastward 1500 feet in a second. If, therefore, the body did not partake of the motion of the earth, and descended in a *straight* line in a second, it would reach the ground 1500 feet westward from the foot of the height whence it fell. But it does partake of the earth's motion,

Fig. 84.

and, moving eastward as fast as the height, it describes the curved line of a projectile. Suppose a ball falls from a height, A, Fig. 84, and in a second of time that height passes to C. The forward or projectile force would tend to carry the ball to C, and the force of gravity would tend to carry it to B. But both forces acting together, it pursues a middle path, and this path is a curved line, because one of the forces is a continued force. (See § 63.) For the same reason, if a ball be dropped from a railway car in motion, and it take a second to fall, at the end of that second it will strike the floor just under that part of the car from which it fell. Although the car may have moved a considerable distance, the dropped ball, partaking of its motion, goes along with it in its fall. For the same reason, a ball dropped from a masthead when a ship is in motion, partakes of the motion of the ship. The ball in each of these cases describes in its fall a curved line.

69. **Motion in Orbits.**—Why is it, let us ask, that a cannon-ball shot horizontally from some great height will not revolve around the earth like the moon. It has the same two forces acting upon it as the moon has —viz., a projectile force and the attraction of the earth—and both ball and moon describe a curve in their motion. But the curve of the ball bends to the earth, while that of the moon ever sweeps around the earth. Why is this? In the first place, the resistance of the air continually retards the velocity of the ball. But, secondly, even if the ball could be projected from an elevation sufficiently high to be outside of the atmosphere, the force of the projection would not be great enough. We know, from the rate of progress of the heavenly bodies in their orbits, that it would require an immense velocity to keep the ball from being brought to the earth by its attraction. The Creator of these worlds, when he launched them into their orbits, gave them precisely that impulse which is needed to balance the centripetal force (§ 73) of attraction, and so they pursue a middle course between the two directions in which these two forces tend to carry them. And as their velocities have never been retarded by the resistance of air or any other substance, they have been ever the same from the beginning.

70. **Newton's Laws of Motion.**—By investigating the

principles of motion, Sir Isaac Newton arrived at three laws, which have been in some measure anticipated in the preceding sections. These laws may be briefly stated as follows:

I. A body free from the interference of external matter or force will, if at rest, remain at rest, and if in motion, will move uniformly in a straight line.

II. A given force always produces the same effect, whether the body on which it acts be in motion or at rest, and whether it be acted upon by one or more forces simultaneously.

III. Action and reaction are equal and opposite.

These laws are far more comprehensive than they at first sight appear, and require some explanation.

The first part of the first law follows from inertia, § 14; a body at rest will remain at rest until some external force causes its motion; and a body has no power in itself to change the rate or direction of its motion. Hence in the communication of motion a certain amount of time elapses before its effects are made evident. Of this we have many examples, some of which are given in the following sections.

71. **Inertia Shown in the Communication of Motion.**— When the sails of a vessel are first spread to the wind, the vessel does not move swiftly at once, for some time is required for the force applied to overcome the inertia of so large a mass, and to put it in rapid motion. Horses make a greater effort to start a load than they do to keep it in motion after it is started. If a person stand up in a carriage, and the horses start off suddenly, he falls backward, because his body, from its inertia, does not readily and at once partake of the motion of the carriage. If a person start forward quickly with a waiter, filled with glasses, in his hands, the glasses will slide backward.

The foregoing illustrations show that it requires some time to communicate motion to any body. We will give some illustrations of this fact of a more striking character. If a ball be thrown against an open door, it will move the whole door, and perhaps shut it; but the same ball fired from a rifle will pass through the door without moving it perceptibly. In the latter case its velocity is so great that there is not time enough to communicate motion to the whole door, and it moves only that part of it with which it comes in contact. A bullet thrown with but little force against a window will crack a whole pane of glass; but if shot from a pistol, it merely makes a round hole. A cannon-ball having a great velocity may pass through the side of a ship, doing perhaps comparatively little damage, while one moving with much less velocity may do vastly more damage by splintering the wood to a considerable extent. For the same reason a rapid ball hitting a person may occasion less suffering and do less harm than a slow ball; for a rapid ball kills merely the parts which it touches, leaving the flesh around in a sound state, while the slow ball bruises over a large space. If a large pitcher filled with some heavy liquid be quickly taken up, the handle will break, leaving the pitcher behind. Large dishes are sometimes broken in this way when heavily loaded.

72. Inertia Shown in the Disposition of Motion to Continue.—As in the case of the ship, in the first illustration in § 71, it takes time to communicate motion to the whole ship, or, in other words, to overcome its inertia, so, when the ship is once in rapid motion, it does not stop suddenly when the sails are taken down, but its inertia tending to keep it moving is gradually overcome by the resistance of the water. If a person stand up in a carriage in motion, and the horses suddenly stop, he will be thrown forward, for his body has a motion in common with the carriage, and from inertia is disposed to go on when the carriage stops. When you strike your foot against anything to get the snow off, you give the foot and the snow a common motion together, then arresting the motion of the foot, the snow through inertia passes on. The same thing is illustrated in striking two books together to remove the dust. If a ship strike upon a rock, everything on board which is loose is dashed forward. The earth, as it revolves on its axis, has a velocity at the equator of about 1000 miles an hour. If this revolution should be suddenly arrested, everything loose on its surface, having acquired the motion of the earth, would be at once thrown eastward, just as the furniture, etc., on board ship are dashed forward when the vessel is stopped by running against a rock. All the houses, monuments, and structures of every kind would fall prostrate eastward.

An Equestrian Feat.—In the feat of jumping over a cord from the back of a galloping horse, represented in Fig. 85,

Fig. 85.

the only exertion made by the rider is to raise himself sufficiently to pass over the cord. He comes down again upon the horse's back, simply because of the motion which he has in common with the horse, his feet going in the path represented by the dotted line. If he should attempt to throw himself forward, as in leaping from the ground, he would go too far, and perhaps strike upon the horse's neck instead of his back. Skill in jumping from a moving carriage consists in making the proper allowance for the forward motion which is had in common with the carriage. Most persons are apt to overdo the matter, and so come to the ground prostrate, and with unnecessary violence.

A Case in Court.—A dashing young man driving a light phaeton ran against a heavy carriage. His father was induced by his son's representations to prosecute the driver of the carriage for driving too fast. A knowledge of motal inertia very readily decided the case. The son and his servant both declared in the witness-stand that the shock of the carriage against the phaeton was so great that they were thrown over the horses' heads. They thus proved themselves guilty of the fast driving, for it was their own rapid motion that threw them out when the phaeton was stopped by running against the carriage. The following case is a parallel one: If two boats—the one, of large size, sailing slowly up stream; the other, a small one, sailing rapidly down—run against each other, a man standing in the bow of the one going down will be thrown much farther forward than one standing in the bow of the other.

73. Centrifugal Force. — The second part of the first law states that if a body in motion be not interfered with, it will move uniformly in a straight line. This disposition of motion to be straight is well illustrated by a consideration of centrifugal force; but, before explaining it, we will briefly mention the nature of *curved motion.*

We have shown in § 60 that when two or more forces act upon a body simultaneously, the motion resulting is in a straight line. If, however, you have understood the explanation of the parabolic path of a projectile (§ 67), you have observed that two forces may also produce curved motion, provided one of them communicate a single impulse and the other a succession of impulses.

Of this we have a familiar example in a ball whirled around at the end of a string. You can give it an impulse, and then, holding it in your hand, let it whirl. Here the impulse given the ball is one force, and the tension of the string is the other, the latter acting continuously. Your hand holding the end of the string is the centre about which the motion revolves; the impulse which you have given the ball tends to make it fly away from the centre in a straight line, and hence is called the *centrifugal* force; the tension of the string keeps it from thus flying off, and is called the *centripetal* force.

This will appear clearer by examining the diagram, Fig. 86. If the hand holding the string be at A, the impulse given the ball B will tend to move it in the direction B C,

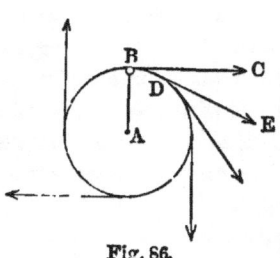

Fig. 86.

but the string A B pulls it towards the centre; and when the ball reaches any other point, as D, it is prevented from pursuing the straight path D E by the same cause; consequently, the ball revolves in a circle. That the ball would take the direction B C but for the resistance of the string can easily be shown by experiment: in place of a ball fastened to a string we may use a stone in a sling, and the instant the stone is set at liberty it will dart off as straight as an arrow (Fig. 87).

When the earth, at the creation, was put in motion, it would have moved in a perfectly straight line were it not constantly drawn towards the sun by attraction, the continuous action of this latter force being the same as the tension of the string in the case of the whirling ball. The force of attraction, then, is the centripetal force of the earth, and the impulse which was given to it by the Creator in the beginning is its centrifugal force; and, balanced between these two forces, the earth and all the heavenly bodies move uniformly onward in their orbits. The centrifugal force in these illustrations is simply the tendency of motion to a straight line, which is constantly counteracted by the centripetal force.

74. Illustrations of Centrifugal Force. — When a wet mop is whirled, the water flies off in every direction by its centrifugal force. On the same principle a dog, coming out of the water, shakes off the water by a semi-rotary motion.—When a suspended bucket of water is revolved swiftly, the water rises high on its sides, and leaves a hollow in the middle. It is the tendency to fly away from the centre of motion that causes this. If the

Fig. 87.

bucket be held firmly by the cord and swung swiftly around the hand as a centre, the centrifugal force of the water against the bottom and sides will prevent its escaping, even when the bucket is upside down, as shown in Fig. 88. —Large wheels, revolving with great velocity, have been broken by the centrifugal force of their particles, and hence the necessity of having such wheels made very strong. The immense grindstones used in gun-factories have sometimes been broken through in the middle, or have burst into pieces with destructive violence from the same cause. — A man riding horseback on turning a sharp corner inclines his body towards the corner, to avoid being thrown off by the centrifugal force. So, in the feats of the circus, a man standing on a horse running at full speed around the ring inclines his body strongly inwards, as shown - in Fig. 89. The horse also instinctively inclines in the same direction for the same reason. If the rider find himself in danger of falling, by making the horse go a little faster, thus adding to the centrifugal force,

Fig. 88.

the difficulty is relieved.—The centrifugal force is made use of in milling. The grain is admitted between two circular stones by a hole in the centre of the upper one, and as the stone revolves

Fig. 89.

it constantly moves towards the circumference, and there escapes as flour.

Bends in Rivers. — We see the operation of centrifugal force in the bends of rivers. When a bend has once commenced in a river, it is apt to increase, for as the water sweeps along the outer bank of the bend it presses strongly against it, just as the water in the whirled bucket presses against its sides, by its centrifugal tendency, or, in other words, its tendency to assume a straight motion. Of course, the result is a wearing-away of this outer bank, and in proportion to the looseness of the material of which it is composed and the velocity of the river's current. And when one bend is formed, another is apt to form below, but on

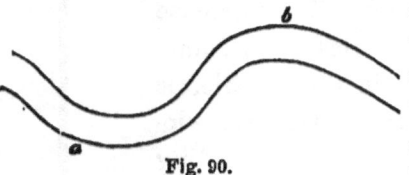

Fig. 90.

the opposite bank. The water, by sweeping along the bend *a*, Fig. 90, is directed by it towards the opposite bank at *b*, and makes a bend there also. It is in this way that a river, running through a loose soil— the Mississippi, for example—acquires a very serpentine course. As the water in the whirled bucket rises around the sides, so in the river the water will be higher against the bank *a* than on the opposite side. Eddies and whirlpools are produced on the same principles, when water is obliged to turn quickly around some projecting point. If a current were moving swiftly along the shore *a* towards the point *b*, Fig. 91, it would be directed outwards by the resistance of this projection, and so a depression would be left at *c*, just behind it, and this depression would be surrounded by a revolving body of water.

Fig. 91.

75. Application of Centrifugal Force in the Arts.—Much use is made of centrifugal force in the arts, and we will mention a few examples. In the art of pottery the clay is made to revolve on a whirling table, the workman at the same time giving the clay such shape as he chooses with his hands and various instruments. In doing this he constantly pays attention to the centrifugal force, giving the table a velocity proportioned to the amount of this force which is needed in each stage of the operation. One of the most beautiful applications of this force is in the manufacture of common window-glass as formerly conducted. The glass-blower gathered up on the end of his iron tube a quantity of the melted glass, and blew it out into a large globe. When it was of sufficient size and thinness, he placed it on a rest, as shown in Fig. 92. A second man then came with a rod having some melted glass on the end, and attached this to the globe at a point opposite to that where the tube of the first man joins it. Then a boy

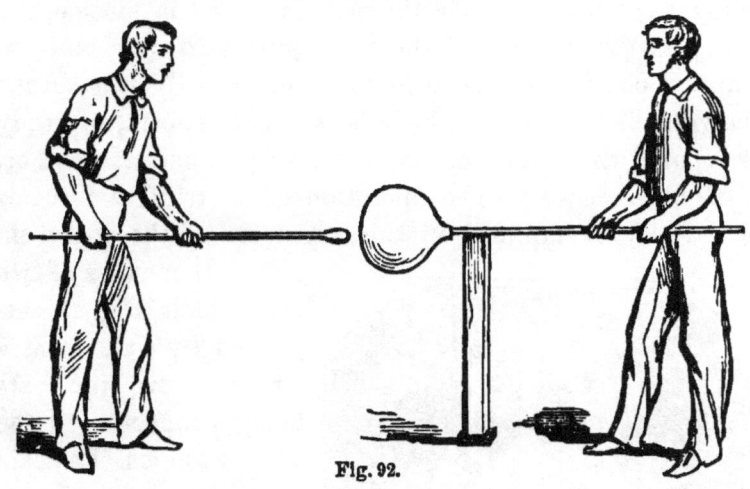

Fig. 92.

gave this tube a quick blow and severed its connection with the globe, leaving a hole in the globe where the glass breaks out. The second man, having the globe attached to his rod, carried it to a blazing furnace, and, resting the rod on a bar at its mouth, put the globe directly into the flame. The glass being soon softened, he whirled the globe continually around. The hole in the globe enlarged by the centrifugal force, and at length by this force the globe was changed into a flat, circular disk, from which were cut panes of glass.

In sugar-refineries the crystallized sugar is freed from

F

the viscid molasses by being placed in a box revolving with great speed; the liquid is thrown off by the centrifugal force, and collected in a suitable manner. This method of drying substances by means of centrifugal motion is frequently adopted in the arts; perhaps the most curious application is to the honey-comb. It has been observed that honey-bees provided with a clean comb will at once proceed to fill it with honey; accordingly, filled combs are carefully shaved to remove the caps on the cells, and placed in a centrifugal machine. When the machine is set in motion, the honey is thrown out of the cells quite perfectly, and the emptied comb is replaced in the beehive. By this means the valuable time of the busy bees is economized, and they are spared the trouble of making fresh wax and new combs.

Steam-Governor.—The operation of centrifugal force is beautifully exemplified in this regulator of the steam-engine. It consists of two heavy balls, Fig. 93, suspended by bars from a vertical axis, the bars being connected to the axis by hinges. The bars have also a hinged connection at their lower ends with two smaller bars, and these latter have a similar connection with a collar that slides up and down on the axis. Now the faster the axis turns, the farther the balls fly out from it, from the centrifugal force, and the higher the collar slides up on the axis. From the collar extends a lever. This is connected with a valve in

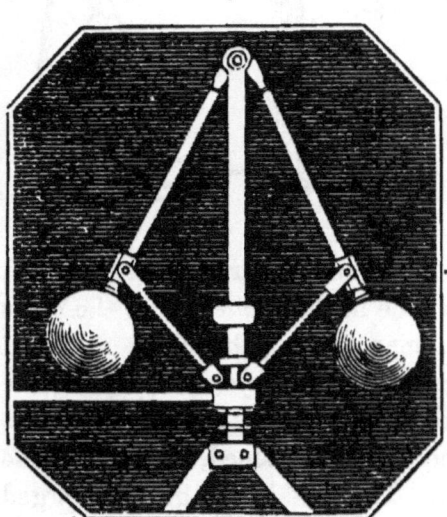

Fig. 93.

the steam-pipe, and so regulates the amount of steam that enters the working part of the engine. The object of this ingenious contrivance is to make the engine regulate its own velocity. When it is not working too fast, the valve in the steam-pipe is wide open. But the moment that the engine begins to move too rapidly, the balls extend out far from the axis, so that the collar rises, and by the lever partly closes the valve. Less steam, therefore, can come to the engine; and the engine working, in consequence, less rapidly, the balls fall again, opening the valve. You see, then, that the regulation of this valve by the governor effectually prevents the engine from running at a danger-ously high speed.

76. Shape of the Earth Influenced by Centrifugal Force.— If the potter should make a ball of soft clay revolve rap-idly around on a stick run through it, the ball would bulge out at the middle, where the centrifugal force is greatest, and would be flattened at the ends where the stick runs through it. This is precisely what has happened to the earth. At the equator, where the centrifugal force is great-est, it has bulged out about thir-teen miles, while it is ·flattened at the poles. This shape was of course assumed before the earth became solid. Fig. 94 represents the shape of the earth, N S being the polar diameter, and E E' the equatorial diameter. The depres-sion at the poles is much exag-gerated in the figure in order to

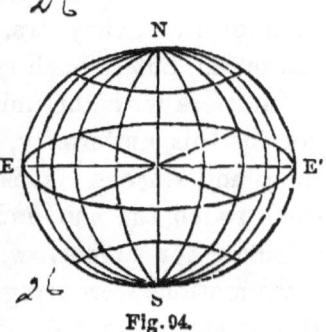

Fig. 94.

make the shape manifest. The tendency to take this shape from the centrifugal force may be illustrated by the in-strument represented in Fig. 95. It consists of two circu-lar hoops of brass connected with an axis, $b\ a$. The hoops

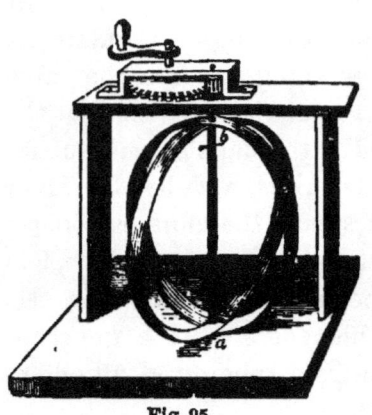

Fig. 95.

are fastened to the axis at *a*, but are left free at *b*. By some simple machinery at the top the hoops can be made to revolve rapidly; and bulging out at the sides by the centrifugal force, they slide down on the axis at *b*.

77. **Uniformity of Motion.**— A third point in the first law refers to the uniformity of motion in the absence of any interfering cause. This uniformity is true both of the direction and of the velocity.

Suppose a body to be set in motion, and to meet with no opposition from friction, or the resistance of air, or attraction, it would move on forever, and with the same velocity with which it began. Now precisely these circumstances exist in the motion of the heavenly bodies in their orbits. They are, it is true, under the influence of attraction, but in such a way, as you will soon see, as not to interfere with the uniformity of their motion. Were it not for this uniformity, we should have no regularity of times and seasons. It is only by the uniform motion of the earth round the sun, and round its own axis, that we can calculate for to-morrow, or next week, or next year. If these motions were irregular, it would throw confusion into all our calculations for the future and all our recollections of the past. We can measure time by nothing else than regular motion; and were there no regular motion, we should have merely the very inaccurate measure furnished by our sensations. To measure time with accuracy, we take some great and extensive uniform motion as our standard. Thus, the revolution of the earth around the sun we take as one division of time, and call it a year. We observe that during this time it whirls around on its own axis 365 times, and the time occupied by each of these revolutions we call a day.

The impossibility of producing on the earth's surface a condition of things similar to that in the empty space through which the heavenly bodies move is an argument against the attainment of *perpetual motion.* It is evidently impossible to annihilate external forces, such as gravity, the resulting friction, etc., and consequently the motion of any object will not be uniform, but continually retarded. Perpetual motion (the dream of visionary philosophers of many centuries) is, then, a mechanical impossibility.

78. The Second Law of Motion.—The second law of motion states that a given force always produces the same effect, whether the body on which it acts be in motion or at rest, and whether it be acted upon by one or more forces simultaneously. This is in reality an expansion of the first law, and its principles have been anticipated in speaking of compound motions (§ 60) and of projectiles (§ 67). Thus, whatever the number of forces acting upon a body, each force may be regarded as producing independently its own change of motion. When a ball is thrown horizontally, the deviation from a straight line leads to the inference that it is affected by another and vertical force—that of gravity.

79. The Third Law of Motion.—The third law of motion states that action and reaction are equal—that is to say, when any of the causes of motion act, the action is met by an opposite and equal reaction. If, for example, a blow be given, an equal blow is received in return. For this reason, if one in running hit his head against the head of another, both are equally hurt. When a child knocks his head against a table, there is sound philosophy in the common saying that he has given the table as good a blow as he has received, though it may afford him no comfort.

Many very interesting illustrations of this law of motion suggest themselves, of which we will give a few.

A swimmer, pressing the water downward and backward with his hands and feet, is carried along forward and upward by the reaction of the water. And in this case, as in every other, the greater the action, the greater is the reaction; in other words, the more strongly he presses with his hands and feet, the more rapidly is he borne along by the reaction of the water against the pressure. A boat advances in proportion to the force with which the oars press against the water. So the rapidity of a steamboat depends on the force with which the paddles drive the water astern. Birds rise in the air by the reaction of the air against their wings as they are pressed downward. A sky-rocket pursues its rapid flight because a large quantity of gaseous matter issues from its lower end, and, being resisted by the air, its pressure throws the rocket upward. If a ship fire guns from the stern, its advance will be accelerated; but if from the bow, it will be retarded. When a broadside is fired, the ship inclines to the other side.

Further Illustrations. — If a spring be compressed between two equal bodies, it will throw them off with equal velocities. If they are unequal, the velocity of the smaller body will be greater than that of the larger, and in proportion to its smallness. For this reason, when a ball issues from a cannon, though the cannon and the ball are equally acted upon by the elastic or expansive force of the gases set free by burning the powder, the gun is moved but very little, because the force is diffused through so large a mass; while the ball, being so much smaller, moves with great velocity. When a volcano throws stones from its crater, the earth may be compared to the cannon, the stones to the ball, and the explosive materials throwing the stones to the exploding powder projecting the ball. Since the cannon is moved as much as the ball, the earth also is moved as much as the stones, the only reason that it does not move so far and so rapidly being that the force is diffused through so large a bulk. These examples illustrate very well the relation of action and reaction; for whenever there is an action of one body upon another, it is as if a spring were between the two bodies, acting equally upon both. When a man jumps from the ground, it is as if a spring were compressed between him and the earth, and this expanding moves the earth exactly as much as it does the man. He really kicks the earth away from him. The motion of the earth is not obvious because it is diffused through so large a mass. The case is parallel to that of the ball and cannon. The same force is exerted upon the man and the earth; but the man, like the ball, moves the most, and in proportion to his small size. So when a bird hops from the ground, the earth moves as really as the bird. If the bird hop from a twig, you perceive that the twig is moved by the pressing-down of the bird as it rises. When it starts from the ground, it exerts the same downward pressure, and moves the earth as really as in the other case it did the twig. Of course, many of these motions are far too small to be perceptible, and too insignificant to take into account under ordinary circumstances.

80. **Communication of Motion in Elastic Bodies.** — Additional proof of the truth of this law is shown in the communication of motion in elastic bodies. Momentum is transferred from one body to another very differently in elastic and non-elastic bodies. As shown in § 61, when one non-elastic body strikes upon another, the momentum is divided between them, and both move on together. Now, if a and b, Fig. 96, were elastic bodies, as ivory balls, and b should be let fall against a, it would give all its momentum to a. Therefore b would stop, and a would move on to the same height from which b came. The reason is, that the velocity lost by b and received by a is just double what it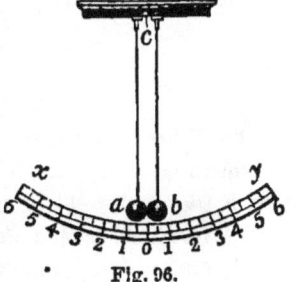

Fig. 96.

would be if the balls were non-elastic. For the same reason, if a and b, being elastic, meet each other from equal heights on the arc, they will both rebound, and return to the same heights from which they came. But if non-elastic, they simply destroy each other's momentum and stop. The effect produced in the former case is just twice as great as in the latter, as you may see by reckoning on the arc. For the same reason, too, if you have a

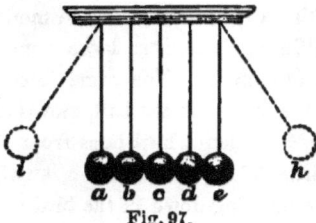

Fig. 97.

row of elastic balls, as in Fig. 97, and let a fall from the point i upon b, it will stop there; and communicating all its momentum to b, this momentum will pass from b to c, and so on through all the row of balls to e, the last one, which will fly off to the point h, at the same height with i, the point from which a fell. If b be held still, and a be let fall upon it, a will rebound to the height from which it fell, for then the compressed elastic spring of each ball, b being immovable, communicates all the motion to a. It is for this reason that an elastic ball, on being thrown against anything fixed, rebounds. If it be thrown against a perfectly elastic body, it rebounds with a force equal to that with which it is thrown. The transmission of sound by the air takes place in a somewhat similar manner, as will be shown in Chapter XIV.

QUESTIONS.

66. What is said of the course of bodies thrown into the air? Give the comparison of a man on a steamboat. Why does the atmosphere move with the earth?—67. Explain the parabolic path of projectiles. What three forces act upon a stone thrown into the air? What is the path of a body projected horizontally? Show that the shape of the curve is modified by the force exerted. Show why a ball dropped from the mouth of a cannon will fall to the ground in the same time as one fired from it.—68. By what two forces is a falling body acted upon? Explain Fig. 84. What is the course of a ball dropped from a railway car or from a mast-head?—69. Give the comparison between the cannon-ball and the moon. What is said of the velocities of the heavenly bodies?—70. What are Newton's laws of motion? From what principle does the first law follow?—71. Illustrate the fact that inertia is shown in the communication of motion.—72. Give

illustrations of inertia as shown in the disposition of motion to continue. Describe and explain the equestrian feat mentioned. What is said of skill in jumping from a moving carriage? Relate the case in court.—73. What is stated in the second part of Newton's first law? What is said of curved motion? Explain the diagram (Fig. 86). What are centrifugal and centripetal forces? What forces correspond to these in the revolution of the earth around the sun?—74. Give illustrations of the various operations of centrifugal force. What is said of the formation of bends in rivers? Show how eddies and whirlpools are formed.—75. How is centrifugal force used in the art of pottery. How in making window-glass? How in sugar-refineries? What is said of the honey-comb? Describe and explain the operation of the steam-governor.—76. What is said of the agency of centrifugal force in shaping the earth? Explain the operation of the apparatus mentioned.—77. What is said of uniformity of motion? What of its uniformity in velocity? State by what means we calculate time. What is said of perpetual motion?—78. What is stated in the second law of motion?—79. What is stated in the third law of motion? Illustrate the law. Give the illustration of the spring. Of the cannon-ball. Of the volcano. What is said of the jumping of a man from the ground? What of the reaction in the case of a hopping bird?—80. Explain the additional proof of the law shown in the communication of motion in elastic bodies.

CHAPTER IX.

THE SIMPLE MACHINES.

81. **Machines not Sources of Power.**—As shown in § 56, the forces of nature at our command are few in number, and under many circumstances only one—gravitation—is available. Nature seldom provides forces in a form directly suited to the accomplishment of work, and we therefore resort to contrivances called *machines*, by which one form or degree of force is transformed into another, and rendered serviceable. From an erroneous idea of the principles involved, six simple machines, named respectively the *Lever*, the *Wheel and Axle*, the *Inclined Plane*, the *Wedge*, the

F 2

Screw, and the *Pulley*, were formerly spoken of as the *Mechanical Powers*. These machines, however, are in no sense powers, but merely means of applying power to advantage, and are not in themselves sources of power. No instrument or machine can create power, and the only use of all the variety of tools and machinery is to enable us to *apply* power in such a manner, with such a velocity, and in such a direction as to effect the objects which we have in view. Excepting old usage, there is no reason why the term mechanical powers should have been confined to the six contrivances above named. Any arrangement of solid or rigid parts, moving with different velocities whereby one manifestation of force is converted into another, equally deserves the appellation mechanical power.

Before proceeding to a consideration of the six simple machines, we will explain a few of the terms employed. *Power* is the force by which a machine or instrument is moved. *Weight* is the resistance to be overcome. If the resistance be in some other form than that of weight, it is called technically by this name. So what is called *power* may be in the form of weight. The *fulcrum* is the point on which the instrument or machine is supported while it is in motion.

82. **Lever of the First Kind.**—A beam or rod of wood, iron, or any other material, resting at one part on a prop, or fulcrum, about which the beam can move is called a *lever*, the name being derived from a Latin word meaning to lift. The lever is the most simple of all simple machines, and is therefore in universal use. Though the savage makes use of few tools in comparison with the civilized man, he uses the lever almost constantly in some form. The wedge is the only one of the other simple machines that he uses to any great extent. Levers are of three kinds, commonly called the *first*, *second*, and *third*

kinds, the difference depending on the relative position of the power, the weight, and the fulcrum.

In the lever of the first kind the fulcrum, or prop, is between the weight and the power. The common crow-bar, or hand-spike, is a familiar example, as seen in Fig. 98—the stone, S, or other heavy body to be moved being the weight, the stone or block of wood, F, on which the bar rests being the fulcrum, and

Fig. 98.

the pressure of the hand, H, the power. The nearer the fulcrum to the weight, or the farther the power from the fulcrum, the greater is the force of the lever. This may

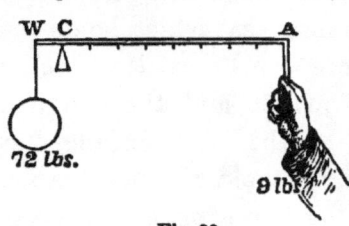

Fig. 99.

be illustrated in Fig. 99. Here the short arm of the lever, as it is called, C W, is one eighth of the length of the long arm, A C. If the weight hanging at the end of the short arm be 72 pounds, a weight of 9 pounds, or the force of a hand equivalent to this, will balance it at the end of the long arm. But if the power should be applied at only four times the distance from the fulcrum at which the weight is, then it would require a force of 18 pounds to balance the 72 pounds on the short arm. Similar variations can be made by altering the length of the short arm. The power and the weight balance each other when the weight multiplied by the length of the short arm, and the power multiplied by the length of the long arm, give equal products.

Steelyards and the Balance.—Examples of levers of this

Fig. 100.

kind are very common. In an ordinary balance, Fig. 100, we have a lever the two arms of which are equal, and therefore equal weights suspended at the ends balance. If they be not exactly equal, a heavier weight will be necessary on the shorter arm. The inequality will injure the buyer if the prop be too near the pan in which the weights are placed, and the seller if it be too near that which holds the article to be sold. Any difference can be easily detected by changing the places of the article and the weights. Whenever cheating is practised by the "false balance," it is of course done in a small way, to avoid any observation by the eye of the inequality of the two arms of the scale-beam, and the weight of the pan hanging from the shorter arm is made a little greater than that of the other, so that they may balance. Balances may be rendered very accurate by making the fulcrum or pivot of hardened steel, and of a wedge shape, with a sharp edge, in order to avoid friction as much as possible.

The steelyard, Fig. 101, differs from the balance-beam in having the arms of different lengths. The principles on which this instrument is constructed were developed in explaining Fig. 99. With either the balance or the steelyard, when two weights balance each other, the centre of the weights and the apparatus taken together is just over the fulcrum. Hence the necessity for placing the prop near the large weight when we wish to balance it by a small one.

In Fig. 101 C is the ful-
crum, P the weight to be
determined, and Q the
power applied in the form
of a hanging weight.

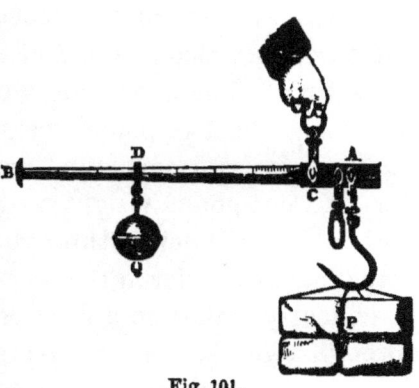

Other Examples. — Scissors
are double levers of the first
kind. The fulcrum is the rivet,
the weight or the resistance to
be overcome is the article to be
cut, and the power is applied to

Fig. 101.

the long arms of the levers by
the fingers. With large shears hard substances can be cut. Even plates
of iron are cut like paper by shears worked by a steam-engine.—Pincers
are double levers. The hinge, or rivet, is the fulcrum.—The common
hammer, as used in drawing nails, is a good example of the power of this
kind of lever. Though crooked, it acts in the same way with a straight
lever. The fulcrum is the point on the board where the hammer rests,
and this is between the resistance to be moved, the nail, and the power,
that is, the hand which grasps the handle.

83. **No Gain of Power in this Lever.**—We will now illustrate the truth that there is no gain or saving of power in
this lever, as might at first thought seem to be the case.

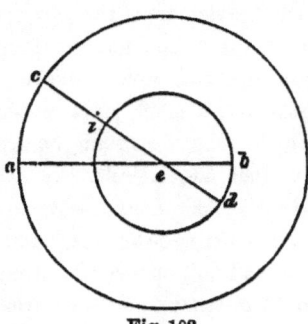

Let $a\,b$, Fig. 102, represent a lever,
and e its fulcrum. Let the arm $a\,e$
be twice as long as $e\,b$. A pound,
then, suspended from a will balance
two pounds at b. · If now, when the
weights are suspended, the long
arm be raised so that the lever occupy the position represented by
the line $c\,d$, and then let go, the

Fig. 102.

one pound at c, balancing the two
pounds at d, will bring the lever again to the position $a\,b$.
It will be perceived that the end of the long arm of the

lever moves through the space *a c*, which is larger than *b d*, through which the end of the short arm moves, in the same time. The one-pound weight, in fact, falls two feet in raising the two-pound weight one foot, and it moves twice as far as a one-pound weight suspended at *i* would do. If a one-pound weight could raise a two-pound weight without thus moving through twice as much space, an actual gain of power in the lever would indeed result. But it evidently makes no difference whether one pound moves through two feet or two pounds through one foot; the force is the same in both cases. For the momentum or force of a moving body is in proportion to its weight and velocity (§ 61); and therefore the pound weight moving through two feet has as much momentum as the two-pound weight moving through one foot in the same time. The small weight does the same amount of work that the larger one would by moving twice as far in the same time, just as a boy who carries a load half as large as a man will do as much work as the man if he carry it twice as fast.

Fig. 103.

The Seesaw.—The same thing is illustrated in the seesaw, Fig. 103. The man, being much heavier than the boy, is nearer the prop; and as they move up and down the boy passes over a much larger space than the man, describing an arc in a much larger circle.

Archimedes's Lever. — Archimedes, a distinguished mathematician and philosopher who lived about 250 years before the Christian era, said that if he could have a lever long enough and a prop strong enough, he could move the world by his own weight. But he did not think how far he himself would have to move to do this, owing to

the vast difference between the weight of his body and that of the earth.
"He would have required," says Dr. Arnott, "to move with the velocity
of a cannon-ball for millions of years to alter the position of the earth by a
small part of an inch." Somewhat analogous to this is the case of the
Hydrostatic Bellows and of Bramah's Press, as will be explained in Chapter X. In all these cases great effects are produced by small power, which
itself has to accomplish extensive motion.

84. Lever of the Second Kind.—In the second kind of
lever the weight is between
the fulcrum and the power, as
shown in Fig. 104. The same
rule of equilibrium applies here
as in the case of the lever of the
first kind. The 72 pounds of
weight can be sustained by 8
pounds of power, because the

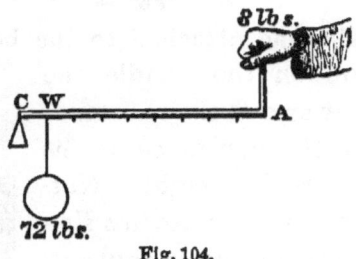

Fig. 104.

power acts on the lever at 9 times the distance from the
fulcrum that the weight does, for $1 \times 72 = 9 \times 8$. The common wheelbarrow, Fig. 105,
is an example of this kind
of lever. The point at
which the wheel presses
on the ground is the ful-
crum, and the weight is
the load, its downward
pressure from its centre
of gravity being indicated

Fig. 105.

at M. Of course, the nearer the load is to the fulcrum, the
easier it is, on starting, to raise the handles. The common
hand-barrow furnishes another illustration (see Fig. 106).
If the load be placed in the centre, each of the men carries
half, for the pole becomes a lever, of which each porter
is a fulcrum as regards the other; if the load be shifted
towards one of the men, he will have to carry a larger

Fig. 106.

share than the other. The crow-bar can be used as a lever of the second kind when its point is placed beyond the weight to be raised. The chipping-knife, Fig. 107, is another example. The end attached to the board is the fulcrum, the pressure on the handle the power, and the resistance of the substance to be cut is the weight. Nut-crackers operate in a similar manner. In shutting a door, by pushing it near its edge we move

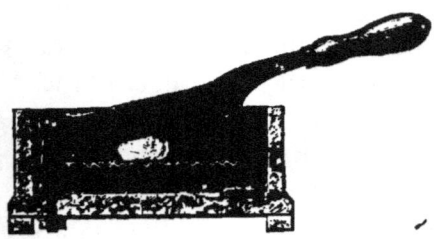

Fig. 107.

a lever of this kind. The hinge is the fulcrum, and the weight is between this and the hand. We see, then, the reason that the slight push of a hand shutting the door may even crush a finger when caught in it at the side

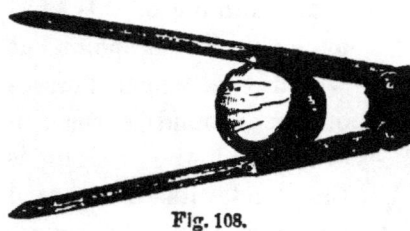

Fig. 108.

where the hinges are. The finger is a resistance so near the fulcrum that the power moving through a great space acts upon it with immense force. The same explanation applies to the severe bite of the finger when it is caught in the hinge of a pair of tongs. The oar of a boat is a lever of this kind, the weight to be moved being the boat, and the fulcrum, singularly enough, being the unstable water.

85. **Lever of the Third Kind.**—In the third kind of lever

the power is between the fulcrum
and the weight, as seen in Fig.
109. In the first two kinds of
lever the power may be less than
the weight, but in this the power
must always be greater than the
weight. The advantage of this
lever consists in the great extent

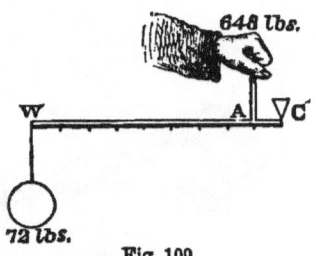

Fig. 109.

of motion obtained. Applying the same rule here as in the
other levers, let us calculate the result. If the weight, as
in Fig. 109, be 9 times as far from the fulcrum as the power, it will require a power equal to a weight of 648 pounds
to sustain a weight of 72 pounds, for $9 \times 72 = 1 \times 648$.

Examples.—When a man puts his foot against the end of a ladder, and
raises it by taking hold of one of the rounds, the ladder is a lever of this
kind. It is evident that he spends his force upon it at a great mechanical
disadvantage, for the power is applied much nearer to the fulcrum than is
the weight of the ladder, taken as a whole. If you push a door to by placing your hand very near the hinges, you do not shut it as easily as when
you take hold of it at its edge. In the first case it is a lever of the third
kind, and the hand moves through a small space, and therefore must exert
a considerable force; while in the latter case the door is a lever of the second kind, and the hand, moving through a greater space, puts forth less
force. When we use a pair of tongs, we use a pair of levers of the third
kind. They are an instrument in which convenience rather than power is
needed. We cannot grasp anything very firmly with them because the
power is so much nearer to the fulcrum than the weight to be lifted. For
this reason, a pinch with the ends of the tongs is of small moment compared with one in the hinge. The anatomical structure of animals furnishes a most beautiful example of this lever. Take, for example, the
principal muscle which bends the elbow, as represented in Fig. 110. This
comes down from the shoulder in front of the bone of the arm, and is inserted just below the elbow-joint into one of the bones of the forearm. It
pulls upon the forearm very near the fulcrum, which is the elbow-joint, and
so acts at a great mechanical disadvantage. The object of this arrangement is to secure quickness of movement, which is here, as in almost all
muscular motions, of more importance than great strength. When great

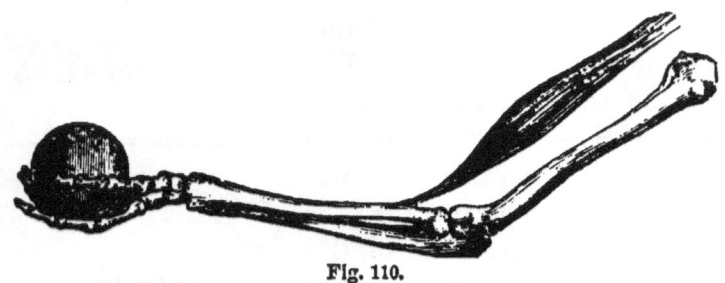

Fig. 110.

weights are lifted, the fact that the muscles act at such mechanical disadvantage makes the exhibition of power wonderful.

86. Compound Levers. — When several levers are connected together, we call the whole apparatus a compound
lever. Let each of
the levers in Fig.
111 be 3 inches long,
the long arms being
2 inches, and the
short ones 1 inch.

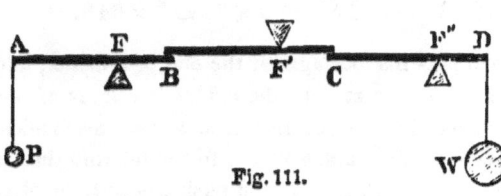

Fig. 111.

One pound at A will, according to the rule, balance 2 at B,
and 2 at B will balance 4 at C, and 4 at C will balance 8 at
D. Therefore 1 pound at A will balance 8 pounds at D.
Hence it is evident that an equilibrium is effected when the
power is to the weight as the product of all the short arms
is to the product of all the long arms. The compound lever
is used in weighing heavy loads—as hay, coal, etc. Fig. 112
shows a represen-
tation of the ar-
rangement. The
load, W, stands
on a platform,
A B, which rests
upon two levers,
E D and E C.
The long arms of

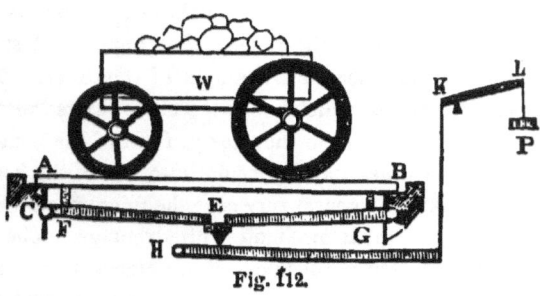

Fig. 112.

these levers are E G and E F, and the short arms are G D and F C. The ends of the long arms press upon the fulcrum of the lever, H I. The pressure is transmitted from the end of the long arm by the rod, I K, to a small lever, K L, where a small weight or power, P, balances the weight of the heavy load, W. The two objects secured by this arrangement are accuracy and the occupation of a small space.

87. **Wheel and Axle.**—The next of the simple machines is the wheel and axle. The most familiar applications of this power are seen in drawing water and in raising heavy articles in stores. The principle of this power is the same as that of the lever, as may be shown in Fig. 113, which represents a section of the wheel and axle. The power, P, hangs by a cord which goes round the wheel, and the weight, W, by a cord around the axle. We may consider the power as pulling on a lever represented by A B, the long arm of which is A C, and the short arm B C. You see that the wheel and axle, then, may be viewed as a constant succession of levers, and it

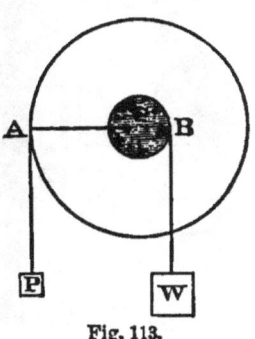

Fig. 113.

is therefore sometimes called the perpetual lever. And the same rule of equilibrium applies here as in the simple lever.

In the common windlass the power is ap-

Fig. 114.

plied to a winch or crank, C F, Fig. 114, instead of a wheel. In estimating the power of this arrangement, F C must be considered the long arm of the lever, and half the diameter of the axle, B B, as its short arm.

The Capstan.—In the capstan, represented in Figs. 115 and 116, the axle is in a vertical position. The top of it is pierced with holes, into which levers are introduced. Fig. 115 shows the instrument as commonly used in moving buildings. Some-

times horse-power is applied at the ends of the levers. Great power is exerted by this instrument; but we have the same fact here as in all other cases where a small force produces a

Fig. 115.

great effect—the effect is slow, and the force passes over a great space in producing it. The moving of a building a foot requires many circuits of the horse around the axle. The capstan, as constructed for use on ships, Fig. 116, has a circular head, with many holes for levers, so that many

men can work together in raising a heavy anchor. The cable, being wound around the capstan several times, is prevented from slipping by friction; and, as one end of the cable unwinds, the other is wound up.

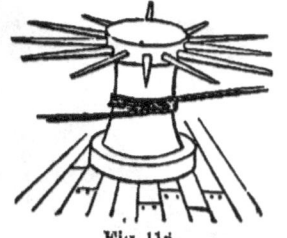

Fig. 116.

Fusee of a Watch.—In the fusee of a watch we have a wheel and axle of a peculiar construction. When we wind up a watch, the chain is wound around the spiral pathway on the fusee, B, Fig. 117, and, at the same time, the spring is coiled up tightly in the round box, A. The spring, in gradually uncoiling itself, turns this round box around, and thus pulls upon the chain, c, making the fusee to revolve, and thus give motion to other parts of the machinery. Now the spring, in its effort to uncoil, acts strongest at first; and,

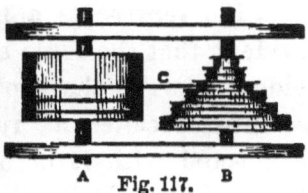

Fig. 117.

therefore, if the fusee were of uniform size, the watch would go fastest when first wound up, and go gradually slower as it ran down. This difficulty is obviated by giving the power a small wheel to pull on at first, and gradually enlarging the wheel as the spring uncoils. Because, in order to produce a certain effect on a given weight, the less the power, the longer must be the arm of the lever on which the power acts.

88. Pulleys.—Another simple machine is the *pulley*, by which masses moving with different velocities may be connected, and thus forces of different degrees of intensity balanced. Pulleys are of two kinds, *fixed* and *movable*. Fig. 118 represents a fixed pulley; the wheel, A B, has a groove in its circumference which prevents the rope from slipping. Its action may be conceived of as that of successive levers of equal arms, and, therefore, equilibrium requires equality of power and weight. Fixed pulleys are used to change the direction of forces, as in hoisting sails on board ship. By a combination of two fixed pulleys a horizontal force may be used to raise weights vertically, as shown in Fig. 119.

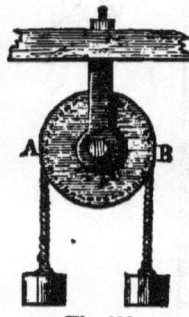

Fig. 118.

Fig. 119.

Fig. 120.

Movable Pulleys.—

Fig. 120 represents a movable pulley. In this case it is evident that the force exerted by the weight is equally divided between the ropes S_2 and S_1. A movable pulley is sometimes called a "runner," and a fixed pulley is often connected with it in order to give the desired direction to the force. Pulleys are often connected in complicated

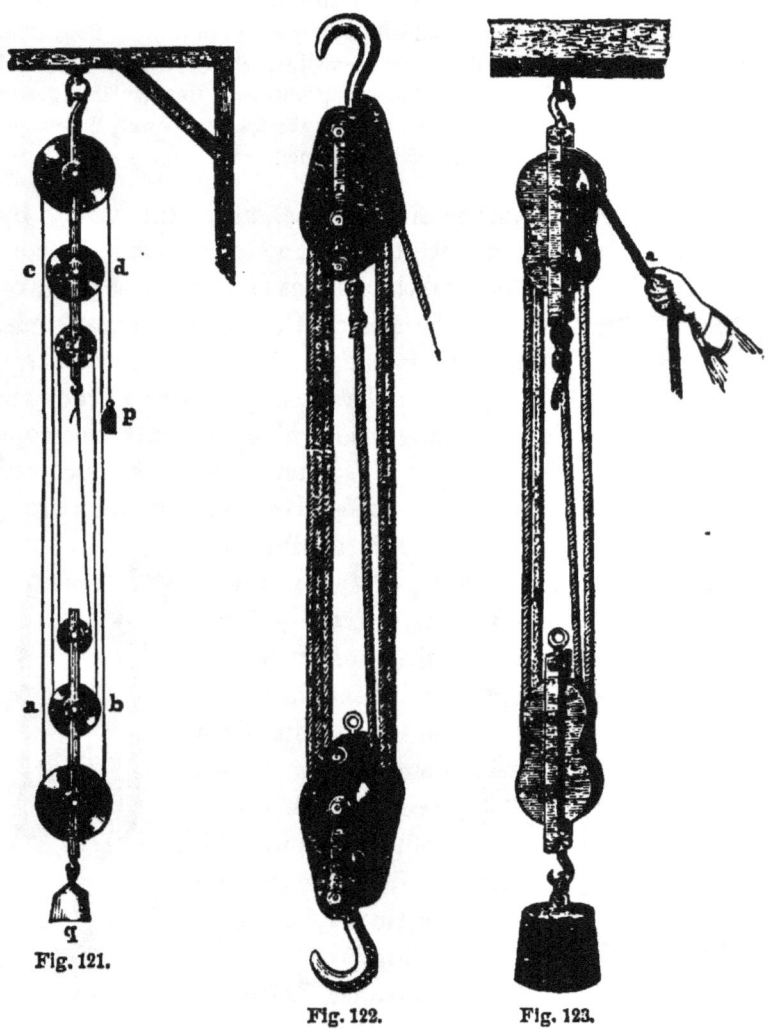

Fig. 121.

Fig. 122.

Fig. 123.

ways. Fig. 121 shows a system of fixed and movable pulleys; the weight, q, is evidently upheld by six cords, which divide the weight equally among them. If q weigh six pounds, equilibrium will be obtained by making the weight of p equal to one pound; at the same time, it must be remembered that p will move six feet while q moves only one. Other arrangements of pulleys are shown in Figs. 122 and 123. The combination of pulleys in one block having a single axle (Fig. 122) is in common use.

89. **The Inclined Plane.**—This, being a very simple contrivance, is much used, especially when heavy bodies are to be raised to only a small height, as in moving large boxes and hogsheads into stores. The advantage of the inclined plane may be illustrated by Fig. 124. The line A c represents an inclined plane. If a weight be drawn up this plane, it is raised only the height B c.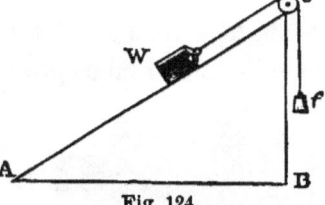

Fig. 124.

A smaller power is requisite to draw the weight up the plane than to raise it perpendicularly; and the power necessary will be the less the longer the plane. A power which would balance a weight on an inclined plane would be to the weight as the height of the plane to its length. Thus, if A c be twice as long as B c, a weight of four pounds on the plane may be balanced by a two-pound weight suspended by a cord passing from the weight over the summit of the plane. A flight of stairs is constructed on the principle of an inclined plane, the projections in it being for the purpose of affording a sure footing in ascending or descending. In like manner hogsheads are let down the steps of a cellar-way by ropes, and it makes no difference in the principle

of the operation whether planks be laid on the steps
or not.

It is supposed that the immense stones in the pyramids and other
massive Egyptian structures were put into their position by means of
inclined planes.   Roads, when not level, are inclined planes; and the
steeper the inclination, the more power is required to draw a load up the
road.   Great mistakes were formerly made in carrying loads straight over
high hills.   Besides failing to take advantage of the principles of the inclined plane, in many cases the horse, in going over a hill, passes over quite
as much space as he would if the road were made to go round the base of
the hill, and sometimes even more.   If the hill were a perfect hemisphere,
a road over it would be just equal in length to a road around its base to
the opposite point.

90. The Wedge.—This simple device may be considered as
two inclined planes placed with their bases together, as seen

Fig. 125.

in Fig. 125.   Indeed, sometimes the
wedge has one side only inclined,
it being but half of the ordinary
wedge.   The difference between the
inclined plane and the wedge in op-
eration is that in the first the inclined plane is fixed, and
the weight is made to move up along its surface; while in
the latter the weight—that is, the resistance—is stationary,
and the surface of the plane is made to move along upon
it.   The power of the wedge is estimated just like that
of the inclined plane; that is, by comparing the thickness
of the wedge with the length of its side.   The less the
thickness of the wedge compared with its length, obviously
the more powerful is the wedge as a penetrating instrument.   The wedge is used for splitting blocks of wood and
stone, for producing great pressures, for raising heavy bodies, etc.   All cutting and piercing instruments—knives, ra-
zors, axes, needles, pins, nails, etc.—act on the principle of
the wedge.

Knives, planes, chisels, etc., are often used somewhat in the manner of a saw, by drawing their edges against the object to be cut, at the same time that the pressure applied exerts the influence of a wedge. The edge of a sharp knife examined under a microscope proves to be serrated. The sharpest razor, it is said, may be pressed directly against the hand with considerable force without cutting the skin, while if drawn ever so little lengthwise it will inflict a wound.

91. **The Screw.**—This is another of the simple machines. Its principle is essentially that of the inclined plane. The "thread" running around the screw is an inclined plane which is spiral instead of straight; and the corresponding part in the nut is an inclined plane running in the opposite direction. In the common screw the nut is fixed, and the screw is made to play up and down in it; but sometimes the screw is fixed, and the nut is made to play around it. The screw acts like a wedge, and has the same relation to a straight wedge that a road winding up a hill has to a straight road of the same length and rise. Especially does the comparison hold when the screw is forced into wood; the wedge goes straight into the wood, but the edge of the screw's thread enters the wood spirally.

To estimate the force of the screw, we compare the length of one turn of the thread around it with the height to which the thread rises in going round. Let *a b*, Fig. 126, represent one turn of the thread, *b c* the height to which it goes. It is clear from the figure that the principle which applies to the inclined plane and to the wedge applies here also. Since the less the height of the plane, the easier it is for a weight to be drawn up it; and since the less the depth of the wedge, the less is it resisted; therefore, the less the height of the turn of the screw's thread, the easier is it to move the screw, and the greater the force which it exerts. Hence the

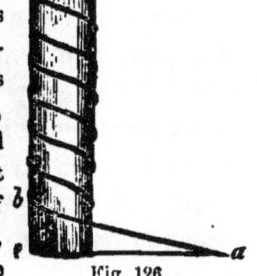

Fig. 126.

G

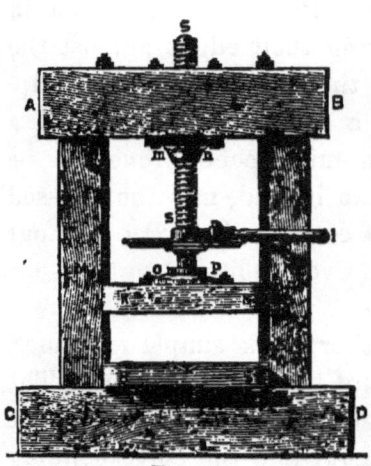

Fig. 127.

prodigious power of a screw with a thread which rises very slowly in its spiral turns. Screws are much used when great pressure is required, as in pressing oils and juices from vegetable substances, in compressing cotton into bales, in bringing together with firm grasp the jaws of the vice, etc. In turning the screw a bar is used, so that we have in this instrument the combined advantages of the screw and the lever (Fig. 127). That you may have some idea of the power of these two instruments acting together, we will state an imaginary case. Suppose it is desired to raise by the screw a weight of 10,000 pounds. Let a turn of the screw be ten inches long, and the rise be but one inch. Then, so far as the screw is concerned, the power requisite to raise the 10,000 pounds will be 1000—the ratio of the height of the thread's turn to its length. But the power of the lever is yet to be estimated. Let the length of the lever, passed through the head of the screw so that it is equal on each side, be thirty inches. The diameter of the screw is about three inches, or one tenth of the diameter of the circle described by the end of the lever. It will take but a power of 100 pounds to raise the weight, the ratio of the radius of the screw to half the length of the lever.

92. **Other Simple Machines.**—When the old idea prevailed that the simple machines were but six in number, it was shown that these could be reduced practically to three, for the wedge and the screw are modifications of the inclined plane, and the wheel and axle is a modification of the lever. As already mentioned, however, there is no necessity for limiting to so small a number the simple machines, provided we regard them as simply modifiers of motive power. The *toggle-joint* is a simple machine, in which the connected parts are arranged so as to move with different velocities. It is used to raise carriage-tops, and, when

made of immense strength, to exert great pressure through a small space, as in shearing and punching iron. Fig. 128 represents, in skeleton form, a toggle-joint. Machines which change the direction of the force applied are very numerous, but a study of them belongs properly to higher mechanics. We see their applications in complicated machines in common use — the printing-press, sewing-machine, steam-engines, locomotives, and many others.

Fig. 128.

93. **The Pendulum.**—Certain mechanical contrivances are designed for the modification of *mere motion*, without any reference to the transmission of forces: such are watches, clocks, and timepieces of every description. These instruments have for their object the production of a perfectly uniform motion with a view to the measurement of time. The motion-regulator by which clocks are controlled—*the pendulum*—demands a somewhat extended notice.

Various modes of measuring time have been adopted by mankind. At first, time was inaccurately divided by merely observing the sun. But after a while man resorted to various contrivances to measure short periods of time with accuracy. All of these depend upon the uniformity of motion alone. The sundial measures time by the uniform movement of the shadow on its face, caused by the uniform movement of the earth in relation to the sun. The hour-glass measures time by the uniform fall of sand produced by the attraction of gravitation. The best measurement of time is by the comparatively modern invention of clocks and watches, in which time is divided into very minute periods by the uniform motion of the pendulum or the balance-wheel. The pendulum furnishes an interesting example of motion sustained by the influence of gravity. It was not till the time of Galileo, less than three centuries ago, that its operation was understood and appropriated to the measurement of time. He observed that a chandelier hanging from the lofty ceiling of a cathedral in Pisa vibrated very long and uniformly when accidentally agitated, and the thought of the philosopher evolved from this phenomenon the most important results. Though it had been before men's eyes in some shape or other since the creation, it was reserved for Galileo to observe its significance, and to pave the way which has led to the use of the pendulum as man's time-keeper over the whole earth.

Explanation of its Operation.—A pendulum commonly consists of a ball or weight at the end of a rod suspended so as to vibrate with little friction at the point of the suspension. Let *a b*, Fig. 129, represent such a pendulum. When it is at rest, it makes a plumb-line hanging towards the centre of the earth. If it be raised to *c* and be left to fall, the force of gravity will not only carry it to *b*, but, by the accumulated momentum acquired in its descent, gravity will carry it to *d*. The same would be true of its return from *d*. And it would vibrate forever in this way if it could be entirely freed from the resistance of the air and from friction. But, as it is, the pendulum, left to itself, gradually loses its motion from these obstacles. In the common clock, the office of the weight is to counteract the influence of these obstacles and keep the pendulum vibrating. In the watch, the mainspring performs the same office to the balance-wheel.

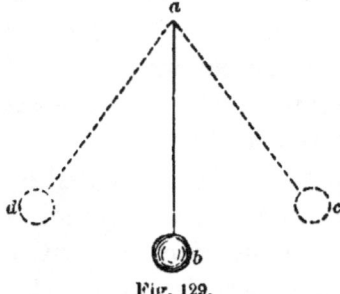

Fig. 129.

The times of the vibrations of a pendulum are nearly equal, whether the arc it describes be great or small; for

when the vibration is a large
one, the velocity which the
pendulum acquires in falling is
greater than when the vibration
is of small extent. The reason
is, that the higher it rises, the
steeper the beginning of its descent. Thus *a c*, Fig. 130, is

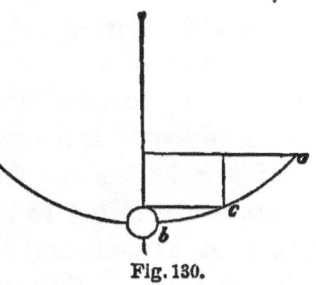

Fig. 130.

steeper than *c b*. The longer a pendulum, the longer time
does its vibration occupy. It requires a pendulum of
the length of a little over thirty-nine inches (39.13″) to
vibrate once every second. Cold weather, by contracting
the pendulum, makes it vibrate quicker than in summer,
and so makes the clock go faster. Various contrivances
have been resorted to in order to counteract the variation
of length in pendulums by heat and cold, one of which we
will describe in the chapter on heat (§ 167).

The popular idea that a heavy body falls quicker than a
light one is dispelled by the fact that pendulums vibrate
equally fast or slow, no matter of what material they are
constructed. Similar pendulums of lead, glass, iron, or
wood, or even a hollow ball, vibrate at the same rate.

94. **Friction.**—Friction is the resistance offered to a body
by the surface on which it moves. It seems to arise from
adhesive attraction between the touching substances and
from the roughness of their surfaces. The rougher the
surfaces brought in contact, the greater the friction, the
little cavities and projections fitting into each other and
necessitating a certain force to raise the projections on one
surface from the cavities in the other. Substances which
appear quite smooth to the naked eye, as polished steel,
nevertheless exhibit inequalities of surface when examined
under the microscope.

Friction acts as an obstacle to motion. When we roll

a ball, the rougher the surface on which it is rolled, the greater the friction and the sooner the ball stops. Machinery, no matter how carefully constructed, suffers a waste of power through friction, even to the extent of a third or a half of the force applied. Hence various expedients are resorted to for the purpose of diminishing friction, such as polishing the rubbing surfaces, and oiling or otherwise lubricating them. Using wheels, as on carriages, effects the same end. A heavy load, which the most powerful horse could not move if placed on a "stoneboat," is readily drawn along in a wheeled vehicle. Casters attached to household furniture prevent the friction arising from dragging it over carpets.

The very fact that rapidity of motion is lessened by friction is in some cases of the greatest importance to us. "But for friction, men walking on the ground or pavement would always be as if walking on ice; and our rivers that now flow so calmly would all be rapid torrents. It is friction which retains all loose objects on earth in the situations in which, for convenience, men choose to place them— the furniture of a house, the contents of libraries, museums, etc. Friction, therefore, is essential to our existence."

Fig. 131 illustrates a simple method of taking advantage of friction. The weight, P, can be lowered gradually, and with comparative ease, by wrapping the rope about the cylinder, A C B; whereas, without the friction of the rope, the force of gravity would be beyond the control of the power, Q. In this way heavy casks which are otherwise unmanageable are lowered into cellars. — Fig. 132 shows us at once the advantages of friction, and a means of overcoming its disadvantages. The block of stone, Q, is supported by roll-

Fig. 131.

Fig. 132.

ers, in order to overcome the friction of its surface on the ground; and by means of the friction between the rope, B, and the axle, O C, the rope is prevented from slipping when power for moving the block, Q, is applied to the lever, A.

The friction of the driving-wheels of a locomotive upon the rails prevents them from slipping. In this case the wheel pushes backward on the rail at each successive point of contact. To make this clear, suppose a common wheel be deprived of its rim and be made to revolve on the ends of its spokes. The end of each spoke gives a backward push as it strikes the ground. Now the rim of a wheel makes the same pushes, but they are more numerous—they are continuous, being made by all the successive points in the rim. Sometimes the rails of a railroad are too smooth, from frost or some other cause, and then sand is thrown upon them to enable the locomotive to start. The sand serves to prevent the wheels from sliding by enabling them to get some hold upon the rails in their backward pushes.

95. The Real Advantages of Machinery.—If there is, then, no saving, but a loss, of power in tools and machinery, what, let us inquire, are their advantages?

If one man can do alone by the aid of some instrument that which would otherwise require the exertion of many men, although slow in accomplishing it, yet it is a great advantage. Thus, one man with a lever can move a stone which it would require perhaps thirty men to move without it; and though it take him thirty times as long, it saves him the trouble of getting a company of men to help him. So if a man can raise his goods by a wheel and axle to the upper loft of his store, though he raise them more slowly than several men would lift them directly by ropes, it is an advantage to him, since it saves the hiring of a number of laborers. It must, however, be remembered that what is gained in amount of work is lost in time, and what is gained in time by using any machine is lost in the amount of work.

Another advantage is that, in applying the force, intervals of rest may be secured without any loss. This is obvious in the case of the pulley, but still more so in the case of the screw. It is friction in both these cases which enables the workman to rest. It saves to him all that he has gained by opposing any tendency to slip back. The same thing is true of the wedge. When this is driven into wood, it remains fixed because prevented from returning by the friction of the wood against its sides. It is the same cause which holds a nail in its place, and opposes any effort to draw it out. In driving the wedge, the workman can have as long intervals as he pleases between his blows, because friction saves all that is gained. This effect is very well exemplified in the capstan, Fig. 115. It requires but little exertion of the man who sits there to hold the rope, because the few turns of it around the axle prevent its slipping easily.

A third advantage which often attends the use of tools and machines is that force may be made to produce motion at various distances, in various directions, and in various degrees of velocity. Thus, as to distance, a man standing on the ground can raise a weight to the top of a house by a pulley. A water-wheel may by the connections of machinery produce motion at considerable distances from it. Then, as to direction, horizontal motion may be converted into vertical, rotary into straight, etc. The velocity of motion is generally varied by cog-wheels. Thus, a wheel of 60 cogs, revolving once in a minute, playing on a wheel of 10 cogs, will make it revolve once in 6 seconds.

Another advantage of tools and machines is that they secure a better mode of applying power than we otherwise could have. Thus, when several men are pulling on a rope, much power is lost by their pulling irregularly, a difficulty which is removed by the pulley. The same can be said of applying pressure by the screw. One man presses more steadily, and therefore more effectually, than fifty men would without the screw. The arrangements of tools and machines are so made as to provide convenient ways of applying our strength. An instrument, for example, for moving a weight by hand is so shaped as to hold the weight well, and also to afford a good handle for the hand to grasp. The common claw-hammer is a very good illustration. We grasp the nail by an iron claw; with the handle we can apply not merely the force of the hand, but that of the whole arm, and then we have the immense lever power of the instrument. We have a good illustration of convenience in an instrument in what is called a Lewis, represented in Fig. 133. It is used for raising blocks of stone in building. It has three parts, A B C. It is used in this way: A hole of the same shape as the instrument is made in the upper part of the block of stone to be raised; then A and C are inserted, and B is pushed in between them. With the ring, D, bolted through the instrument, the stone is raised to its place by the ordinary machinery. The principle of the instrument, you see, is that of the wedge.

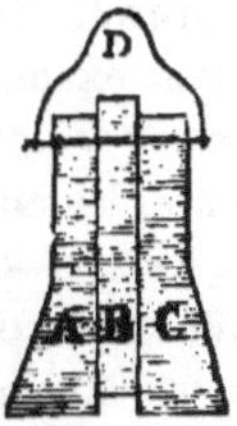

Fig. 133.

96. **Man a Tool-making Animal.**—Though there is no actual saving of power in the tools and machines which man uses, yet so great are the advantages which he reaps from them that, more than two thousand years ago, a philosopher thought that man could not be better distinguished from brutes than by calling him a tool-making animal. If the distinction was so striking in the time of Aristotle, when tools and machines were so few 'in number and so rudely contrived, and so few of the sources of power were appropriated by man to his use, how much more striking is it now, with all the variety and perfection of instruments and machinery, and with the ever-extending appropriation of the sources of power furnished by the elements! The power which air and water and gravitation supply is applied constantly with more and more variety and effect;

G 2

and the appropriation of that mighty source of power, steam, is wholly a modern invention.

QUESTIONS.

81. What is said of the forces of nature and of machines? Name the six simple machines. Why is the term mechanical powers, formerly applied to them, erroneous? Explain the terms power, weight, and fulcrum. —82. How are levers classified? What is the lever of the first kind? Illustrate its uses. What is said of steelyards and the balance? Give other examples of this kind of lever.—83. Show that there is no gain of power in this lever. What is said of the seesaw? What boast did Archimedes make?—84. What is a lever of the second kind? Give examples of its use.—85. What is a lever of the third kind? Give examples.—86. What constitutes a compound lever? Explain its use in platform scales. —87. Explain the action of the wheel and axle. What is said of the windlass? What of the capstan? Explain the construction of the fusee of a watch.—88. What is the advantage of the pulley? Describe the fixed pulley and illustrate its uses. Describe a movable pulley. Explain some of the arrangements of pulleys.—89. Illustrate the mechanical advantage of the inclined plane. What is said of railways?—90. What is a wedge? Give examples of its uses.—91. Upon what principle does the screw work? How is its force estimated? Give an example.—92. Describe some other simple machines.—93. What is said of the pendulum? Who first made use of it for the measurement of time? Explain its operation.—94. What is meant by friction? What is its effect in machinery? How may it be overcome? Illustrate the uses made of friction.—95. What is the first advantage of the simple machines which is mentioned? Give the illustrations. What is the second advantage? Give the illustrations. What is the third advantage? Give examples. How is the velocity of motion in machinery usually varied? What is the fourth advantage? Mention examples. Describe the instrument called a Lewis.—96. What is said of the title by which Aristotle distinguished man from other animals?

CHAPTER X.

HYDROSTATICS.

97. What Hydrostatics Teaches. — A single substance may, as we have seen, exist in three forms—solid, liquid, and gaseous—and these forms are distinguished one from another by the difference in the mobility, or *flow*, of the particles composing the substance. In a solid the particles are comparatively rigid, the force of cohesion being strong; in a liquid or a gas they are not in such close contact, and are freer to move about each other. A body in either the liquid or the gaseous state is called a *fluid*, on account of the *flow* of the particles. A very important branch of Natural Philosophy relates to the pressure, motion, and other phenomena of fluids, which for convenience is considered under three heads, viz.: *Hydrostatics*, which treats of the pressure of liquids; *Hydraulics*, which treats of their motion; and *Pneumatics*, which treats of the same phenomena in air and gases.

In order to understand the phenomena of Hydrostatics, you must bear in mind that they result from the influence of the attraction of the earth upon liquids, and that they depend upon the two great characteristics of liquids, their mobility and incompressibility (§ 17).

98. Level Surface of Liquids.—Owing to the perfect mobility of the particles among each other, and their being equally attracted towards the centre of the earth, liquids at rest assume a level surface. The particles forming the surface may be regarded as the tops of so many columns of particles supported by a uniform resistance or pressure below; for no particle below can be at rest unless urged equally in all directions, and therefore all the particles, at any one level, which, by equally urging one another, keep themselves at rest, must be bearing the weight of equal columns. Thus, a higher column, however produced, must sink and a lower one must rise until just balanced by those around; that is, until all become alike. The particles of water may be compared to shot; if you place shot in a box and heap them up in any portion of the surface, on shaking the box those that are highest will roll down, and a level surface will result. They would do this without agitation if they were as free to move among themselves as are the particles of water. If a microscope could be made strong enough to distinguish the shape of the particles of water, the surface might possibly appear like the level surface of shot in a vessel. But the particles of water are so exceedingly minute that the surface of water, when entirely free from agitation, is so smooth as to constitute a perfect mirror, often feasting our eyes with another world of beauty as we look down into its quiet depths. Strictly speaking, the surface of a liquid is not level or horizontal; being parallel to the surface of the earth, it forms a curve, but this curved surface is of so great a radius that it cannot be perceived unless we take into view a very large surface, as the ocean. Here it is very manifest; for whenever a ship sails out of port, the topmost sail is the last thing seen from the shore, the rest of the ship being concealed by the water rounded up between it and the observer. This is illustrated in Fig. 134.

Fig. 134.

If the earth had no elevations of land, or if there were water enough to cover them, the water would make a perfectly globular covering for the earth, being held to it by the force of attraction. The reason for this is precisely the same as was given in § 31 for the disposition of a drop of liquid to take the globular form. As in that case, so in this, it can be demonstrated that each particle is attracted towards a common centre, and that this will produce in the freely moving particles a uniformly rounded surface. What could thus be shown to be true if the earth were wholly covered with water is true of the portions of water which now fill up the depressions in the earth's crust.

Spirit-Level.—What is commonly called a perfectly level surface is, then, one in which every point is equally distant from the centre of the earth, and is therefore really a spherical surface. But the sphere is so large that any very small portion of it may be considered, for all practical purposes, a perfect plane. A hoop surrounding the earth would bend about four inches in every mile. In cutting a canal, therefore, there is a variation in this proportion from a straight level line. Since the variation is but an inch in a fourth of a mile, it is of no account in taking the level for buildings. Levels are ascertained by what is called a spirit-level. This consists of a closed glass tube, Fig. 135, nearly filled with alcohol. This liquid is used in pref-

Fig. 135.

erence to water because it never freezes and is more mobile. The space not filled with alcohol is occupied by air. The tube is placed in a wooden box for convenience and security, there being an opening in the box at *a*. Now, when the box with its glass tube is perfectly level, the bubble of air will be seen in the middle at *a*; but if one end be higher than the other, the bubble will be at or towards that end. This simple instrument is used by masons and carpenters for the purpose of levelling walls or floors of buildings, by engineers in surveying, and by others.

99. **Flow of Rivers.**—If a trough be exactly level, the

water will be of the same depth at one end as at the other, for the surface of the water at both ends will be at the same distance from the centre of the earth. But if one end be raised, the water will become deeper at the other end. If it were not so, the surface at the two ends would not be at the same distance from the centre of the earth. Now, if water run in at the upper end of an inclined trough and out at the lower, we have an illustration of what takes place in all rivers—the water is in constant motion. A very slight incline gives a flow to water, for the particles are so mobile that, in obedience to the force of gravitation, they descend the inclined plane to seek a level. Three inches per mile in a smooth straight channel gives a velocity of about three miles an hour. The Ganges, which gathers the waters of the lofty Himalaya Mountains, in running 1800 miles falls only 800 feet; and to fall gradually these 800 feet in its long course, the water takes nearly a month. The gigantic Rio de la Plata has so gentle a descent to the ocean that, in Paraguay, 1500 miles from its mouth, large ships arrive which have sailed against the current all the way by the force of the wind alone.

100. **How some Rivers have been Made.** — Changes are constantly produced in the earth by the disposition of water to seek a level. In doing this the water carries solid substances of various kinds from elevated places into depressed ones, tending to fill up the latter. New channels are also sometimes made by the water. The boy who makes a little pond with his mud-dam, and lets the water overflow from it into another pond on a lower level, as he sees a channel worked by the water between the two ponds becoming larger and larger, witnesses a fair representation on a small scale of some extensive changes which have, in ages past, taken place in some parts of the earth. It is supposed, and with good reason, that many rivers had their origin in the way above indicated. For example, where the Danube runs its long course there was once a chain of lakes. These becoming connected together by their overflow, the channels cut between them by the water continually became larger, until at length there was one long, deep, and broad channel, the river; while the lakes became

dry, and constituted the fertile valley through which that noble river runs to empty into the Black Sea. It is said that a similar process is manifestly going on in the Lake of Geneva, the outlet of it becoming continually broader, while the washing from the neighboring hills and mountains is filling up the lake. Towns that a century ago lay directly upon the borders of the lake now have gardens and fields between them and the shore; and Dr. Arnott says, "If the town of Geneva last long enough, its inhabitants will have to speak of the river threading the neighboring valley, instead of the picturesque lake which now fills it."

101. **Canals.**—The management of the locks of a canal is in conformity with the disposition of water to seek a level. A ground view of one lock and a part of two adjacent locks is given in Fig. 136. The lock C has two pairs of flood-

Fig. 136.

gates, D D and E E. The water in A is higher than in C, but the level is the same in C and B, because the gates, E E, are open. Suppose there is a boat in the lock B that you wish to get into the lock A. It must be floated into the lock C, and the gates E E must be closed. The water may now be made to flow from 'the higher level, A, into C, till the level is the same in both A and C. But this cannot be done by opening the gates D D, for the pressure of such a height of water in the lock A would make it difficult, perhaps impossible, to do this; and, besides, if it could be done, the rapid rush of water into C would flood the boat lying there. The discharge is therefore effected by openings in the lower part of the gates

D D. These openings are covered by sliding shutters, which are raised by racks and pinions, as represented in Fig. 137. When the water has become of the same level

Fig. 137.

in A and C, the gates D D can be easily opened, and the boat may be floated from C into A. If a boat is to pass downward in the locks, the process described must be reversed.

Canals are also extensively used for supplying water by side openings to turn water-wheels for the working of machinery. The water turns the wheel by the force which gravitation gives it as it descends from the level of the canal to the level of the river. One of the grandest canals in the world is that cutting through the Isthmus of Suez, constructed in 1864–69 under the direction of a French engineer, M. De Lesseps. It is about one hundred miles in length, and connects the waters of the Mediterranean with those of the Red Sea. The importance of canals as a means of transportation of heavy goods is evident from the fact that a horse which can draw but one ton on our best roads can draw thirty with the same speed in a canal-boat.

102. **Other Illustrations.**—We see the tendency of liquids to rise to the same level in other ways. In a coffee-pot

the liquid has the same level in the spout as in the vessel itself, whatever may be its position, as seen in Fig. 138.

Fig. 138.

If it be turned up so far that the level of the fluid in the vessel is higher than the outlet of the spout, the fluid runs out. If two reservoirs of water be connected, the water will stand at the same height in both, whatever the distance between them may be. In the aqueduct pipes that extend from a reservoir, the water will rise as high as the surface of the water in the reservoir itself. If the outlets of the pipes be lower than this level, the water will run from them, as in the case with the coffee. The cause of these and similar facts is the same as that of the level surface in vessels and reservoirs—the action of gravitation. This may be made plain by Fig. 139. Let the figure represent the section of a vessel with divisions of different degrees of thickness, these divisions, however, not extending to the bottom of the vessel. Water in this will stand at the same level in the different compartments, just as it would if the vessel had no such divisions. This is simply because the attraction of the earth acts upon the water in the same way with the divisions as

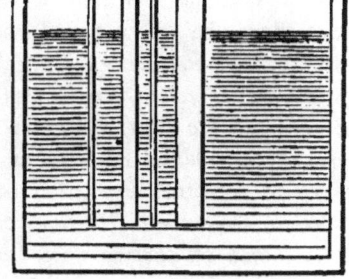

Fig. 139.

. without them. And you can see that it will make no difference whether these divisions be thick or thin, or whether
the apartments be near together or far apart, as in the case
when branch pipes extend from a reservoir. A branch
pipe may be considered as having the same relation to the
reservoir as one of the narrow compartments in the figure

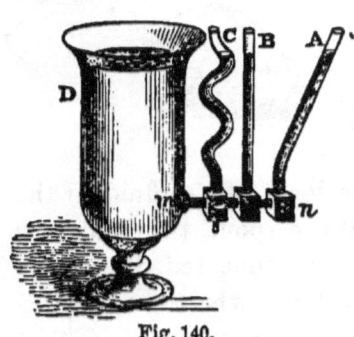

has to the rest of the vessel.
The result is not at all affected
by either the size or form of the
tubes that may be connected
with a common reservoir: a fluid
will stand at the same height in
all. Thus we have, in Fig. 140,
tubes of various size and shape,
A, B, C, connected by the pipe,
m n, with a reservoir, D; and if

Fig. 140.

water be poured into the latter, it will rise to the same
height in all, just as in the different compartments of the
vessel represented in Fig. 139.

A man once thought that he had solved the chimerical problem, perpetual motion, by means of a vessel constructed as in Fig. 141. He rea-
soned in this way: If the vessel contain a pound
of water and the tube only an ounce, since an
ounce cannot balance a pound, the water in the
vessel must be constantly forcing that in the tube
upward. It therefore must constantly run out of
the outlet of the tube, and as it flows into the
vessel the circulation must go on, the only hindrance to perpetual circulation being the evaporation of the water. He was confounded when

Fig. 141.

he discovered, on pouring water into the vessel, that it stood at precisely
the same level in the vessel and in the tube. He forgot that a common
teapot is nearly such a vessel, and yet does not overflow.—A glass tube on
the outside of a cistern or a boiler, and connected with it at the bottom,
shows at once the level of the water within.—Another illustration of the
fact that water is always seeking its level is found in the water-pipes which

distribute water to the inhabitants of large cities: the water pumped into a reservoir situated on an elevation rises by the action of gravity and its perfect mobility to the height of every cistern not above the level of the reservoir, no matter how much lower may be the depression crossed by the pipes.—We are not to suppose that it was ignorance of this law of liquids that led the ancients to build aqueducts of stone at immense expense, in some cases spanning valleys at great heights; but this enormous labor was necessitated by the lack of a suitable material such as iron.

103. Springs and Artesian Wells.—The principles developed in the previous paragraphs will explain the phenomena of springs, common wells, and Artesian wells. The crust of the earth is largely made up of layers of different materials, as clay, sand, gravel, limestone, etc. When these were formed they were undoubtedly horizontal, but they have been thrown up by convulsions of nature in such a way that they present every variety of arrangement. Since some of these layers are much more pervious to water than others, the rain which falls and sinks into the ground often makes its way through one layer lying between two others which are impervious to water, and so may make its appearance at a great distance from the place of its entrance, and at a very different height. How this explains the phenomena of springs, common wells, and Artesian wells is made clear by Fig. 142. A A and B B B are designed to represent porous layers of earth lying between other layers which are impervious to water.

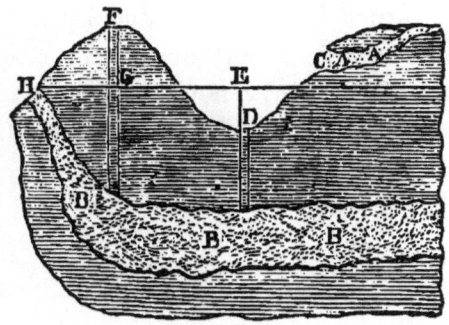

Fig. 142.

The water in A A will flow out at C, making what is commonly called a spring. If we dig a well at F, going down to the porous layer, B B B, the water will rise to G, because this is on a level with the surface of the ground, H, where the supply of water enters. From this point it may be raised by a pump. If the well be dug at D, the water will rise not only to the surface, but to E, because this is on a level with H. Water is sometimes obtained under such circumstances from very great depths. In this case the porous stratum containing the water is reached by boring, and then we have what is termed an Artesian well. The name comes from the province of Artois, in France, where this operation was first executed. There is a celebrated well of this sort at Grenelle, a suburb of Paris, where the water rises from a depth of nearly 1800 feet below the surface, and is further carried to a height of a hundred feet above it, furnishing a supply of beautifully clear water at the rate of 800,000 gallons a day.

104. Pressure of Liquids is in Proportion to Depth.—The pressure of a fluid is in exact proportion to its depth. For, all the particles being under the influence of gravity, the upper layer of them must be supported by the second, and these two layers together by the third, and every layer must bear the weight of all the layers above it. This pressure, being occasioned by the weight acting vertically downward, is not dependent on the amount of the surrounding liquid, nor on the shape or size of the containing vessel. In a vessel having the shape A, Fig. 143, it is evi-

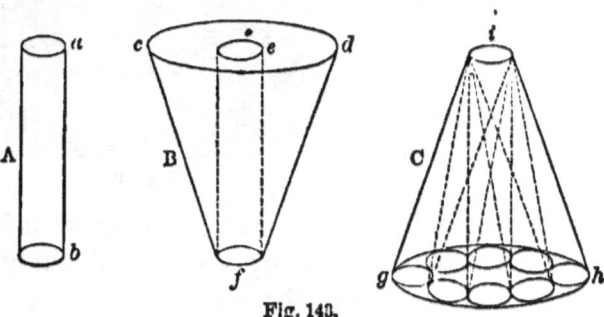

Fig. 143.

dent that the pressure of the liquid within upon the bottom *b* depends upon the height of the column *a b*; in the vessel B, which widens at its mouth, the pressure on the bottom *f* is equal to that of the weight of the column *e f*, for the rest of the liquid exerts pressure upon the sides *c f, d f*, and balances the column *e f* on all sides round about.

In the case of a vessel tapering at its mouth, C, let us suppose the bottom *g h* divided into a number of portions of the same size as the mouth *i*, and that there are eight of these portions. The pressure on the bottom *g h* is the same as that of eight columns of liquid of the same size as *i*. An inspection of Fig. 144 will perhaps facilitate comprehension of this law. The pressure on the bottom *k l* of the vessel

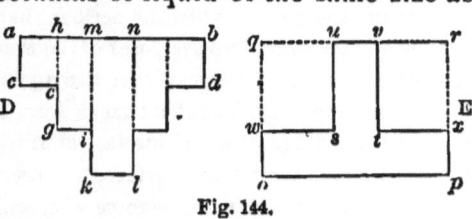

Fig. 144.

D is equivalent to the weight on the liquid column *m n k l*; the pressure on *g i* depends upon the height of column *h m g i*, etc. In the case of a vessel having the form E, the bottom *o p* has to bear the weight of the liquid column *w x o p*; this, however, has itself to sustain the pressure of the column *u v s t* exerted upon the liquid at *s t*, and this pressure produces an effect equivalent to that of a column whose base is *o p*, and whose height is *o q*. It is evident from the foregoing that the total pressure of a liquid on the bottom or side of a vessel depends on the area and the depth. A tube two feet long and a square inch in section holds nearly a pound of water; hence the pressure of water at any depth, whether on the bottom of a vessel or on its side, is little less than *one pound on the square inch for every two feet of depth*—a general truth worth remembering.

Illustrations.—The increase of pressure at great depths produces the most striking effects. Thus, if an empty corked bottle be let down very deep at sea, either the cork will be driven in or the bottle will be crushed before it reaches a depth of ten fathoms. A gentleman tried the following experiment: He made a pine-wood cork, so shaped that it projected over the mouth all around. He then covered this with pitch, and fastened over the whole several pieces of tarpaulin. The bottle, thus prepared, he let down to a great depth by attaching to it a weight. On raising it up he found that it contained about half a pint of water strongly impregnated with pitch, showing that the pressure of the water forced water through the several pieces of tarpaulin, the pitch, and the pores of the wooden cork. When a ship founders near land, the pieces of the wreck, as it breaks up, float to the shore; but when the accident happens in deep water, the great pressure forces water into the pores of the wood, and thus makes it so heavy that no part of the vessel can ever rise again to reveal her fate. When a man dives very deep, he suffers much from the pressure on his chest. If we watch a bubble of air rising in water, it is small at first, but it grows larger as it approaches the surface, because it sustains less pressure than when deep in the water. The force with which a fluid is discharged from an opening in a vessel depends on the height of the fluid above the opening. The difference in this respect between a full barrel and one nearly empty is very obvious.—It is not known whether there is a limit to the pressure which fishes can bear with impunity, but they abound chiefly in the shallow waters on coasts, or on banks in the midst of the ocean, such as the banks of Newfoundland, or the bank of Lagullas, off the Cape of Good Hope.

105. **Sluice - Gates, Dams, etc.**—The application of the above principles in the construction of sluice-gates, dams, etc., is a matter of great practical importance. Pressure in a fluid is always in proportion to the height of the fluid above the point of pressure. The pressure upon any portion of the side of a vessel containing a fluid must be in proportion to its distance from the surface; or, in other words, it is the weight of a column of water extending from this portion to the surface. Let A B C D (Fig. 145) represent a *section* of a cubical vessel—that is, one in which each side is of the same size with the bottom. The press-

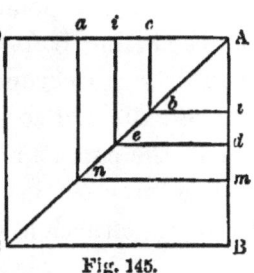

Fig. 145.

ure on the point *a*, in the line A B, is that of a column of particles, A *a*. But A *a* is equal to *c b*, and *c b* is equal to *b a*. Therefore *b a* may represent the pressure on *a*. In the same way, it can be shown that *e d* represents the pressure on *d*, *n m* the pressure on *m*, C B that on B. Therefore the pressure on all the points in A B will be represented by lines filling up all the triangular space A B C, and this is half of A B C D, which represents the pressure on the line C B. It is clear, then, that since the pressure on a vertical line in the side is half that on a line at right angles to it in the bottom, the pressure on the whole side is half that on the whole bottom.

We see from the above demonstration why a dam is

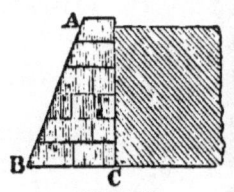

Fig. 146.

built in the form represented in Fig. 146. We learn, also, why the hoops and other securities at the lower part of the monstrous vats used in breweries (some of them holding many thousand barrels) require to be made of very great strength. It is manifest, also, that if a sluice-gate is to be kept shut by a single support, this must be applied at one third of the distance from the bottom, there being as much pressure, as shown in Fig. 145, on the lower third as on the upper two thirds of the gate.

106. **Lateral Pressure in Fluids.**—The pressure of a liquid on the side of a vessel, just mentioned, is a *lateral* pressure, and it is caused by the downward pressure of gravitation in the liquid. But how? The particles of a fluid are freely movable among each other, and therefore are ready to escape from pressure in any direction. The particles at *a*, Fig. 145, pressed upon by the column of particles extending above them to the surface, are ready to escape

laterally, and would do so if an opening were made in the vessel at that point. But if the vessel contained a block of ice, fitting it as accurately as the body of water, there would be no escape at the opening, because the particles of the solid are so held together that the downward pressure of the earth's attraction occasions no lateral pressure.

The manner in which the downward pressure of the earth's attraction causes lateral pressure may be made

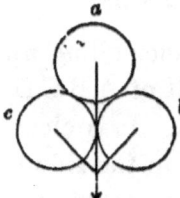

Fig. 147.

clear by Figs. 147 and 148. We will suppose that the particles of solids and liquids are alike round, and that a solid differs from a liquid only in having its particles firmly united by attraction. Let *a*, *b*, and *c*, in Fig. 147, represent three particles of a solid. Since they are united firmly, they will have

a united pressure from the centre of gravity directly towards the centre of the earth, as represented by the arrow. Let now *d*, *e*, and *f*, Fig. 148, represent three particles of water. These being but very slightly coherent, will make each an independent pressure towards the earth's centre, as indicated by the arrows. It is plain that *d* tends to separate *e* and *f*, and will do so if they are left free to move in a

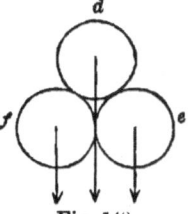

Fig. 148.

lateral direction. For example, if *e* be at the side of a vessel, and an opening be made there, the downward pressure of *d* will give *e* a lateral movement, forcing it out of the opening.

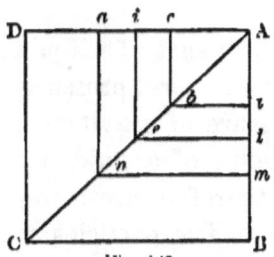

Fig. 149.

Another View.—Referring again to Fig. 145, here repeated, observe that the lateral pressure at any point in the side of a vessel, as *a*, is occasioned *wholly* by the downward pressure of a vertical column of particles extending from that point to the surface. The neighboring columns of particles have nothing to do with it. The same thing is true in regard to any other point either in the line A B or another line drawn on the side of the vessel. And therefore the pressure upon the whole side is occasioned solely by the columns of particles in close proximity to it, and not at all by the other columns of particles in the vessel.' The number of these columns, therefore, in the vessel—or, in other words, the breadth of the body of water in it—makes no difference in the pressure on its side. For this reason two flood-gates so near together that a few hogsheads or even pails of water fill up the space between them sustain as much pressure as they would if a lake or an ocean of water lay between them. The project of digging a ship-canal between the Red Sea and the Mediterranean, now so triumphantly accomplished, was objected to on the ground that the water in the former being twenty feet higher than in the latter, it would burst through the flood-gates with such force as to produce most disastrous results. But according to the principle just illustrated, there is no more danger of this than there would be if two ponds were united by a canal, in one of which the water is twenty feet higher than in the other.

107. **Pressure in Liquids Equal in all Directions.**—We are now prepared to go a step farther. The pressure occasioned by gravitation in fluids operates equally in all directions when the fluid is at rest. That is, any particle of a liquid is pressed equally in all directions. If it were not so, it would not remain at rest, but would be moved in the direction in which the superior pressure operates. Suppose that a, Fig. 150, is a stratum of particles in a vessel containing water at rest. The upward pressure on it being equal to the downward pressure, the stratum neither rises nor falls. If a body of liquid be disturbed by wind or any other cause, those particles which are raised above the common level in waves are pressed downward more than upward or laterally, in obedience to the action of gravitation. They therefore move downward, pushing laterally and upward the neighboring particles, until the liquid regains its level surface and its state of rest. In like

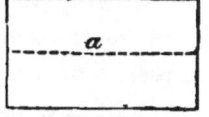

Fig. 150.

II

manner, if any particles are heated they become lighter than their neighboring particles; and the latter, being more strongly attracted than the former, push them upward in order to take their places. When all the liquid acquires the same temperature, it is at rest, each particle having an equal pressure upon it in all directions.

Illustrations.—If a bladder filled with water be compressed by the hand, the water is pressed no more immediately under the hand than in any other part of the bladder, and wherever an opening be made the water will

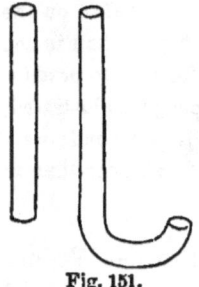

rush out with equal readiness. A hose-pipe as readily bursts upward as in any other direction. A large cork, sunk in very deep water, will be uniformly reduced in its dimensions, showing that it has been pressed equally on all sides. In the experiments with the closed bottles (§ 104), the result is the same if the bottle be sunk with its mouth downward. If two tubes, shaped as in Fig. 151, be thrust down into water, the water will rise with equal facility in both, although in the straight one the pressure which carries up the water is wholly upward, while in the bent one it is at first downward.

Fig. 151.

108. Upward Pressure as the Depth.—It has been shown that the downward and the lateral pressure are in proportion to the depth. The same is true of the upward pressure, and owing to the same cause—the attraction of the earth. Let us examine this case. Why is any particle of a fluid pressed upward at all? It is owing to the struggle on the part of the neighboring particles to get below it. And why this struggle? It results from the attraction of gravitation, and the greater this attraction the greater the upward pressure. The upward pressure therefore, as well as the downward pressure, differs at different depths. Thus, in Fig. 152, the upward pressure against the layer or stratum of particles, *b*, is greater than

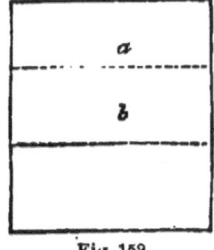

Fig. 152.

that against a, for the same reason that the downward pressure on b is greater than that on a. But the two pressures at b are equal, and so are they at a, and therefore each stratum remains at rest.

Experiments.—Some very neat experiments show that the upward pressure varies with the depth. Take a large glass tube, A B C D, Fig. 153, fitted at one end with a circular plate of brass, which may be held there by a string, F. Thus arranged, plunge it quite deep into water, and you will find it unnecessary to hold on to the string, for the brass disk will be held tight to the tube by the upward pressure of the water. Now draw up the tube slowly, and at length the disk will fall from the end of the tube. Why? Because the end of the tube has come to a

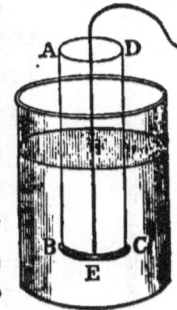

Fig. 153.

point where the upward pressure of the water is less than the downward pressure of the disk. To succeed with this experiment, the end of the tube where the disk is applied must be very even and smooth.

Another experiment may be tried in this way: Tie to one end of a glass tube a piece of thin India-rubber or of a bladder, and fill the tube partly with water. The India-rubber will of course bulge out or become convex from the weight of the water. Press the closed end down a little way in a vessel of water, so that the level in the tube shall be above the level in the vessel. The India-rubber is still somewhat convex, for the upward pressure, being in proportion to its distance from the surface of the water outside of the tube, is not so great as the downward pressure of the higher water within the tube. Lower the tube so that the level in the tube is the same with that in the vessel. The India-rubber then becomes flat, because the downward and upward pressures upon it are equal, just as would be the case with a stratum of water in the same place. But press the tube lower down, and the India-rubber bulges upward into the tube, because the upward pressure is then greater than the downward.

109. Illustrations of Liquid Pressure.—You are now prepared to understand the explanation of some very striking phenomena in the pressure of liquids. If you take a perfectly tight cask filled with water, and screw into its top a long tube, you can burst the cask by pouring water into the tube. To understand this you must bear in mind two facts—that the fluid in the cask is not compressible, and that its particles move freely among each other.

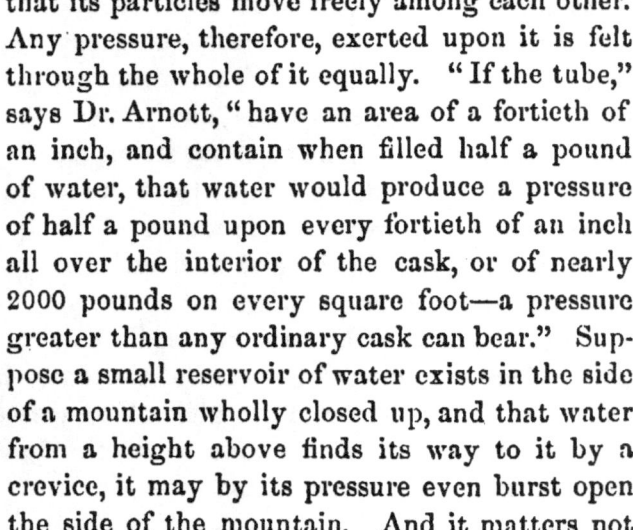

Any pressure, therefore, exerted upon it is felt through the whole of it equally. "If the tube," says Dr. Arnott, "have an area of a fortieth of an inch, and contain when filled half a pound of water, that water would produce a pressure of half a pound upon every fortieth of an inch all over the interior of the cask, or of nearly 2000 pounds on every square foot—a pressure greater than any ordinary cask can bear." Suppose a small reservoir of water exists in the side of a mountain wholly closed up, and that water from a height above finds its way to it by a crevice, it may by its pressure even burst open the side of the mountain. And it matters not how large or small the crevice may be, for pressure in a liquid is only as the height. If the reservoir be ten yards square and an inch deep, and the fissure leading to it be but an inch in diameter and two hundred feet in height, it is calculated that the pressure of the water in the fissure would be equal in force to the weight of 5000 tons.

Fig. 154.

The manner in which these effects are produced may be made clear by Fig. 155. Let A be a close vessel filled with water, and let a tube, b, be made fast in it, with a movable plug or piston at c. If the surface of the water be pressed upon by this piston with the force of a pound, the water being incompressible and its particles freely movable among each other, the pressure will be extended equally through

all the water, and every
portion of the vessel A
of equal surface with *c*
will bear a pressure to
the extent of one pound.
If another tube, *d*, of the
same size were inserted
with a piston, *i*, the force
of a pound applied to the

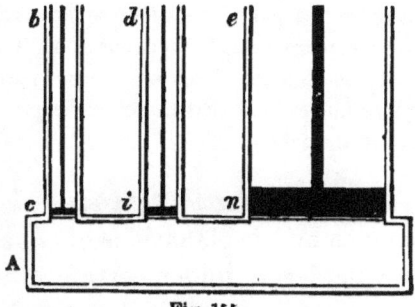

Fig. 155.

piston *c* would push upward the piston *i* with the same
force. And if there were several pistons of the same size,
by pushing upon one with the force of a pound they would
all be pressed upward with exactly this force. Further, if
e be a tube five times as large as *b*, its piston, *n*, will be
forced upward with a pressure of five pounds by the
downward pressure of a pound upon *c*. Suppose, now,
that a pound of water were substituted for the piston *c*,
the other pistons would be pressed upward as before.
And if all the pistons be removed, the pound of water in
b will press the water up the tube *d* with the force of a
pound, and up the tube *e* with the force of five pounds.

To make this still clearer, we will present it in a little different form.
Let B, Fig. 156, be a close vessel with two tubes, one of which is five

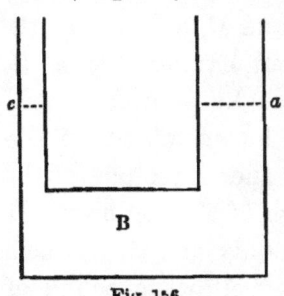

Fig. 156.

times as large as the other. If sufficient water
be poured into the vessel to occupy a part of
the tubes, it will stand at the same height in
both tubes, as indicated. If there be a pound
of water, then, in the tube *c*, there will be five
pounds in *a*. Now, if the five pounds of water
in *a* pressed any heavier on the whole body of
water in B than the pound of water in *c*, it
would force the water in *c* to a greater height.
But this is impossible, as has been shown in
§ 102. Observe that the pressure of five pounds in *a* is spread over five
times the area or extent of surface as the pressure of one pound in *c*. If
the tube *c* have an area of an inch square, the water in it will exert a

pressure of a pound on every square inch in the vessel. The water in *a* exerts a pressure of five pounds; but it must be remembered that it does not press with this force on every square inch, but on each space of five square inches, and that therefore its pressure on each inch is the same as that in the tube *c*.

110. Hydrostatic Paradox.—It is evident from the phenomena and explanations given above that a small quantity of a fluid can, under certain circumstances, exert an enormous pressure. This fact has been called the Hydrostatic Paradox. It does seem, at first view, incredible or paradoxical when one asserts that a few ounces of water can be made to raise weights of hundreds or even thousands of pounds. But the explanations given show you that there is no unexplainable mystery in the fact. The cause of it is the same as that which gives a level surface to liquids; viz., the force of gravitation acting upon a substance whose particles are freely movable among each other. In fact, there is nothing more paradoxical in it than that one pound at the long end of a lever should balance ten pounds at the short end, an explanation of which is given in the chapter on the Simple Machines.

Hydrostatic Bellows.—The instrument called the Hydrostatic Bellows is represented in Fig. 157. It consists of 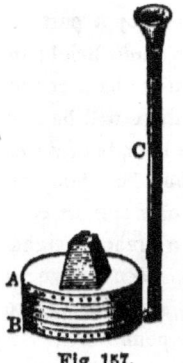 two circular boards, A and B, united together by strong leather, and having a tube, C, through which water can be poured into it. The weight which can be sustained on the bellows without forcing the water out of the tube depends on the size of the bellows. If the area of the tube be only one thousandth of that of the top of the bellows, a pound of water in the tube will balance a thousand pounds' weight on the bellows. It is for the same reason that one pound of water in the

Fig. 157.

tube *c* balances five pounds in *a* (see Fig. 156). As the weight presses upon the top as a whole, it is just as if a vessel of the same size with the bellows were resting upon it and containing a thousand pounds of water. The water, in that case, would stand at the same height in the vessel and in the tube. This shows that the Hydrostatic Paradox is only one illustration of the great fact that a liquid, under the influence of gravitation, seeks its level.

When the weight on the bellows is less than is required to balance the water in the tube, the weight can be raised continually by pouring water into the tube. But observe that although the lifting force be so strong, it is very slow in its operation. If the comparative areas of the tube and the bellows be as above supposed, the water must fall in the tube ten inches in raising the weight the one hundredth part of an inch.

111. **Bramah's Hydrostatic Press.**—These principles have been applied by Mr. Bramah in his Hydrostatic Press.

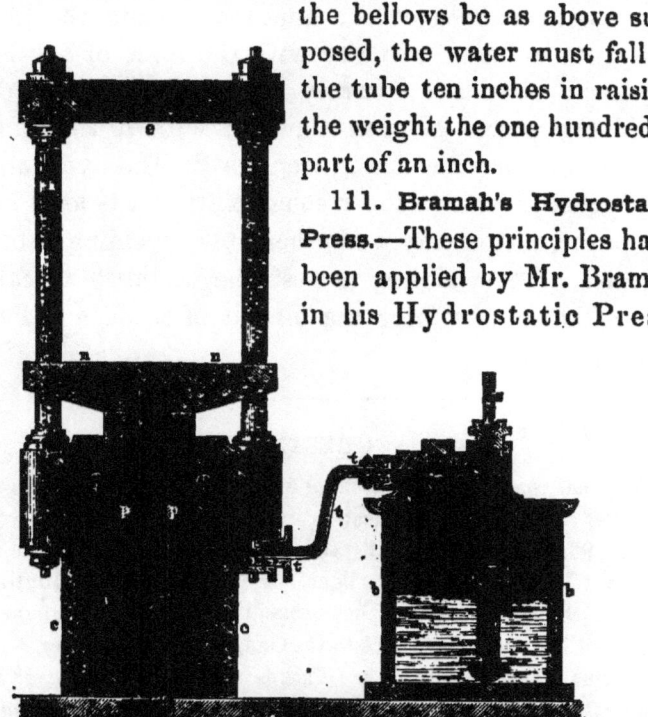

Fig. 158.

This consists of a small metallic forcing-pump, Fig. 158, in which the water in the reservoir *b b* is pumped up by the piston, *s*, worked by a lever not shown in the cut, and forced into a strong and large cylinder, *c c*. In this cylinder is a stout piston, P, having a flat head, *n n*, above. Between this plate and another, *e*, is placed the body to be compressed. It is obvious that the pressure exerted will be in proportion to the difference between the size of the pump and the cylinder, *c c*, just as in the case of the bellows it depended on the difference between the areas of the tube and of the top of the bellows. "If the pump have only the one thousandth the area of the large cylinder, and if a man by means of its lever-handle press its piston down with a force of one hundred pounds, the piston of the great cylinder will rise with the force of one hundred thousand pounds. Scarcely any resistance could withstand the power of such a press; with it the hand of a child might break a strong iron bar." The hydraulic press is of great service in the mechanic arts; it is used in pressing paper, cotton, hay, and other bulky yielding substances, to raise great weights, to test the strength of cables, to launch vessels, to force the oil out of seeds, and for many other purposes.

√

QUESTIONS.

97. What constitutes the essential difference between the different forms of matter? Give the classification of that branch ·of Physics relating to fluids.—98. Why do liquids at rest assume a level surface? Give the comparison. Is the surface of a liquid strictly horizontal? Illustrate this. Describe the spirit-level and its uses.—99. What is said of the flow of rivers?—Illustrate by reference to the Ganges and other rivers.—100. How have some rivers been made? What is stated about the river Danube? What of Lake Geneva?—101. Describe the arrangement of canal locks. Why are canals so useful?—102. Illustrate the tendency of liquids to rise to the same level by reference to a coffee-pot. Describe a foolish man's

plan for perpetual motion, and give the reason of its failure. What is said of ancient and modern aqueducts?—103. Explain the operation of springs and Artesian wells. Whence comes the name Artesian? What is stated of a well in Paris?—104. Why is the pressure of a liquid in proportion to its depth? Give the illustrations of this.—105. Explain the application of these principles to the construction of dams. What is said about the construction of dams and brewers' vats?—106. Explain the lateral pressure of liquids. Show the difference between a liquid and a solid in this respect. Show how the earth's attraction causes the lateral pressure. Give another view. What is said of the proposed ship-canal between the Mediterranean and the Red Sea?—107. Show that pressure in liquids is equal in all directions. Give the illustrations. —108. Show that the upward pressure in a liquid is as the depth, and that this is produced by gravitation. Describe the experiment represented in Fig. 152. Give the experiment with the tube and India-rubber.—109. State the examples given of great effects produced by small quantities of a fluid. Explain these effects by reference to the diagram. Explain Fig. 156.—110. What is the Hydrostatic Paradox, and why is it so called? Describe and explain the Hydrostatic Bellows.—111. Describe and explain Bramah's Hydrostatic Press. Mention some of its uses.

CHAPTER XI.

SPECIFIC GRAVITY.

112. **Nature of the Subject.**—We now reach a very interesting subject, intimately connected with Hydrostatics, and the principles which have been developed in relation to liquids are to be here applied to various kinds of substances. As we proceed you will see that all the phenomena brought to view in this chapter are to be referred to the same cause as those of the previous chapter—viz., the attraction of gravitation.

Before proceeding with the study, we will define Specific Gravity. The specific gravity of any substance is its weight

II 2

as compared with that of an equal volume of another substance taken as a standard. For solids and liquids distilled
water is the standard of comparison, and its specific gravity
is for convenience called 1. Mercury, for example, is thir-
teen and a half times as heavy as an equal volume of water,
and is said to have a specific gravity of 13.5. For gases air
is usually taken as the standard, though hydrogen gas is
sometimes so employed. .

We shall postpone explaining the methods of determining the specific gravities of bodies until we have more fully
detailed the principles involved. (See § 116.)

113. **Action of Gravity on Solids in a Liquid.**—The reason
that a very heavy substance—a stone, for example—sinks in
water is simply because the earth attracts it more strongly
than it does the water, and drags the stone down through
it. If the stone lay upon a bladder filled with water, it
would press upon it with the force with which it is attracted by the earth. But where water is not thus confined, the
stone thrusts its particles to the one side and the other till
it gets to the bottom.

It is the attraction of gravity, also, that makes light substances, as wood and cork, rise in water. In this case the
water is attracted by the earth more strongly than the
wood or cork, and so gets below it, and in so doing pushes
up the lighter substance.

But you will observe that the wood, on rising in the water, does not come completely out of it and lie upon the
surface, but a part of it remains im-
mersed in the water. The explanation
of this will furnish you with the key to
many very interesting facts. Suppose
that half a block of wood, A, Fig. 159,
weighing a pound, is above the surface
of water. As it is attracted to the

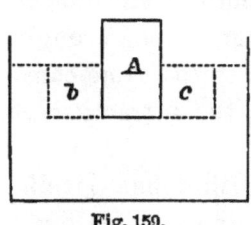

Fig. 159.

earth with the force of a pound, it has pushed to the one side and the other just a pound of water, and taken its place. It is drawn down towards the earth with the same force as the pound of water on either side of it, *b* or *c*. If it were attracted more strongly—that is, if it weighed more than a pound—it would displace more than a pound of water. If it had just the same weight as the same volume of water, it would displace a volume of water equal to its own bulk; it would be wholly immersed, and would stay in the water wherever you placed it, because it is attracted by the earth with the same force as an equal bulk of water.

Imagine water in a vessel divided into equal portions of a pound each, as represented in Fig. 160. Now, suppose that the portion *a* should at once change into solid ice without at all altering its bulk or weight. It would not move from its position, because it is attracted by the earth with precisely the same force as when it was water, and as strongly as each of the equal portions of water around it. But

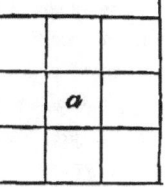

Fig. 160.

since water on-becoming ice does really increase in bulk and therefore become lighter, this block of ice would rise so that a part of it would be above the surface.

The lighter a substance immersed in water, the more of it will there be above the surface. Consider the case of two blocks of wood having different weights, though of the same size. Suppose the heavier one, A, Fig. 161, is one third lighter than the same bulk of water. One third of it will be above the surface. If the other, B, be half the weight of water, half of it will be above the surface. We would say, then, that the specific gravity of

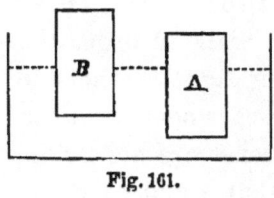

Fig. 161.

the wood in the first block is two thirds that of water, and the specific gravity of the wood in the second is one half that of water.

Illustrations.—There are many interesting facts illustrating the principles here developed. A stone is lifted much more easily in water than in air because of the support afforded by the upward pressure of the water. A boy often wonders why he can lift a very heavy stone to the surface, but can raise it no farther. When a bucket of water is drawn up a well, much less exertion is required to raise it through the water than through the air after it emerges from the water. While it is in the water you raise only the bucket itself; the water in it exerts no pressure on the rope, being sustained by the water around it. But when it reaches the air, you have to raise the weight of the water added to that of the bucket. When a person lies in a bath for some time, on raising his arm from the water it seems to be very heavy. This is because the arm has had for so long a time the support of the water that when it is lifted into the air the want of this support is sensibly felt. It is said that Archimedes conceived of the principles of specific gravity as his limbs felt the liquid support of a bath, and so overjoyed was he with the discovery that he ran home, crying out all the way, "Εὕρηκα! εὕρηκα!"—I have found it! I have found it! It was a rational joy, for he had found a principle of immense value to science and to the world.

Boats and Life-boats.—A boat of iron will float as high out of water as one of wood of the same size, provided the iron be made so thin that the boat is not heavier than the wooden one. For what is it that floats? Not the iron or wood, but a wooden or iron boat filled with air. If it were filled with water instead of air, it would sink, the specific gravity of the materials of which it is built being, on the whole, of greater specific gravity than water. Life-boats have in their structure either a large quantity of cork or air-tight vessels of tin or copper, and consequently are so light that they will float even when filled with water.

114. Specific Gravity of Animals.—Birds have a much less specific gravity than animals that walk, in order that they may rise easily in the air. Their light feathers increase greatly their bulk, as you may see whenever a bird is stripped of them. Besides this, the bones are hollow and communicate with the lungs. Birds that swim, as ducks, swans, etc., have so small a specific gravity—that is, are so large in proportion to their weight—that but a small part of the body is under water, and the motion of their feet is not required to sustain them, but serves, like oars, to propel them along. Insects are of small specific gravity, those that fly the most swiftly being the lightest. Fishes have very nearly the same specific gravity as water, and hence require but little muscular effort to move about in their element. They are assisted much in rising and falling by a contrivance by which they can instantaneously alter their specific gravity. They have an air-bladder, which they can dilate or contract at pleasure. When dilated, the bulk of the fish is increased and his specific gravity lessened, and he rises easily and at once. By compressing it he as readily sinks.

The human body, when the chest is filled with air, is so much lighter than water that it will float with about half the head above the surface. A knowledge of this fact, with proper presence of mind, might ordinarily save persons from drowning; for if the body be put in the proper position, the feet downward and the head thrown backward, the nose and mouth will be out of the water. So little is required in the way of support to keep the whole head out of water, that persons who cannot swim are often saved from drowning by grasping very small pieces of wood. An oar would support half a dozen men if they would be satisfied with keeping only the head out of water; but if each one struggle to get his whole body upon the oar, they may all be lost.

115. Avoidable Causes of Drowning.—The reasons that in water-accidents so many people are drowned who might easily be saved are thus summarized by Dr. Arnott:

1st. They believe that the body is heavier than water; and, therefore, that unless continued exertion be made, they must sink. Hence, instead of lying quietly and a little on the back, with the face only out of the water, they generally assume the position of a swimmer, in which the face is downward, and the whole head has to be kept out of the water to allow of breathing. To do this requires practice; and if a person cannot swim, the first attempt at floating in this position will prove a disastrous failure.

2d. The body raised for a moment by any exertion above the floating level sinks as far below that when the exertion ceases, and the plunge terrifies the unpractised and renders them easier victims to their fate.

3d. They make a wasteful exertion of strength to prevent water entering the ears, not thinking that it can only fill the outer ear, as far as the drum, and that this is of no consequence.

4th. They generally attempt, in their struggle, to keep their hands free above the surface, forgetting that any part of the body held out of the water in addition to the face (which must be out) requires an additional effort to support it. The tendency of the body to sink diminishes just in proportion to the quantity immersed; because all those parts which are out of water, not being supported by the water, become so much additional absolute weight to the portion immersed. This is, indeed, one of the most frequent causes of death by drowning.

5th. If the accident occur at sea, they cannot, like the practised swimmer, choose the proper interval for breathing, which is when the crest of a wave has passed over, and the head is for an instant above water.

6th. The chest should be kept as full of air as possible, which without other effort will cause nearly the whole head to remain above water. If the chest be once emptied while the face is under water, and the person cannot inhale again, the body remains specifically heavier than water, and will sink.

A life-preserver is a great aid in preservation from drowning, for it diminishes the specific gravity of the body. It is commonly an air-tight bag fastened round the upper part of the body, which can be filled when required by blowing into it through a tube fitted with a valve. "On the great rivers of China," says Dr. Arnott, "where thousands of people find it more convenient to live in covered boats upon the water than in houses on the shore, the younger male children have a hollow ball of some light material attached constantly to their necks, so that in their frequent falls overboard they are not in danger."

In wading a river the feet press upon the bottom with a force equal to half the weight of the person's head, this being the difference between the weight of the body and the weight of the same bulk of water. Now, this pressure is not sufficient to give a sure footing against even a moderate current. Many persons have been drowned from ignorance of this fact. A man carrying a load may often ford a river safely where without a load to press him down, and thus give him a sure footing, he would be carried down the stream.

116. **How to Determine the Specific Gravity of Solids.**— From the principles explained in the preceding pages it is evident that, owing to the upward pressure of water, a body weighs less in water than in air; hence we determine its specific gravity by comparing its weight in water with its weight in air—water, you remember, being the standard.

In determining the specific gravity of solid bodies, several cases may arise: the solid may be (I.) heavier than water and insoluble in it, (II.) heavier than water and soluble in it, (III.) lighter than water and insoluble in it, and (IV.) lighter than water and soluble in it. For examples of the first case we have gold, lead, and the other heavy metals, also rocks and minerals of various kinds; as an example of the second case we have sugar, salt, saltpetre, and many other substances; cork, the varieties of wood, and the bodies of animals are examples of the third case; the fourth case is very rare.

(I.) We will consider the simplest first, and take for an example a piece of lead. Suspend the lead by means of a hair from one of the pans of a balance, as shown in Fig. 162, and weigh it carefully. Then

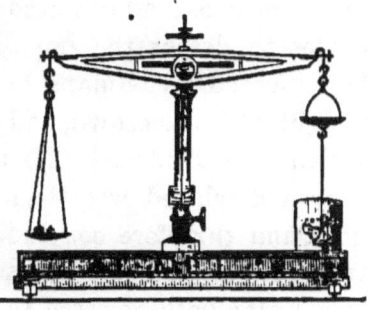

Fig. 162.

introduce the lead into a cup of water, and you will find that a portion of the weight must be removed from the opposite pan to preserve the equilibrium. The weight which you take from the pan will be the weight of a quantity of water equal in bulk to the piece of lead, for the immersed body is supported with a force equal to the weight of the water it displaces (§ 113). Thus if a piece of lead weighing eleven grammes weigh only ten grammes in water, it will prove that lead is eleven times as heavy as water. And if a lump of copper weigh nine grammes in air and eight in water, it is nine times as heavy as water. Calling, therefore, water=1, the specific gravity of lead is 11 and of copper 9. It is obvious that a body of the same specific gravity with water would weigh nothing when immersed in water, for it would be supported with an upward pressure precisely equal to its own weight, just as the same bulk of water is. A hundred grammes of water, therefore, will weigh nothing in water. The experiment can easily be tried. Weigh an empty glass bottle, suspended from one arm of the scale-beam, and then put a hundred grammes of water in it. On immersing it in water it will be balanced.

Archimedes and the Crown.—Hiero, King of Syracuse,[*] stipulated for a crown of pure gold. But suspecting the maker of it had adulterated the gold, he called upon Archimedes to detect the imposture. He did it in this way: he procured two lumps of gold and silver of the same weight with the crown, and observed the quantity of water which each displaced. He then tried the crown, and found that it displaced less than the silver and more than the gold, and therefore concluded that it was an alloy of the two metals. All this was suggested to him by his experience in the bath, referred to in § 113.

---

[*] Hiero II., died 216 B.C.

117. More about Determining Specific Gravity.—The process explained in § 116 is called the method by direct weighing. We may condense the instructions into the following rule : To find the specific gravity of a solid heavier than water and insoluble in it, weigh it in the air, and again in water; then divide the first weight by the difference between the first and the second weight. Suppose we represent the weight of the solid by the letter W, its weight in water by Z, and the words specific gravity by the abbreviation sp. gr., then we may write the rule in the following manner :

$$Sp.\ gr. = \frac{W}{W - Z}$$

Observe that $W - Z$ gives us the weight of the water displaced. An examination of the following proportion will show how the above equation is obtained :

$$\left\{ \begin{array}{c} \text{Weight of} \\ \text{water displaced,} \\ or \\ W - Z \end{array} \right\} : \left\{ \begin{array}{c} \text{Sp. gr.} \\ \text{of water,} \\ or \\ 1 \end{array} \right\} :: \left\{ \begin{array}{c} \text{Weight of} \\ \text{body in air,} \\ or \\ W \end{array} \right\} : \left\{ \begin{array}{c} \text{Sp. gr.} \\ \text{of} \\ \text{the} \\ \text{body.} \end{array} \right\}$$

$$\text{Whence,}\quad Sp.\ gr. = \frac{1 \times W}{W - Z} = \frac{W}{W - Z}$$

Example.   A piece of lead weighs in the air 8.19 grammes.
"   "   "   in water 7.47 "
Difference (or W−Z),   .72 "

$$.72 : 1 :: 8.19 : sp.\ gr.$$

$$Sp.\ gr. = \frac{8.19 \times 1}{.72} = 11.37,\ sp.\ gr.\ of\ lead.$$

(I. A.) Another way of ascertaining the specific gravity of a body heavier than water and insoluble in it is known as the method by the flask; it is particularly applicable to fragments of minerals or substances in powder. For this method small bottles of peculiar construction are used; they are provided with a stopper ground to fit the neck of the bottle well, and pierced by a small hole running ver-

Fig. 163.

tically through it.' When the flask is filled with water and the stopper inserted, that portion in the neck escapes through the stopper, permitting the flask to be completely filled with precisely the same weight of liquid each time it is used.

Suppose you want to ascertain the specific gravity of a certain kind of sand: weigh a portion first in the air; then weigh the flask filled with water; next introduce the sand into the flask, and allow it to force out the water (equal in bulk to the solid); insert the stopper carefully, wipe the flask dry, and weigh it again. The specific gravity may then be calculated from the following equation:

Let weight of sand $= W$
　"　of flask and water $= W'$
Let weight of flask, water, and
　sand　$= W''$

Then,

$$\text{Sp. gr.} = \frac{W}{(W + W') - W''}$$

(II.) To determine the specific gravity of a substance soluble in water, a known weight of it is weighed in oil or some liquid which does not dissolve it, and the specific gravity of the oil having been determined by some one of the methods explained in § 118, the specific gravity of the substance is calculated by means of the following formula:

If the weight of the substance in air $= W$
and　"　"　"　in oil $= W'$
Sp. gr. of the oil　$= A$
Sp. gr. of water being　$= 1$
then $W - W' = W'' =$ the liquid displaced;
and　$A : 1 = W'' : W'''$
Whence sp. gr. of substance $= \dfrac{W}{W'''}$

(III.) To determine the specific gravity of a body lighter than water and insoluble in it, weigh it first in the air, then attach to it a piece of lead sufficiently heavy to sink it, and weigh the two together in water; lastly, weigh the lead alone in water, then calculate from this formula:

Weight of cork in air　$= W$
　"　lead in water　$= W'$
Weight of lead and cork in
　water　$= W''$

Then,

$$\text{Sp. gr.} = \frac{W}{W' - W'' + W}$$

(IV.) The fourth case, that of substances lighter than water and soluble in it, is of comparatively rare occurrence; examples are found, however, in the case of the alkaline metals, sodium, potassium, etc. Weigh the body in the air, then in some liquid of low specific gravity in which the body is not soluble—naphtha, for example—and calculate as below:

$$\text{Weight of body in air} = W$$
$$\text{"} \qquad \text{"} \qquad \text{naphtha} = W'$$
$$W - W' = W''$$
$$\text{Sp. gr. of naphtha} = A$$
$$\text{"} \quad \text{water} = 1$$
$$A : W'' :: 1 : W'''$$
$$\text{Sp. gr.} = \frac{W}{W'''}$$

Before proceeding to the determination of the specific gravity of liquids, we will give one more formula which enables us to find the weight of each of two substances when combined in one mass. Some such formula must have been used by Archimedes in ascertaining the proportion by weight of the gold and silver forming the alloy of which Hiero's crown was made (§ 116).

$$\text{Sp. gr. of the alloy} = \text{Sp. gr.}$$
$$\text{Weight "} \quad \text{alloy} = W$$
$$\text{Sp. gr. of one constituent} = s'$$
$$\text{Sp. gr. of second "} = s''$$
$$\text{Weight of one "} = w'$$
$$\text{Weight of second "} = w''$$

Then,
$$w' = W \frac{(\text{Sp. gr.} - s'') \, s'}{(s' - s'') \, \text{sp. gr.}}$$
And
$$w'' = W - w' *$$

To insure accuracy, all determinations of specific gravity should be made at one and the same standard temperature. This fixed temperature is 4° C. —that at which water has its greatest density.

118. Determination of the Specific Gravity of Liquids.—
Several methods may be employed for ascertaining the

---

* For proofs of this formula see "Galloway's First Step in Chemistry," p. 74.

specific gravities of different liquids. We will describe three of them: I. By the flask; II. By weighing a substance in it; III. By the hydrometer.

I. The method by the flask is exceedingly simple. Having selected a specific-gravity flask, determine its weight, fill it with water, and weigh again. Then empty it, dry it carefully, and, filling it with the liquid of which the specific gravity is desired, weigh again.

Let weight of flask $\qquad = F$
" " " and water $= W$
" " " " liquid $= W'$

Then, Sp. gr. of liquid $= \dfrac{W' - F}{W - F}$

II. Take a body of known specific gravity, and insoluble in the liquid to be examined; weigh the body in air, and then in the liquid; if

Weight of body $\qquad = W$
" " in liquid $= W'$
Sp. gr. of body $\qquad = A$

$W : (W - W') :: A : \text{Sp. gr.}$

Or, Sp. gr. $= \dfrac{(W - W')A}{W}$

III. The most expeditious method of ascertaining the specific gravity of liquids is by means of an instrument called a hydrometer. This instrument consists of a glass tube widened into a large and a small bulb at one end, the smaller bulb containing a few shot or a little mercury to cause the centre of gravity of the instrument to fall in the lower part; the narrow portion of the tube, called the stem, is furnished with a scale for reading the depth to which the instrument sinks when plunged in any liquid. The lighter the liquid to be tested, the deeper will the hydrometer sink in it. The manner of using a hydrometer is obvious: it is simply floated in the liquid to be tested, and the figure on the scale at the point where it touches the upper surface of the liquid is accurately noted. The graduation of the hydrometer varies for each liquid; or, if the scale indicates specific gravity, and not arbitrary degrees, the relation between specific gravity and the strength of the liquid examined is ascertained by reference to tables printed for the purpose. Hydrometers receive different names according to the liquids for which they are constructed: that for testing alcohol is called an *alcoholometer;* for solutions of sugar, *saccharometer;* for milk, a *lactometer.*

Both in Europe and America the lactometer is used to test the quality of milk. In large cities the adulteration of milk with water has become so common a fraud that the police (in some instances) are authorized to collect samples of suspected milk for examination with the lactometer. In New York City the Board of Health has recommended a certain lactometer as a standard. Besides the forms of hydrometer mentioned, there are many others; Fig. 164. as, for example, the *salimeter* (for salt solutions), the *vinometer* (for wines), the *acidometer* (for acids), etc. In determining the specific gravity of liquids, attention must be paid to the temperature at which the observation is made, for bodies increase in volume with a rise of temperature, and this increase is not uniform for all substances.

119. **Tables of Specific Gravity.**—Use is made of a knowledge of the specific gravity of certain substances to identify them; especially is this the case with precious stones and minerals. We give below two tables—one of the specific gravity of solids, and the other of liquids. Observe that the specific gravity of living men being 0.89, or lighter

than water, they should float if the precautions mentioned in § 114 were properly taken.

TABLES OF SPECIFIC GRAVITY.

Solids.		Liquids.	
Cork	0.24	Gasoline	0.66
Oak-wood	0.84	"B." Naphtha	0.72
Living men	0.89	Ether	0.72
Starch	1.50	Kerosene oil	0.80
Alum	1.70	Alcohol (absolute)	0.80
Charcoal	1.85	Oil of turpentine	0.86
Roll sulphur	2.00	Ammonia (solution)	0.87
Saltpetre	2.10	Olive-oil	0.92
Quartz	2.65	Distilled water	1.00
Marble	2.83	Sea water	1.02
Glass (flint)	3.33	Milk (cow)	1.03
Diamond	3.52	Human blood	1.06
Iron pyrites	5.00	Water of Dead Sea	1.16
Tin	7.29	Glycerin	1.27
Iron (malleable)	7.84	Chloroform	1.49
Copper (cast)	8.78	Nitric acid	1.51
Silver (fused)	10.50	Sulphuric acid	1.84
Lead	11.34	Bromine	2.98
Gold	19.50	Thallium ethylate	3.55
Platinum	21.50	Mercury	13.59

Specific Gravity of Gases.—The specific gravity of gases is determined by a process much like that for liquids, mentioned in § 118. Air (sometimes hydrogen) is assumed as the standard. A large glass globe filled with air is weighed, then exhausted by an air-pump, and weighed again, the access of air being prevented by a stop-cock. The difference between the weights gives the weight, A, of a certain volume of air; the globe is then filled with the gas under examination, and weighed a third time; by subtracting the weight of the empty globe the weight, B, of the gas is obtained. And $\frac{B}{A}$, or the weight of the gas divided by the weight of an equal volume of air, gives the specific gravity of the gas. Corrections must of course be made for temperature.

QUESTIONS.

112. Explain what is meant by specific gravity. What are the standards of comparison for the three forms of matter?—113. Explain the sinking of heavy substances in water. Explain diagrams Figs. 159 and 160. Give the illustrations: lifting a stone; raising a bucket; raising the arm in a bath. . Relate the anecdote of Archimedes. What is said of iron boats? —114. What is said of the specific gravity of birds? Of insects? Of fishes? What of the specific gravity of the human body?—115. State the principal avoidable causes of drowning. What is narrated about children in China? Why is wading in deep rivers sometimes dangerous?—116. What four cases may arise in determining the specific gravity of substances? Explain the manner in which the specific gravity of a solid is obtained. Describe the experiment of weighing water. What is stated of Archimedes and the crown?—117. Give a condensed rule for finding the specific gravity of a body heavier than water. Illustrate by an example. Give what is known as the method by the flask. How is the specific gravity of a substance soluble in water determined? How that of a body lighter than water and insoluble in it? How that of substances lighter than water and soluble in it? How can you find the weight of two metals in an alloy? At what temperature should accurate determinations be made?—118. Describe the first method for determining the specific gravity of a liquid. The second method. What is a hydrometer? How is it used? Name some of the varieties of hydrometers. For what is the lactometer used?—119. What is said of tables of specific gravity? Give a few examples from the table of solids. Give examples from the table of liquids. What is said of the process for determining the specific gravity of gases?

√

CHAPTER XII.

HYDRAULICS.

120. **Hydraulics.** — Hydraulics teaches about liquids in motion, whether issuing from vessels or moving in channels, of the employment of water as a source of work-power, and of machines used for raising water to a height. If an opening be made in the bottom or side of a tank filled with water, the liquid will flow through the orifice in obedience to gravitation, the particles of liquid near the orifice being pushed out by the pressure of those around and above them.

Let us examine more carefully some of the phenomena connected with this flow of liquids through an opening.

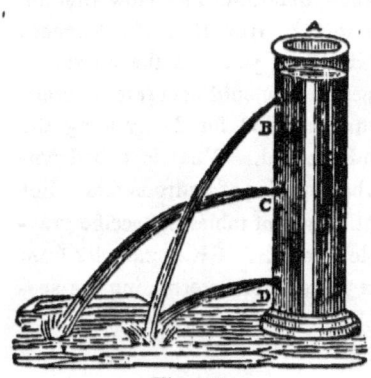

Fig. 165.

Let A, Fig. 165, represent a vessel of water having three openings, B, C, and D, C being equidistant from B and D. Suppose B and D are closed and water flows from C; it is plain that the rapidity with which it issues must depend upon the pressure, and consequently upon the height of the liquid above the opening; and since this level continually falls, the pressure of the liquid and the velocity of the flow diminish also. If the level be not maintained by replenishing a vessel, it takes twice as long to empty it as it otherwise would do.

Again, suppose the orifice B is one foot below the surface of the water, and that the pressure there causes a certain quantity, say a litre, to flow out in one minute; if we want the water to issue twice as fast, say two litres a minute, we must make the pressure four times as great; or, what is the same thing, another opening, C, of the same size must be made four feet below the level of the water. For the discharge of three litres a minute the pressure must be nine times as great; for a flow of four litres a minute the force must be sixteen times as great, and so forth, in the proportion of squares. The reason for this is that to move double the number of water particles would require double the force if they moved with only the same velocity; but because twice as many have to press through the same-sized opening in the same time, each must move with double speed, and hence the force must again be doubled; but two doublings are equivalent to a fourfold increase.

When a liquid descends from an opening in the side of a vessel, it follows the path of a projectile (§ 67). In Fig. 165 the water is represented as spouting farthest horizontally from the orifice C, in accordance with the law that a stream will spout to the greatest distance from an opening half-way between the surface and the bottom of the liquid. If B and D are equidistant from C, the water issuing from them will strike the ground at the same distance from the foot of the vessel, A.

The amount of the water discharged depends upon the size of the orifice and the velocity of the stream. For any given time the rule for finding the quantity discharged is as follows: Multiply the area of the orifice by the velocity per second, and this product by the

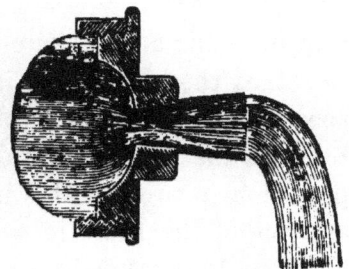

Fig. 166.

I

number of seconds. The shape of the aperture through
which the water flows has also a marked influence on the
volume discharged. A funnel-shaped tube having a circular section, Fig. 166, discharges more liquid in a given time
than an opening of any other shape.

121. **Water-Clocks.**—The ancients took advantage of this regular
flow of water through openings to measure time before the invention of
clocks and watches. The water-clocks, or *clepsydra*, as they were called,
were analogous in principle to the common sand-glass. Ctesibus, a celebrated Greek philosopher of Alexandria, about 250 B.C., contrived a most
ingenious form of this instrument. Water flowed as tears from the eyes
of a statuette which seemed to be deploring the passage of time; the tears
gradually filled a reservoir, and raised a floating figure which pointed to
the hours marked on a scale. This reservoir emptied itself by means of a
siphon arranged, as in the cup of Tantalus (§ 143), once every twenty-
four hours, and the discharge of the water worked mechanism which indicated the day and the month.

122. **Flow of Liquids through Tubes.**—The flow of liquids
through long tubes and pipes is considerably affected by
friction. An inch tube 200 feet long, connected horizontally with a reservoir, will discharge water only one quarter
as fast as an inch orifice in the side of the reservoir.
Sudden turns in a pipe should be avoided, because they occasion so much friction against the sides of the pipe and
among the particles of water by disturbing the regularity
of the current. In the entrance of the arteries into the
brain, in order to prevent the blood from flowing too rapidly into this organ, there are sudden turns in the arteries
to retard the blood; and in grazing animals, since there is
special danger that the blood will flow too freely to the
brain as the head is held down in eating, there is a special
provision to prevent this in a net-work of arteries. If the
arteries of the brain in such animals were straight tubes,
they would continually be dying of congestion of the brain
or of apoplexy.

Friction of liquids in a small pipe is greater in proportion to its size than in a large pipe. In a pipe an inch in diameter water moves only one fifth as fast as in a tube two inches in diameter. This may be made clear by Fig. 167, which represents the area of a small tube inside of the area of a tube having twice its diameter. Suppose the effect of the friction in the large tube to extend in to a. In the small one it will extend in as far—that is, to b. But $e\ a$ is about five times as long as $e\ b$, so that there is fully five times more water uninfluenced by friction in the large tube than in the smaller one.

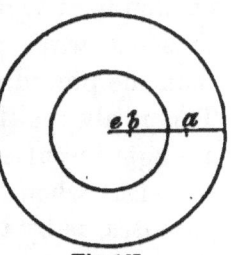

Fig. 167.

Friction in Streams.—The retarding effect of friction is very obvious in brooks and rivers. The water in the middle of a stream runs much more rapidly than it does near its banks. When a river is very shallow at its sides, the water there scarcely moves, though in the middle the water may be running at a rapid rate. A tide, therefore, flowing up a river, moves more freely near its banks than it does in the middle of the stream, because it there meets with less resistance from the downward current. Water moves less rapidly at the bottom of a river than at the surface. For this reason, if a stick be so loaded at one end as to stand upright in water, in the current of a river its upper end will be carried along faster than its lower end, and therefore it will incline forward, as in Fig. 168. As the sea rolls in over a beach, each wave at length pours over its crest and breaks, because the lower part of the wave is retarded by friction on the beach. Were it not for the constant retardation of friction at the sides and bottom of rivers, and at their bends, those rivers which have their rise at a considerable height above the level of the sea would acquire an immense velocity. Thus the Rhone, drawing its waters from 1000 feet above the level of the ocean, would pour them forth with the velocity of water which had fallen perpendicularly the same height—that is, at the rate of 170 miles an hour—did not friction continually diminish the velocity.

Fig. 168.

123. Waves.—Waves are generally formed by the friction of air upon water. As soon as any portion of water is raised above the general surface, it tends by gravity to fall to a level with the water around it, and in so doing the portion next to it is forced upward, forming another wave; thus one wave produces another, each one being smaller than the preceding, till at length the motion is wholly lost. This is always the process when the cause of the motion is a single impulse, as when a stone is dropped into the water. But when the waves are produced by a succession of impulses, as by the wind, they are mostly of the same size. It is quite a common notion that the water moves forward as rapidly as the waves appear to do; but the water really remains nearly stationary, rising and falling, while merely the form of the wave advances. The same wave is made up continually of a succession of different portions of water, or rather it is a succession of different waves. This is very well illustrated by the waving of a rope or carpet. In an open sea a wave slopes regularly on either side; but when it comes near the shore, for the reason given in § 122, it grows more and more nearly perpendicular on the side toward the shore, till at length it falls over; and if it be very large, the roar thus caused by its breaking is heard to a great distance.

Height of Waves.—"So awful," says Dr. Arnott, "is the spectacle of a storm at sea that it is generally viewed through a medium which biases the judgment; and, lofty as waves really are, imagination pictures them loftier still. Few waves rise more than fifteen feet above the ordinary sealevel, which, with the fifteen feet that its surface afterwards descends below this, gives thirty feet for the whole height from the bottom of any water-valley to an adjoining summit. This proposition is easily verified by observing at what height on a ship's mast the horizon remains always in sight over the top of the near waves at the time when she reaches the bottom of the hollow between two waves. Allowance must of course be made for accidental inclinations of the vessel, and for her sinking in the water to

much below her water-line. The spray of the sea, driven along by the
violence of the wind, is of course much higher than the summit of the
liquid wave; and a wave coming against an obstacle may dash to an elevation much greater still. At the Eddystone Light-house, reared on a soli-
tary rock ten miles from the land, a wave which has been growing from far
across the Atlantic often dashes above the lantern at the summit, which is
about ninety feet high."

124. **The Tides.**—The rise and fall of the water of the
ocean, called tide, result from the attraction of the moon.
The moon actually lifts the water towards itself. The
attraction of the sun sometimes increases and sometimes
diminishes the tides, according to its position in relation
to the moon and the earth. If the land were as movable
as the water, or, in other words, if its particles were held
together by no stronger attraction than those of water,
there would be the same motion over the surface of the
earth, when in its revolution successive portions of it present themselves towards the moon.

When the flood-tide returning from the sea meets the outward current of a river flowing into a gradually narrow-
ing arm of the sea, the immensely powerful mass of the
ocean moves inland, like an almost vertical wall, with irre-
sistible force. Such a heaping-up of the waters where the
two currents meet is called the *bore*. This phenomenon is
seen to a remarkable degree in the branches of the Ganges;
its roaring is heard long before its arrival, and all small
vessels seek positions of safety on shore, while even large
ships are occasionally damaged by its resistless sweep.
At Calcutta the water sometimes rises five feet instantaneously, and the huge wave rolls on at the rate of fifteen
miles an hour. The effects of a strong tide are also seen
in certain places where the configuration of the coast com-
pels the incoming water to rise to great heights. In the
Bay of Fundy the returning tide advances with such ra-

pidity that a person on horseback who incautiously ven-
tures too near can scarce escape being overwhelmed.

125. **Relation of Bulk to the Resistance of Liquids and
Gases.**—You have already seen, in § 64, that the greater
the surface of a body in proportion to its weight, the
greater the resistance of the air to its motion. This truth,
which applies to liquids as well as to gaseous substances,
explains the fact that small bodies meet with proportion-
ately more resistance than large ones.
The body B, Fig. 169, is made up of eight
cubes of the size of the cube a, that is, it
has eight times the quantity of matter.
Now, if B were moving through air or
water, any one of its sides pushing the
water before it would meet with only four times as much
resistance as a would, for its surface is only four times as
large, although the body is eight times as large as a. And
the greater the difference of size, the greater is the difference of resistance. If B were a cube twenty-seven times
as large as a, it would meet with only nine times as much
resistance. This explains why shells and cannon-balls can
be thrown much farther than bullets and small shot. The
sportsman does not throw away his shot by foolishly aim-
ing at birds at great distances, and yet shells and large
cannon-balls can be thrown a distance of several miles.
The difference is not in the degree of velocity which the
powder produces, but in the resistance of the air. For
the same reason rain falls with greater rapidity than driz-
zling mist.

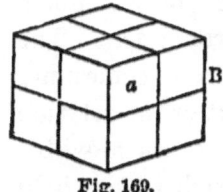

Fig. 169.

Since liquids and aeriform substances resist solids in motion in proportion to the amount of surface which the solids present to them, when
they strike against solids they cause motion in them in proportion to the
amount of surface acted upon. Thus a violent wind which could not
move a lump of tin could, nevertheless, raise a sheet of it, or tear up a

roofing of it if permitted to get beneath. Clouds of sand are raised into the air in the deserts of Africa, although the particles are of the same material as stones, and therefore have the same specific gravity. For the same reason dust, feathers, the down and pollen of flowers, etc., are blown about, although they are heavier than the air. A pebble is moved more easily by a current of water than a stone, because it has a larger surface, in proportion to its weight, to be acted upon by the water. For the same reason sand is moved more easily than pebbles, and fine mud than sand, though stones, pebbles, sand, and mud may all be of the same material. This explains why you find mud where the current is slow, sand where it is faster, pebbles and stones where it is still faster, and where the current is exceedingly rapid you will find nothing but large rocks—sand, pebbles, and stones not being able to resist its force. For the same reason, in the process of winnowing, the chaff is carried away by the wind; while the grain, presenting less surface in proportion to its weight to be acted upon by the air, falls to the floor.

Influence of Shape on Resistance of Liquids to Solids.— The resistance of air or water to a flat surface is greater than to a convex one, because the latter readily turns the particles aside. Thus, a concave surface is resisted much more than a flat one, because the particles of the air or water cannot so easily escape sideways. Fishes are of a spindle-like and slender shape, that they may offer as little resistance as possible to the water. It is for this reason that a fish has no neck, otherwise the upper portion of its body would, from the resistance of the water striking against it, prove a serious impediment to rapidity of motion. Mankind has in some measure imitated the shape of fishes in their boats and ships. Boats which are intended to bear light burdens and go swiftly are made very long and narrow. The webbed feet of water-fowls, when they are moved forward, are folded up so as to meet with as little resistance as possible; but when they are moved backward they are spread out so as to press against the water a broad concave surface. For the same reason the wings of a bird are made convex upward and concave downward; and when it moves its wing upward it cuts the air somewhat edgewise, but in moving it downward it presses directly with the whole concave surface.

126. **Machines for Raising Water.**—A great variety of contrivances for raising water from a lower to a higher level have been devised, some of which are based on a simple application of one or more of the six simple machines described in Chapter VI. Such are the well-sweep, acting on the principle of the lever, and the rope and bucket suspended from a wheel and axle. An old system of raising water is by means of a succession of buckets attached to an endless rope passing over two wheels, so that the buckets fill as they are carried over the lower wheel and discharge as they pass over the top wheel. The *chain pump,* used in many parts of this country, is somewhat similar; but the buckets are replaced by flat disks of metal, which are drawn up through a long tube or barrel, like loose-fitting pistons, and raise an abundant stream of water. The celebrated philosopher Archimedes invented a simple machine, known as *Archimedes's screw,* by means of which water may be readily raised to a moderate elevation. It consists of a tube open at both ends, wound spirally around an inclined cylinder as represented

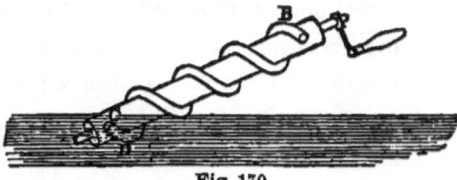

Fig. 170.

in Fig. 170. The lower end of the tube dips below the water; on revolving the cylinder the open end scoops up water, and when it has turned half-way around, the point D is lower than the end C, and, in obedience to gravitation, the water descends to D. On continuing to revolve the screw, the water rises to the top, B, as if drawn up an inclined plane. Archimedean screws are still used in Holland for draining, and are generally driven by windmills.

The various kinds of pumps used for raising water, being dependent upon the principles of pneumatics, will be described in the chapter treating of that topic.

127. **Water-wheels.** — Water flowing in streams having considerable descent affords motive power of first importance. It can be made to perform work through the agency of water-wheels. These wheels are of three principal kinds—the Undershot wheel, the Overshot wheel, and the Turbine. The undershot wheel consists of a wheel revolving on an axle, and having a number of *float-boards* attached to its circumference, Fig. 171. These float-boards

Fig. 171.

dip into the water, which, by its momentum, drives the wheel around, the velocity depending upon the height of the fall of water. In overshot wheels the float-boards are shut in by flat sides, so as to form buckets round the wheel into which the water is allowed to fall at the top of the wheel, Fig. 172. In this wheel the water acts almost solely by its weight; as the wheel revolves, the buckets, filled at the top, descend, and discharge the water, so that by the

I 2

Fig. 172.

time they begin to rise on the opposite side they are empty.

When the water is received half-way up the wheel, or higher, the arrangement is called a Breast Wheel.

The Turbine presents a very different appearance: it consists of a horizontal wheel divided into compartments by curved lines, as shown in that portion of the cut (Fig. 173) without the heavy circle. Within this is fitted a fixed cylinder, also divided into compartments similar to those in the wheel, but running in the opposite direction. Water, from a height, enters a tube connected with this cylinder, and, following the course given by the curved lines, strikes against the partitions of the wheel, causing it to

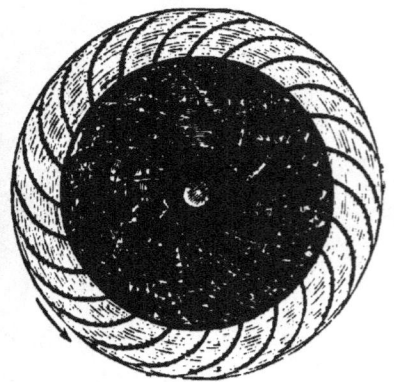

Fig. 173.

revolve about a vertical axis. Owing to the pressure of the water within the tube, and to its striking the partitions nearly at right angles, turbines turn to account a larger proportion of the motive power (four fifths) than any other wheel.

Somewhat similar in principle is the so-called Barker's Mill; it consists of a vertical cylinder arranged in a frame in such a way that it can revolve upon the point upon which it rests. Water running into the cylinder escapes by two arms having holes on the alternate sides; by this arrangement the reaction upon the issuing water makes the cylinder

Fig. 174,

revolve rapidly, causing the ends of the arms to revolve as represented in the figure (Fig. 174).

QUESTIONS.

120. Of what does Hydraulics teach? Describe some of the phenomena connected with the flow of liquids through an opening. What is the path of a liquid issuing from a lateral opening? Upon what depends the amount of water discharged?—121. What is said of water-clocks?—122. How does friction affect the flow of liquids through long tubes? What is said of the effect of friction in brooks and rivers? In what part of a stream does the water move most rapidly? Explain the formation and breaking of the crest of waves rolling over a beach. What is said of the velocity of rivers as affected by friction?—123. Explain the formation of waves. What is it that really advances in the forward movement of a wave? Give the comparison mentioned. What is said of the height of waves?—124. What causes tides? What is a bore? Mention some places where its effects are noteworthy.—125. Illustrate the relation of bulk to the motion of solids produced by moving gases and liquids. What is said of the opposition of gravitation to water and air in moving solids? What difference does the presence of obstacles make in the relation of force to velocity? What is said of the relation of shape to velocity? What is said of the shape of fishes? What is said of the shape of boats? What of

the management of the webbed feet of water-fowls? What of the wings of birds?—126. What is said of machines for raising water? Describe Archimedes's screw.—127. Name the principal kinds of water-wheels. Describe the undershot wheel, and explain its action. Also the other forms. The turbine. Describe Barker's Mill.

CHAPTER XIII.

PNEUMATICS.

128. What Pneumatics Teaches.—Hydrostatics, as you have learned in the preceding chapters, treats of the pressure and equilibrium of liquids, and Hydraulics of the laws governing their motion. *Pneumatics* is that branch of physics which treats of the same phenomena in air and other aeriform bodies. The name is derived from a Greek word signifying *breath* or *air*, just as the term hydrostatics comes from the Greek for *water*. In explaining the laws of "liquid level," "equal pressure in all directions," and of "pressure varying with the depth," we have studied the phenomena with reference only to water as the most convenient liquid; but these laws hold good with all other liquids. In like manner, the laws which we are about to teach concerning common air are equally applicable in the case of all other gases under similar circumstances.

Air Material and has Weight.—That air is a material substance has been shown in § 12, where its impenetrability was demonstrated. It is much less dense than water by reason of a greater separation and repulsion of its particles; but analogous phenomena are observed with both these fluids.

For example, if you fill an India-rubber bag with water and tie its mouth, you cannot flatten it by pressure; and if you blow into it until it is distended and again fasten its mouth, it remains bulky, forming what is known as an air-cushion. Life-preservers and foot-balls are examples of such air-cushions.

Then, again, the resistance offered by air to motion, as in fanning, the power possessed by currents of air to move light as well as heavy objects, and the flight of birds in the air, all prove the material nature of air. That air has weight can be proved by weighing it as you would any other substance. Let a hollow globe, A, Fig. 175, having a neck with a stop-cock, B, be emptied of air and weighed. When you open the stop-cock, and let in the air, the other beam of the scale will rise, because the globe is heavier than

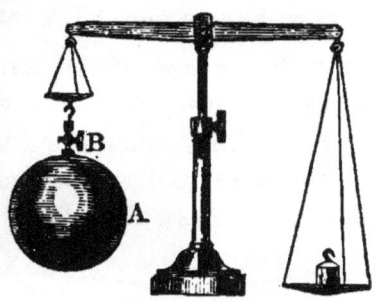

Fig. 175.

it was before. The additional weight required to make the scales balance will indicate the weight of the air which the globe contains. It is about one eight-hundredth ($\frac{1}{773}$) of the weight of the same volume of water. How the globe can be emptied of the air will be shown in another part of this chapter (see § 134).

129. **Air Attracted by the Earth.**—The weight of the air is simply the result of the attraction of the earth (§ 27). Air is attracted by the earth in the same manner as water; and the water takes its place below air because it is attracted more strongly than the air. If you put into a bottle mercury, water, and oil, the mercury will lie at the bottom, because it is more strongly attracted by the earth than the other fluids. The water will be next, then the oil, and lastly, over all, the air, that being less attracted than any of the other substances. This attraction of the air by

the earth is the origin of the chief phenomena of Pneumatics.

Why Some Things Fall and Others Rise in Air.—Most substances fall in air for the same reason that very heavy substances sink in water. They fall because the earth attracts them more strongly than it does the air. The reason that some substances rise in air is precisely the same as that given in § 113 for the rising of substances in water. The air, being attracted more strongly, pushes them up to get below them, as cork or wood is pushed up by water. Thus a balloon filled with hydrogen gas rises in air for the same reason that a bladder filled with air rises in water.

130. **Thickness of the Earth's Air-Covering.**—The air makes a covering for the earth about fifty miles deep. If the earth were represented by a globe a foot in diameter, the air might be represented by a covering a tenth of an inch in thickness. The line *a*, Fig. 176, shows the curve of

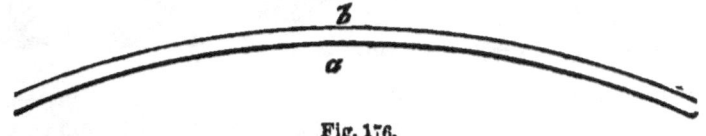

Fig. 176.

the surface of such a globe, and the space between *a* and *b* represents the comparative thickness of the covering of air. This is ascertained by calculation from the pressure of the air upon the earth in the same manner as the depth of water is calculated from the pressure which it exerts.

The earth flies on in its yearly journey around the sun at the rate of 1100 miles per minute, and yet it holds on to this loose airy robe by its attractive force, so that not a particle of it escapes into the surrounding ether. Of itself it is disposed to escape; and it would do so, and be diffused through space, if the attraction of the earth for it were suspended.

131. **Compressibility of Air.**—In considering the influence of gravitation upon air, it must be remembered that air is very compressible, while water is very nearly incompressible (§ 17). While, therefore, in a body of water the particles

are very little nearer together at the bottom than at the surface, the particles of the air are much nearer together at the surface of the earth than at a distance from it. All the particles of the air being attracted or drawn towards the earth, those below are pressed together by the weight of those above. The air therefore becomes more rarefied as we leave the surface of the earth, and in the outer regions of the sea of air it is too rare to support life. Even at the tops of very high mountains, or the heights sometimes reached by balloons, disagreeable effects are often experienced from the rarity of the air. The air has been compared, in regard to its varying density at different heights, to a heap of some loose compressible substance; as, for example, cotton-wool, which is quite light at the top, but is pressed more and more compactly as you go towards the bottom. Hydrogen gas is only one fifteenth as heavy as air at the surface of the earth; and therefore the hydrogen balloon rises till it reaches a height where the air is so rare that the balloon is of the same weight with an equal bulk of air, and there it stops.

132. **Similarity of Aeriform Substances and Liquids.** — You have learned in § 17 in what the air and gases differ from liquids. But in one very important respect they are alike—viz., the mobility of their particles. Hence pressure in air, as well as in water, is equal in all directions, so that in the experiment with the bladder, in § 107, it makes no difference in the result whether it be filled with water or air. For the same reason, pressure is in proportion to the depth in aeriform substances as well as in liquids, and the laws of specific gravity apply to the one as well as to the other.

You are now prepared to understand the results of *the action of gravitation upon air and the gases;* or, in other words, the principal phenomenon of Pneumatics.

133. **Pressure of the Atmosphere.**—The amount of the pressure of the atmosphere is very readily estimated, by a process which we will explain in another part of this chapter. It has been ascertained that the atmosphere presses with a weight of fifteen pounds on every square inch. When you extend your outspread hand horizontally in the air, you feel no pressure upon it, notwithstanding it sustains a pressure of some two or three hundred pounds. If your hand be five inches long and three broad, it presents a surface of fifteen square inches, on every one of which the atmosphere is pressing with the weight of fifteen pounds; that is, there is a pressure on the upper surface of your hand of a column of air weighing 225 pounds, and on the lid of a box only thirty inches square there is a pressure of 13,500 pounds. The whole pressure on the body of a man of common size is about fifteen tons. But why is it that the lid of the box is not broken in, your hand not borne down, and your body not crushed? It is simply from the fact, shown in the previous chapter in regard to liquids, and in this one as to aeriform substances, that the pressure is equal in all directions. The lid and the outspread hand are therefore balanced by an upward pressure equal to the downward, and the body sustains an equal pressure on all sides. If the air could be removed from within the box, the lid would be crushed in; if from under the hand, that would be borne down; and if from one side of the body, the body would be forced violently in that direction till it met with an opposing pressure.

But besides this equal pressure of the air on all sides, air exists within the pores and interstices of all bodies that are not very dense, and its particles are subject to the same laws as are those on the outside.

All this can be made clear to you by experiments with the air-pump.

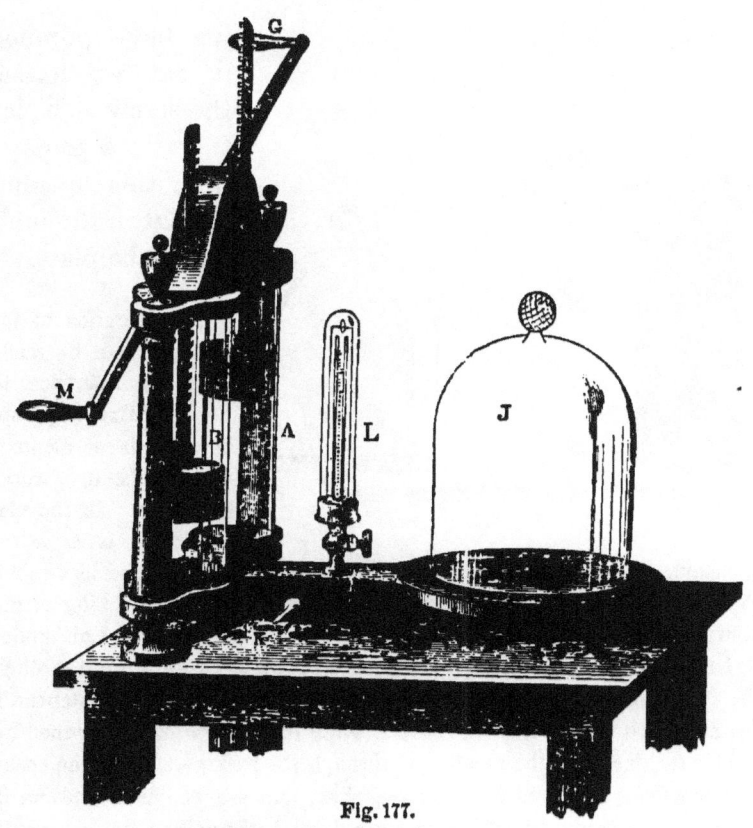

Fig. 177.

134. Air-pump.—Fig. 177 represents an air-pump as commonly arranged. A, B, are two pump-barrels, the pistons in which are worked by means of the handles, G and M. These pumps are very nicely made, and the frame-work to which they are attached is very strong and firm, so that the pumps may work evenly. J is a bell-shaped glass vessel, called a receiver, closed at the top, but open at the bottom, the edge of which is ground very true, so that it may fit exactly on the large, smooth metallic plate. In the middle of the plate is an opening which leads to the pump-barrels, and it is through this that the air is pumped out of the glass receiver, J. If we wish to let the air in after

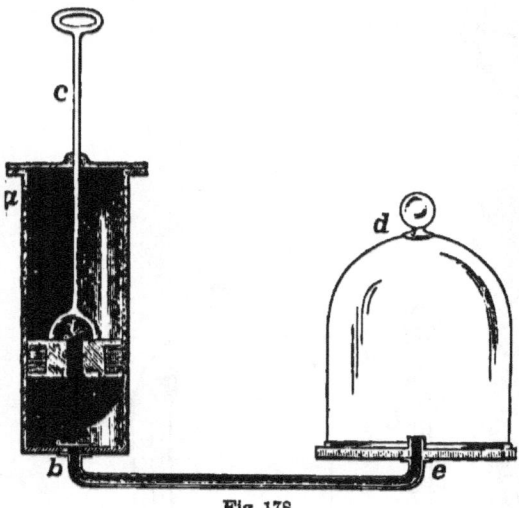

Fig. 178.

we have pumped it out, we loosen the screw at K, for there is a passage from this opening to that in the middle of the plate.

The operation of the air-pump can be made clear by reference to Fig. 178. But one pump-barrel, *a*, is represented, with a piston, *c*, working in it. In the piston there is a valve, *i*, opening upward, and also one at *b*, at the end of the tube leading to the centre of the plate on which is the receiver, *d*. The working of the instrument is as follows: If the piston, *c*, be forced down, the air under it, being compressed, will close the valve at *b*, and will rush upward through the valve *i* in the piston. Let the piston now be raised; the resistance of the air above it will close the valve *i*, while the valve *b* will be opened by the air rushing from the receiver, *d*, through the passage, *e*, to fill the space between the piston and *b*. You see, then, that every time the piston is drawn up air passes out of the receiver through the valve *b* into the space between this valve and the piston. None of this air which has passed out can return; for the moment you press upon it by forcing downward the piston, the valve *b* closes and the air escapes through the valve *i*. Each time, therefore, that you work the piston up and down, you pump some of the air out of the receiver; and after some time exceedingly little air will be left in it, and that, of course, will be diffused throughout the receiver. It will be rarefied like that in the upper regions of the atmosphere. With the double-barrelled air-pump, shown in Fig. 177, the operation is similar but more rapid, because when one piston is raised the other is lowered, and the action is continuous. L is a gauge to indicate the completeness of the exhaustion, which acts on the principle of the barometer.

135. Experiments.—When the receiver J (Fig. 177) is full of air, it can be moved about on the plate easily, and can be lifted from it. But if you work the pumps a few strokes, the receiver will be firmly fastened to the plate, since the air within, being rarefied, presses with little force compared with the air outside. If the pumps be worked for some time, it will be very difficult to release the receiver from the pressure without breaking it. But turn the screw, K, admitting the air, and the equality of the pressure within and without is at once restored. Remove this large receiver, and place a small glass jar, open at both ends, on the plate, with the hand covering the upper opening, as represented in Fig. 179. On exhausting the air, the hand is so firmly pressed into the glass that it requires considerable force to disengage it from the pressure. If we tie a piece of bladder or India-rubber over this jar, as in Fig. 180, and then pump out the

Fig. 179.   Fig. 180.

air, the bladder is at first pressed in; and if we continue

Fig. 181.

to pump, it at length bursts inward with a loud report. It would make no difference in the result of the experiment if the jar were shaped as in Fig. 181, for the pressure is the same in all directions. The resemblance between air and liquids in this respect may be illustrated thus: Suppose that a flat fish rests against the tube of a pump so as to cover the end with one of his sides. He feels no uncomfortable pressure, because the water in the pump and that below it press equally upon him. If, however, the pressure of the water in the pump be suddenly removed by the piston, the fish would be pressed upward into the tube, just as the bladder is pressed upward in Fig. 181, or downward in Fig. 180. The so-called "Magdeburg Hemispheres," Fig. 182, illustrate very strikingly

Fig. 182.

the pressure of the atmosphere. They consist of two hollow half-globes of metal whose edges fit very accurately upon each other. The air being exhausted through the stem and a handle screwed on, great force must be exerted to pull the hemispheres apart. The force required depends upon the extent of their surface. In the famous experiment at Magdeburg, in 1654, by Otto von Guericke, the inventor of the air-pump, two strong hemispheres of brass three feet in diameter were employed; and when he exhausted them on the occasion of a public exhibition, it is said that twenty coach-horses of the emperor were unable to pull them asunder!

In the so-called "Mercury Shower," we have another example of the immense pressure of the atmosphere. Fig. 183 represents a receiver with an opening at the top. Cemented in this opening is a wooden cup, a, terminating in a cylindrical piece, b. If mercury be poured into the cup and the air within the receiver be exhausted, the mercury will be forced through the pores of the wood by the external air, and will fall in a silver shower. A tall jar, c, is placed there to receive it, to prevent any of it from entering the opening in the metallic plate.

Fig. 183.

Fig. 184.

The boy's sucker illustrates the pressure of the air. It is simply a circular piece of leather with a string fastened to its centre, as shown in Fig. 184. When the leather is moistened and pressed upon a smooth stone, it adheres by its edges to the stone, just as the receiver adheres to the plate of the air-pump when the air is pumped out. Many animals have contrivances of a similar character to enable them to walk in all positions, to seize their prey, etc. The gecko and the cuttle-fish furnish interesting examples, as noticed in Hooker's Natural History. Snails, limpets, etc., adhere to rocks by a like arrangement. Some fishes do the same; one, called the remora, attaches itself by suckers to the side of some large fish or a ship, and thus enjoys a fine ride through the water without any exertion on his part. In all such cases it is water instead of air that makes the pressure, but the principle is the same. Flies and some other insects can walk up a smooth pane of glass, or along the ceiling, because their feet have contrivances similar in principle to the boy's sucker. The hind-feet of the walrus are constructed somewhat like the feet of the fly, enabling this huge animal to climb smooth walls of ice.

136. Density of the Air Dependent upon Pressure.—The fact that the degree of the density of the air is dependent on pressure has been already shown in § 131. The same thing can be shown in various ways by experiments with an air-pump. If a small bladder partly filled with air, Fig. 185, and loaded with a weight so as to sink in water, be placed in a jar of water, and the whole be set under the receiver of the air-pump, on exhausting the air the bladder will swell out, owing to the expansion of the air, and will rise. The reason is, that the pressure being removed from the surface of the water, the bladder bears only the pressure of the water, and not that of the air *plus* the water; hence the air within
expands and becomes less dense. If an India-rubber bag be partly filled with air, Fig. 186, and put under the receiver, when the air is exhausted the bag is

Fig. 185.

relieved of pressure, and the air in it becomes expanded—that is, rarefied. For the same reason, if a vessel with soap-bubbles in it be placed under the receiver, on pumping out the air the bubbles will become much enlarged. A very pretty experiment illustrates the same principle. Let an

Fig. 186.

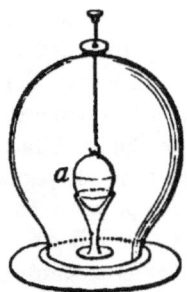
Fig. 187.

egg with a hole in its small end be suspended in a receiver, as represented in Fig. 187, a wine-glass being placed beneath it. On exhausting the air, the egg will run out of the shell into the wine-glass, and then, on admitting the air, the larger part of it will run back again into the shell. This may be explained as follows: The large end of the egg contains air. As soon as the pressure of air is removed from the egg, the air in the egg expands, forcing out the contents; but when the air is admitted into the receiver, the air in the egg is at once condensed to its former small bulk by the surrounding pressure.

Hydrostatic Balloon.—The philosophical toy represented in Fig. 188 illustrates very beautifully the influence of pressure upon the density of the air. The balloon in the jar of water is constructed of glass, having a small orifice at its lower part. Water is introduced into the balloon, care being taken to put in just enough to make the balloon of a little less specific gravity than water. In that case it will rise to the top of the jar, with a very little of its top above the surface of the water. Now tie a piece of India-rubber cloth over the top of the jar, and the apparatus is complete. On pressing upon the India-rubber the balloon will descend in the jar, and on removing the pressure it will rise. The explanation is as follows: The pressure upon the India-rubber is felt through the whole body of the water in the jar, and forces a little more water into the orifice of the balloon, condensing the air within it. The balloon consequently becomes heavier, and, having a greater specific gravity than water, sinks. But when the pressure is removed, the condensed air in the balloon, by its elasticity,

Fig. 188.

returns to its former bulk, expelling the surplus water just introduced; and the balloon, becoming therefore as light as before, rises.

137. **Pores of Substances Contain Air.** — We have said that the pores and interstices of wood, flesh, and a great variety of substances contain air. In all these cases the presence of the air can be made manifest by removing the pressure of the surrounding air, and thus allowing the air in these substances to expand. If an egg be placed in a jar of water, Fig. 189, under the receiver of an air-pump, on exhaustion being made, air-bubbles will constantly rise in Fig. 189.

the water from the egg. In like manner, the surface of a glass of ale, Fig. 190, will be covered with foam, the carbonic-acid gas in it escaping freely when the pressure of the air upon it is removed. The same thing may be seen to some extent even in water, for it always contains some
Fig. 190. air. For a similar reason a shrivelled apple will become plump and fair when the pressure of the external air is removed, but will shrink at once to its shrivelled state when the air is admitted into the receiver.

138. **Elasticity of the Air.** — All the phenomena mentioned in § 136 and § 137 exhibit the elasticity of the air. Owing to this property it is always disposed to expand when pressure is removed from it. This is most strikingly exhibited when the air is much condensed by pressure; the greater the condensation, the stronger the expansive or elastic force. Fig. 191, page 220, represents an instrument called the condenser. In the cylinder, A B, moves the piston, P. Air is admitted to the cylinder at F, and into the receiver, V, at G. The valve at F prevents any air from escaping from the cylinder, and the valve at G prevents it from escaping from the receiver. The instrument operates thus: If the

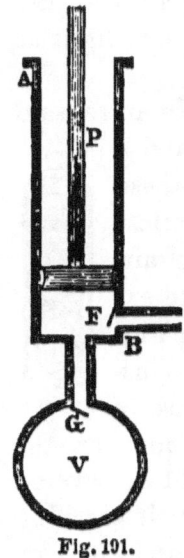

Fig. 191.

piston be pressed downward, the compressed air in the cylinder shuts the valve F and opens G, and so enters the receiver, V. If the piston be raised, air rushes in at F to fill the space in the cylinder. It cannot come from V, because the valve G is shut by the pressure of the air within. By working the piston for some time, you can force a quantity of air into V of very great density. It is evident that this instrument is the very opposite of the air-pump. The receiver, V, contains condensed air, while the receiver of the air-pump contains rarefied air. If you compare the two instruments, you will see that the opposite results are owing to a different arrangement of the valves.

Until quite recently air had never been condensed to the liquid state. This was accomplished by Messrs. Pictet and Cailletet, who subjected it to enormous pressure and a very low temperature. The term *permanent* gas formerly applied to air must now be abandoned.

The elasticity of the air and other gases results from an incessant commotion of their particles. We must picture to our minds the molecules of a gas as moving in all directions, constantly striking against each other, and thus producing pressure on the sides of an enclosing vessel. We have already referred to this motion of the molecules of a gas in § 8. The force with which the molecules strike against the confining walls will be greater the smaller the space through which they are allowed to move, a consideration which explains the fundamental principle known as Marriotte's law—viz., the pressure of any quantity of gas is inversely proportional to its volume. That is to say, the greater the pressure to which a gas is subjected, the less space it occupies. Thus a body of air which under a certain pressure occupies six cubic feet will be condensed to three cubic feet by twice the pressure, and to two cubic feet by three times the pressure, etc.

Illustrations.—Air-guns and pop-guns illustrate the elasticity of condensed air. The air-gun is constructed in this way: A receiver like V, Fig. 191, is so made that you can screw it on and off the instrument. After being charged with condensed air, it is screwed upon the gun, its stem communicating with the barrel. In order to discharge the gun there is a contrivance connected with the trigger for raising the valve, G, so that some of the condensed air may enter the barrel. On doing so, its sudden expansion rapidly forces out the contents. The principle on which the common pop-gun operates is similar. Air is confined between the two corks, P and P', Fig. 192. As the rod, S, is pushed quickly

Fig. 192.

in, the cork P' is carried nearer to P, so that the air between them is condensed. With the condensation the expansive force is increased; and when it becomes so great that the cork P can no longer resist it, it throws the cork out, and so quickly as to occasion the popping sound.

The explosion of powder furnishes a good illustration of the expansive force of condensed air or gases. These gases are produced so suddenly from the powder that at the instant they are in a very condensed state, and therefore expand powerfully. The power of steam is in proportion to its condensation. When formed under the confinement of a boiler, on being allowed to escape it expands with great force. The application of the expansive power of steam will be treated of particularly in § 182.

139. **Pressure of the Air on Liquids.** — If you plunge a tumbler into a vessel of water, and, turning it over, hold it so that its open part is just under the surface, it will remain full. This is because the weight of the air pressing upon the surface of the water in the vessel prevents the water in the tumbler from passing downward. Now,

K

Fig. 193.

if you introduce a bent tube under the tumbler, as shown in Fig. 193, and blow through it, the air forced up into the tumbler presses the water down, taking its place. That is, the pressure of the air within the tumbler acts in opposition to the pressure of the air upon the surface of the water. If instead of a tumbler you take a tall jar, as represented in Fig. 194, and, filling it with water, invert it upon a small shelf placed beneath the surface of the water, you will have a representation of the *pneumatic trough* used by the chemist in collecting gases. To fill the jar *a* with gas he puts beneath it the mouth of the retort from which the gas issues, and the gas pass-

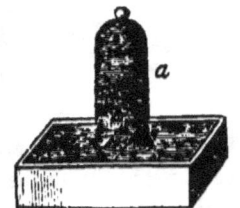

Fig. 194.

ing upward expels the water. In Fig. 195 is represented an experiment which shows not only that the pressure of the air sustains the column of water in the cases cited above, but also that it makes no difference in what direction this pressure is exerted. Take a glass, fill it even full with water, and, placing a piece of writing-paper over its mouth, carefully invert it, as shown in the figure. The paper will remain, and the water will not run out. It is the pressure of the air

Fig. 195.

that sustains the water, and the paper only serves to maintain the surface of the water unbroken. If the paper were not there the particles of the air would insinuate themselves among those of the water, and pass upward in the glass. This explains why a liquid will not run from a barrel when it is tapped, if there be no vent-hole above, unless so large an opening be made as to let the air work its way in bubbles among portions of the liquid. It is this entrance of the air that causes the gurgling sound heard in pouring a liquid from a bottle.

140. **Amount of Atmospheric Pressure.**—If, instead of the glass jar in Fig. 194, you use a tube thirty-four feet long, and closed at the top, it will remain full of water. If the tube be longer, the water will stand only at thirty-four feet, leaving an empty space, or *vacuum*, above it. It makes no difference what the size of the tube is; the result will be the same in all cases.* That is, a column of water thirty-four feet high can be sustained by the pressure of the atmosphere. It is easy, therefore, to estimate the weight or pressure of the air. The pressure of the column of water is found to be fifteen pounds to the square inch of its base, and this, of course, is the amount of pressure or weight of the atmosphere which it balances. Mercury is thirteen and a half times as heavy as water, and therefore the air will sustain a column of it only about thirty inches in height (76 cm.).

141. **Barometer.**—The weight of the atmosphere varies to some extent at different times, and the barometer is an instrument for measuring these variations. It is constructed on the principles developed in the previous paragraphs. Fig. 196, on the following page, represents a very simple form of the instrument. A glass tube about 35 inches

---

* This is true except when the tube is so small that capillary attraction exerts considerable influence.

(88.8 centimetres) long, closed at one end, is filled with mercury, and then inverted in a cup of the same liquid, *n n*. The vacuum produced by the falling of the mercury is called the Torricellian vacuum, from Torricelli, an Italian, who first developed the principles of the instrument in 1642. Fig. 197 shows another form of the instrument, with a scale attached. The mercury generally stands at the height of about 30 inches. But it varies with the weather. When the weather is bright and clear, the air is heavier, and, pressing upon the mercury in the vessel, forces it up higher in the tube. But when a storm approaches, the air is apt to be lighter, and therefore, pressing less strongly on the mercury in the vessel, the mercury in the tube falls.

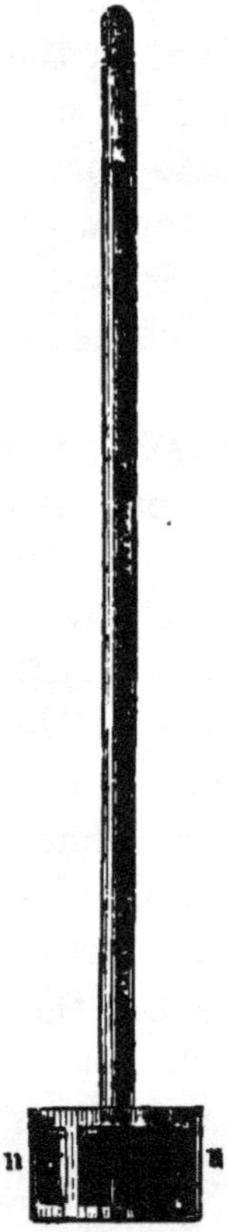

The barometer is of great service, especially at sea, in affording the sailor warning of an approaching storm. An incident is related by Dr. Arnott which strikingly illustrates its value in this respect. He was at sea in a southern latitude. As the sun set after a beautiful afternoon the captain foresaw danger, although the weather was perfectly calm, for the mercury in the barometer had suddenly fallen to a remarkable degree. He gave hurried orders to the wondering sailors to prepare the ship for a storm. Scarcely had the preparations been made when a tremendous hurricane burst upon the ship, tearing the furled sails to tatters, and disabling the masts and yards. If the barometer had not been observed, the ship would have been wholly unprepared, and shipwreck, with the loss of all on board, would in all probability have resulted.

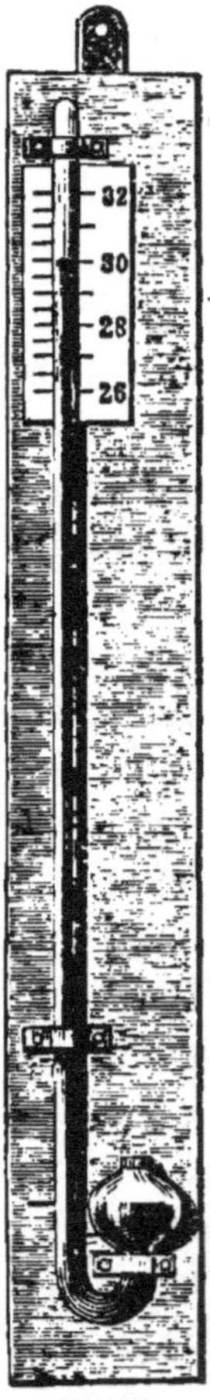

Fig. 196.

Fig. 197.

A water-barometer could be made, but it would be very unwieldy, for the tube must needs be more than 34 feet long. Besides, it would not answer in very cold weather, for the water would freeze. So short a column of the heavy fluid mercury balances the weight of the atmosphere that a barometer made with this is of very convenient size; and then there is no danger of the mercury's freezing, except in the extreme cold of the arctic regions.

The Barometer a Measurer of Heights.—The atmosphere, as stated in § 131, diminishes regularly in density as we go upward. The rate of this diminution has been accurately ascertained, and therefore we can estimate heights by the amount of pressure on the mercury in the barometer. At a height of 500 feet the barometer will be half an inch lower than in the valley below. At the summit of Mont Blanc it stands but half as high as at its foot, indicating a height of 15,000 feet. Du Luc, in his famous balloon ascension from Paris, saw the barometer at one time standing at about twelve inches, showing an elevation of 21,000 feet.

The Aneroid Barometer.—The inconvenience of travelling in mountainous regions with a long tube filled with mercury is very great, and has led to the invention of another form of barometer which is called an *Aneroid.* The principle involved in its construction may be explained by reference to Fig. 198. The curved tube *a b*, when exhausted of air and hermetically closed, is sensitive to the variations in the pressure of the atmosphere, the ends of the tube

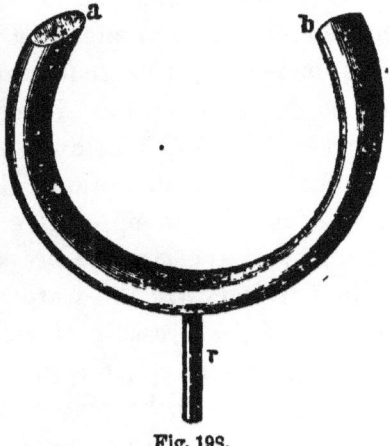

Fig. 198.

approaching with increased pressure, and receding with reduced pressure.

Now, if a similar tube be inserted in a case, and the
curved ends be connected by means of a mechanical contriv-

ance with a hand like
that of a watch, we will
have the simplest possi-
ble form of the aneroid
barometer. The hand
points to figures around
the dial-plate of the in-
strument corresponding
to the height of the mer-
curial barometer. The
general appearance of
such an instrument is
that of a watch, and it
is but little larger (Fig.
199).

Fig. 199.

142. Relation of the Air's Pressure to the Boiling-point.
—Water heated to 212 degrees Fahrenheit (100° Centigrade) boils — that is, it becomes vapor. If water be
heated on the summit of a high mountain, it boils before it reaches this temperature. On the top of Mont
Blanc it boils at 180 degrees (82.2° C.) — that is, 32
degrees (17.8° C.) below the boiling-point of water at
the foot of the mountain. This is because the pressure
of the air acts in opposition to the change of water into
vapor; and the less the pressure, the less heat will be required to vaporize the water. We may illustrate this influence of the pressure of air upon boiling by the following experiment. Let a cup of ether, which boils at 95
degrees (35° C.), be placed under the receiver of an air-
pump. On rarefying the air by the pump, the ether will

boil. The general effect of pressure upon boiling may be prettily illustrated by another experiment. Boil some water in a thin flask over a spirit-lamp. While the steam is still issuing cork the flask tightly, invert it, and let the boiling cease. If, now, you pour some cold water over the flask, the boiling will commence again with considerable energy.

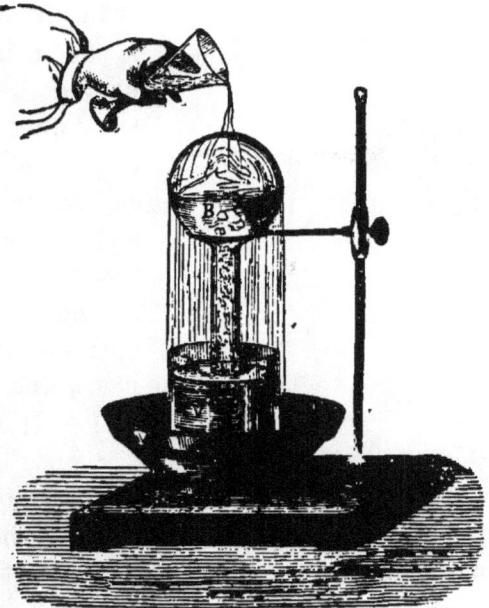

Fig. 200.

Why? Because you condense the steam above the water by the application of cold, and thus remove the pressure. Then, again, if you pour hot water over the flask while the water is boiling, the boiling ceases, because the heat favors the accumulation of steam, and therefore renews the pressure on the surface of the water.

It is evident from what has been stated that most liquids have that form owing to the pressure of the atmosphere upon them. If there were no atmosphere, ether, alcohol, the volatile oils, and even water, would fly off in vapor; and the earth would be enveloped in a gaseous robe, for the particles of the vapors would be held to the earth by attraction, just as the particles of the air now are.

143. **Siphon.**—The pressure of air upon fluids is beautifully exemplified in the operation of the siphon. This instrument is simply a bent tube having one branch longer

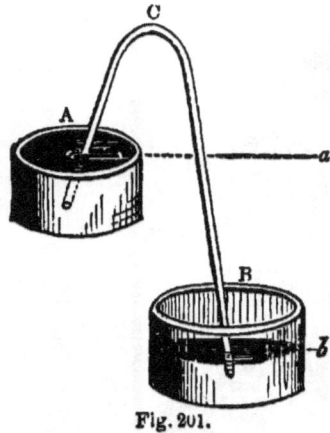

Fig. 201.

than the other. Its operation is shown in Fig. 201. The tube having been first filled with the liquid, its shorter branch is placed in the liquid of the vessel A, which is to be emptied, and beneath the other is held the vessel B, which is to receive the liquid. As shown here, the opening of the long branch is below the surface of the liquid in B. It is manifest, therefore, that the air presses equally upon the surfaces in both vessels, tending to support the fluid in the tube, just as the water is supported in the jar in Fig. 194. But, notwithstanding these equal pressures, the liquid runs up the tube from A, and down its longer branch into B. Why is this? Since the pressure of a column of fluid is in proportion to its height, there is greater pressure or weight in the longer branch than in the other; and it is this difference in weight that causes the flow from A into B through the siphon. The difference in the columns in the two branches is not the difference in length of these branches, but the distance between the levels of the fluid in A and B—that is, the distance from *a* to *b*. The operation, then, of the instrument is this: there is a constant tendency to a vacuum at C, the bend of the tube, from the influence of gravitation on the excess of fluid in the long branch over that in the short one. This tendency is constantly counteracted by the rise of fluid in the short branch, it being forced up by the pressure of the air upon the surface of the fluid in A.

If the siphon were so placed that the surface of the liquid in A were precisely on a level with that in B, as repre-

sented in Fig. 202, the liquid
would remain at rest, for,
since pressure is in proportion to the height, and the
pressures on the two sur-
faces are equal, there would
be an exact balance. But

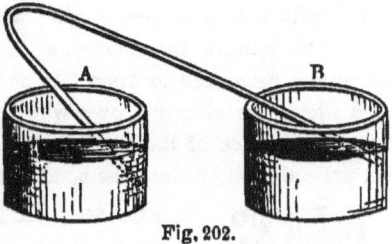

Fig. 202.

let the surface in B be ever so little lower than in A, and
the flow will begin. And the greater the distance between
the two levels, the more rapid will
be the flow, for the greater will be
the influence of gravitation in the
long branch.

Again, if the end of the long
branch of the siphon be free, as in
Fig. 203, the siphon will operate in
the same way, for the air, pressing
in all directions equally, tends to
support the column of fluid in the
long branch by a direct upward
pressure, but is prevented from doing so by the excess of fluid in it
above that in the shorter one. The
operation of the siphon is commonly represented in this
way; but we have given first the arrangement in Fig. 202, in order that you
might more clearly see the principle of
the instrument.

Fig. 203.

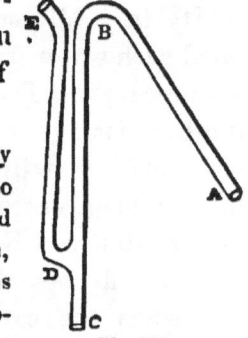

Fig. 204.

Uses of the Siphon.—The siphon is used chiefly
for discharging liquids from one barrel or vessel into
another. For convenience, it is often constructed
after the plan of Fig. 204. To the long branch,
B C, is attached the tube E D. It is used in this
way: The end of the short branch, A, being introduced into the liquid to be drawn off, you close the

K 2

end C with a cork or your finger; and after filling the siphon by suction at E, you remove the finger and let the liquid run. The siphon has sometimes been used to drain pits and mines. It of course can never be used where the.elevation over which the tube is to bend is over 34 feet from the surface of the water to be discharged, for then the air would not press the water up to the bend of the siphon.

The so-called *cup of Tantalus* is a pretty toy; it consists simply of a goblet containing a siphon which is concealed by a human figure. In Fig. 205 the figure is omitted to show the position of the siphon. On pouring water into the cup, it will remain there until you pour in enough to cover the bend of the siphon; as soon as this is done, the siphon fills, and the water flows out through the long branch which passes through the bottom of the cup. The lips of the human figure being on a level with the bend of the siphon, it is apparently prevented from drinking in a tantalizing way.

Fig. 205.

144. **Intermitting Springs.**—The operation of an intermitting spring is essentially the same with that of the cup of Tantalus. Fig. 206 represents such a spring. There is a cavity in a hill, supplied with water from a source above. There is also a passage from the cavity which takes a bend upward

Fig. 206.

like a siphon. Now, when the water in the cavity is low, it will not run out from the siphon - like channel; but when the cavity becomes filled above the level of the bend, the water will at once flow out, just as it does from the cup of Tantalus as soon as the bend of its siphon is covered.

145. **Pumps.**—The accompanying cut represents a common form of pump. A tube extends down into the well, B. Above this is the barrel of the pump, *a b*, in which the piston works up and down. There is a valve in the piston, and another at the bottom of the barrel. Both of them open upward. We will suppose that the pump is entirely empty of water. If the piston descend, the piston valve shuts down, and the lower valve opens, letting the air between pass upward. When the piston rises, the air above the piston cannot get below, for its pressure will shut the valve in the piston. But there will be a tendency to a vacuum below the piston as it rises, and the air will pass up through the valve in the barrel to fill up the space. But why does the air rise?

Fig. 207.

Because of the pressure of the air upon the surface of the water in the well. This forces up in the pump the water and the air above it, just in proportion as the downward pressure in the pump is lessened. If the pumping be continued, all the air will soon be expelled, the water following it and flowing out at the opening, *r*. It is obvious that the pump will be useless if the valve in the barrel be over 34 feet above the surface of the water in the well, because the pressure of the atmosphere will not sustain a higher column of water.

In common language, the operation of the pump is attributed to what is called a principle of suction, as if there were a drawing-up of the water. But that water, you see, is not drawn, but forced up. So it is with all operations of a similar character. When we apply the mouth to suck up a fluid through a tube, the fluid is forced up because the pressure downward in the tube is removed. But how is it removed? It is done by a movement of the tongue downward from the roof of the mouth; thus removing the pressure of the air, in the same manner as the upward movement of the piston in the pump. To fill the space made by the movement of the tongue, the air is forced up the tube, the liquid following; and, as in the case of the pump when the air is all expelled, the liquid will begin to discharge into the mouth.

Forcing-Pump.—The forcing-pump is constructed differently from the common pump. Its plan

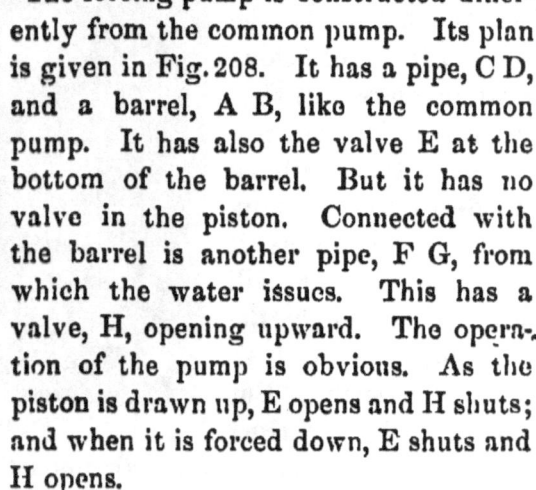

Fig. 208.

is given in Fig. 208. It has a pipe, C D, and a barrel, A B, like the common pump. It has also the valve E at the bottom of the barrel. But it has no valve in the piston. Connected with the barrel is another pipe, F G, from which the water issues. This has a valve, H, opening upward. The operation of the pump is obvious. As the piston is drawn up, E opens and H shuts; and when it is forced down, E shuts and H opens.

146. **Fire-Engine.**—The fire-engine has commonly two

forcing-pumps, with a contrivance for making the water issue in a uniform stream. This contrivance can be explained by reference to Fig. 209. The discharging-pipe, *h g*, extends down into a large vessel, *a*, which is filled with air. The uniformity of the stream depends upon the elastic force of compressed air, as will appear from an explanation of the operation of the machine. When the wa-

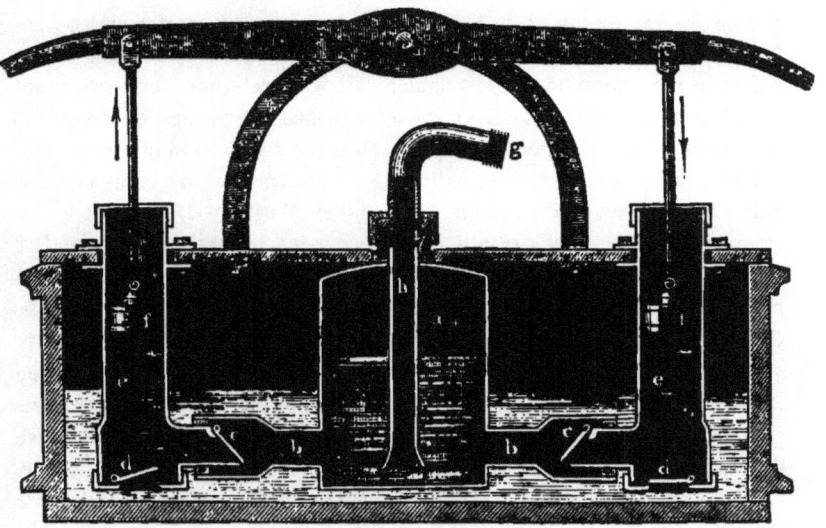

Fig. 209.

ter is forced through the openings *c b*, it compresses the air in *a*, for the tube *h g* is too small to allow all the water to escape that comes from the larger tubes, *b b*. Now, the moment that the piston is raised it ceases to force the water through *c*, and the elastic force of the compressed air operates, shutting down the valve *c* and forcing the water up *h g*. The result is a continuous rise of the water in this tube, and therefore a uniform stream. The valves *d d* permit the water in the reservoir surrounding the cylinders to enter when the pressure in *e* is relieved. By having two cylinders and pistons communicating with one air-chamber, *a*, as in the figure, the continuity of the stream of water is doubly insured.

QUESTIONS.

128. What does pneumatics teach? How can you show that air is material? How that it has weight? Describe the experiment. What is its weight compared with that of water?—129. What is said of the attraction of the air by the earth? Explain why some things rise and others fall in air.—130. How thick is the earth's air-covering? How is the height of the atmosphere ascertained? At what rate does the earth move round the sun?—131. State the influence which gravitation has upon the density of the air at different heights. Give the comparison of air to wool. What is said of hydrogen and balloons?—132. In what are gases and liquids alike, and what are the results of the similarity?—133. What is the amount of pressure of the atmosphere on each square inch of surface? Give the calculations in regard to this pressure. Show why the great pressure of the air does not produce injurious effects. — 134. Describe the air-pump. Explain by Fig. 178 the plan and working of the air-pump. —135. State some of the experiments with the air-pump. How can you prove that air, like water, presses equally in all directions? State the comparison about the fish. What is said of the Magdeburg hemispheres? Give the experiment with mercury. Explain the operation of the boy's sucker. Give the statements about sucker-like arrangements in animals.—136. State the experiment of the bladder and weight. Give the experiment with the India-rubber bag. State the experiment with the egg. Explain the operation of the hydrostatic balloon.—137. What is said of the presence of air in various substances?—138. What is said of the elasticity of air? Describe and explain the condenser. What is meant by a permanent gas? To what is the elasticity of the air due? What is Mariotte's law? Show how the air-gun operates. Explain the pop-gun. Explain the operation of gunpowder. Explain that of steam.—139. Describe and explain what is represented in Fig. 193. Explain the collection of gases in the pneumatic trough. Explain the experiment represented in Fig. 194. What is said of tapping a barrel? What causes the gurgling sound when a liquid is poured from a bottle?—140. How high a column of water will the pressure of the atmosphere sustain? How do you find from this the pressure of the air on every square inch of surface? How high a column of mercury will the atmosphere sustain?—141. Explain the barometer. Relate

the incident given by Dr. Arnott. Why would not a water-barometer answer? What is said of the barometer as a measurer of heights? Describe the aneroid barometer.—142. How is the boiling-point influenced by the amount of the air's pressure? Give the experiment with ether. State the experiment with the flask. What would happen to liquids if the atmosphere were removed from the earth?—143. Explain the operation of the siphon. Explain what happens if the siphon be placed as shown in Fig. 203. Explain the uses of the siphon. Explain the operation of the cup of Tantalus.—144. How are intermitting springs accounted for?—145. Explain the operation of the common pump. Why does the water rise in the pump? How is sucking done? Explain the forcing-pump.—146. Explain the working of a fire-engine.

CHAPTER XIV.

SOUND.

147. That branch of natural philosophy, or physics, which treats of the phenomena of sound is called *Acoustics*, the name being derived from a Greek word meaning "I hear." Acoustics deals mainly with the production, transmission, and comparison of sounds, leaving the question of the pleasurable feelings they may arouse to the science of music. Sound may be defined as a sensation excited in the organs of hearing resulting from the vibratory motion of bodies, which motion is usually transmitted by the air.

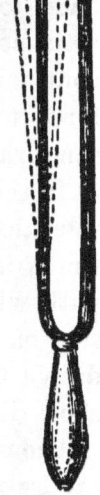

Bodies which emit clear and regular sounds are said to be sonorous. That the production of sound is due to their vibrations may be made manifest to the senses in many ways. If we place the hand upon a large bell that has been struck, we can feel the vibration. If we strike one of the ends of a tuning-fork upon some hard body, we can see the vibration, as represented in Fig. 210 by the dotted

Fig. 210.

lines. If we examine the strings of a piano while it is played, the vibration of the larger strings is very noticeable. If we rub the edge of a drinking-glass with a moistened finger so as to produce a musical sound, the water within it will be thrown into waves by the vibration of the glass.

In wind instruments, as the flute, horn, etc., the sound is caused by the vibration of the body of air within the instrument. In the common tin whistle or bird-call, Fig. 211, the sound is produced by the vibration imparted to the contained air by the impulse of the breath through the orifice, B.

Fig. 211.

148. **An Analogy.**—The vibration of a sonorous body is much like that of a pendulum. The end of the tuning-fork, Fig. 210, on being struck passes to *b*, and in returning passes by the point of rest, A, just as a pendulum does, and reaches *a*. So, also, if a string, tightly stretched between two points, A B, Fig. 212, be drawn aside to D, as it flies back to C it will by its inertia pass on to E, and will continue to vibrate back and forth for some time. The

Fig. 212.

same rule also applies to the extent of the vibrations here as in the case of the pendulum, § 93. The quickness of the vibration is not at all affected by its width. The farther the string, A B, is drawn to one side, the greater the force with which it will return, and hence it will reach its position on the other side of the middle line as quickly when drawn far away from this line as it would if drawn but a short distance.

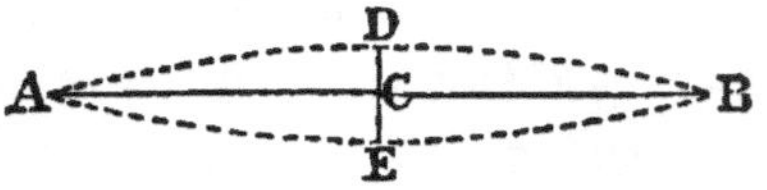

The vibrations, however produced, are transmitted to the ear by means of the intervening air. The latter is set in motion by the impact of the vibrating body, much as motion is communicated through a series of elastic balls (§ 80). The particles of air swing to and fro through a short distance, being condensed at one point and thinner at another. A succession of pulses or waves ensues, each consisting of a pulse of condensation and a pulse of rarefaction. The to-and-fro or wave motion is in the line of the propagation of the sound. The manner in which these sound-pulses act upon the ear will be explained in the next section.

Only a limited number of vibrations produce the sensation of sound; those which are either very slow or very quick will not do it. Thus if a plate of metal or a string make less than 16 or more than 38,000 vibrations in a second, no effect is produced upon the ear. The capacity of hearing differs, however, in different persons, so that although few can hear vibrations which are beyond the range mentioned, there are many whose capacity falls much within it either at one end or both ends of the scale. The range for animals is not the same as that for man. Thus the lion and the elephant can hear a sound when the vibrations are too infrequent to make any impression upon our ears; while small animals have a susceptibility in the organ of hearing for vibrations so rapid that we cannot hear them, and at the same time are not susceptible to the slower vibrations. How far the range varies in different animals has not been ascertained to any extent.

149. How the Sensation of Sound is Produced.—The vibration of a sounding body is transmitted to the ear ordinarily through the air, and there strikes upon a little drum, a membrane at the bottom of the external cavity of the ear somewhat like a common drum-head. There the vibration of the air is communicated to this drum, and from this to a chain of very small bones. From the last of these bones it is transmitted to another very small drum, and from this to a fluid in some very complicated passages in the most solid bone in the body. These may be called the *halls of audience.* In the fluid contained in them are spread out the branches of the nerve of hearing, which receive the impres-

sion of the vibration, and transmit it to the brain, where the mind takes knowledge of it. Observe that the vibration, transmitted first through the air, then through the drum, then the chain of bones, then another drum to a fluid, stops at the fluid. What is transmitted from this to the brain by the nerve we know not, and so we call it an impression.

Sound Transmitted through Various Substances. — In ordinary hearing, sound, as you have seen, is transmitted through various substances before the vibration arrives at the liquid in the halls of audience. But sound need not take this course in all cases to arrive at the nerve of hearing. If, for example, you place a watch between your teeth, the sound will go through the solid teeth and the bones of the jaw directly to the halls of audience by a short cut, instead of going round through the outer ear-passage to the drum, and so through the chain of bones. Fishes in hearing receive the vibration through water. If you place your ear at the end of a timber, while some one scratches with a pin at the other end, you hear the sound distinctly, for the vibration is transmitted through the timber; as in the case of the watch between the teeth, it goes through the solid bone.

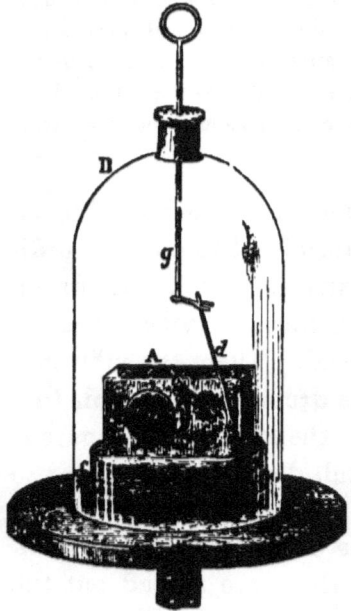

150. Sound not Transmitted through a Vacuum.—As sound is a vibration of some substance, it cannot be transmitted through empty space. This can be proved by an experiment with the air-pump, as represented in Fig. 213. Place under the receiver a clock-work furnished with a bell which can be made to ring by pressing down a sliding rod. If it be struck before the air is exhausted, the sound is heard

Fig. 213.

through the glass. But the more you exhaust the air, the fainter will be the sound; and at length, if you keep on pumping, it cannot be heard at all. A similar experiment can be tried with a music-box. It is owing to the rarity of the air on high mountains, and at the great heights reached by balloons, that all sounds are so faint. The report of a pistol fired off on the top of Mont Blanc is no louder than the snapping of a whip, and trifling compared with its report when fired in the valley below.

151. **Sound Caused by the Resistance of the Atmosphere.** —Sound is often heard at a very great distance on the earth. The sound of an eruption of a volcano has been heard in one case at the distance of 370 miles. But suppose that the same sound should occur at the same distance from the earth—that is, over 300 miles beyond the atmosphere that enrobes the earth—no inhabitant of our world could hear it, for the same reason that you do not hear the bell ringing in an exhausted receiver. If, therefore, any sound, however loud, should be given forth by any of the heavenly bodies, we could not hear it. The course of these bodies in their orbits is noiseless, because they meet with no resistance from any substance. Bodies passing rapidly through our atmosphere cause sound, from the resistance which the air gives to their passage. The whizzing of a ball is an example of this. It is the passage of the electric fluid through the air which produces the thunder. But the heavenly bodies, meeting with no such resistance, make no sound in their course, though their velocity be so immense. In the expressive language of the Bible, "their voice is not heard."

152. **Velocity of Sound.**—The velocity of sound varies in different media. Thus it passes through water four times as rapidly as through air. Dr. Franklin, having placed his head under water, heard distinctly the sound of two stones

struck together in the water at the distance of more than half a mile. Sound passes through solids much more easily, and therefore more rapidly, than through liquids. Thus its velocity through copper is twelve times and through glass seventeen times greater than through air. If you place your ear against a long brick wall at one end, and let some one strike upon the other end, you will hear two reports—the first through the wall and the second through the air. Indians are in the habit of ascertaining the approach of their enemies by putting the ear to the ground. When the eruption of a volcano is heard at a great distance, the sound comes through the solid earth rather than through the air. The ready transmission of sound through solids furnishes us with a very valuable means of examining diseases of the lungs and heart. The sounds occasioned by the movement of the air in the lungs and by the action of the heart are very distinctly heard through the solid walls of the chest.

153. **Measurement of Distances by Sound.** — Whether sound be loud or weak makes no difference in its velocity. Thus the sounds of a band of music at a distance all reach your ear at the same time, the sounds of the instruments that can scarcely be heard keeping exact pace in the air with the sounds of the loudest. The velocity of sound is uniform throughout its whole course, being just as rapid when it is about to die away as it was when it began. This uniformity in the velocity of sound enables us to estimate the distance of the object by which any sound is made. We do it by a comparison between light and sound. Sound moves at the rate of 1120 feet in a second. Now, light moves 192,000 miles a second, and therefore, for all ordinary distances on the earth, we need make no allowance of time for light in comparison with sound. If we see, then, the operation by which a sound

is produced, we can estimate its distance from us by the
length of time which elapses between what we see and
what we hear. In this way we can estimate very accurately the distance of a cannon that we see fired, or the distance of a flash of lightning. When the interval between
the flash and the peal of thunder is about four and a half
seconds, the distance of the cloud is about one mile.

154. **Loudness of Sound.**—The loudness of sound depends
upon the width of the vibrations producing it. The harder
you strike the end of the tuning-fork, Fig. 210, the farther
will it vibrate the one way and the other, and the louder
will be the sound. The same thing is true of the strings
of a piano. A round bell, when struck, tends in its vibration to take an oval form, and the extent of its vibration
back and forth determines the loudness of the sound. As
sound passes from the sounding body the vibration gradually lessens, and at length dies away. It is like the successive vibrations or waves of water produced by dropping a stone in it. The louder the sound, the larger are
the first vibrations and the farther will the vibrations extend; as a large stone dropped into water will produce
larger waves than a small one, and the waves will extend
over a greater space.

The loudness of sound decreases very rapidly as the distance from its source increases; at twice the distance the
intensity is only one fourth; at three times the distance,
one ninth, etc.; the law being the same as that of gravitation—viz., the intensity is inversely as the square of the
distance. (See § 28.) When there is no hindrance, sound
spreads equally in all directions. In this respect the vibrations or waves of air resemble waves of water. Light is
also diffused in the same manner, as you will learn in
another chapter.

155. **Reflection of Sound.**—If an elastic body be thrown

perpendicularly upon a surface, it rebounds in the same path in which it is thrown. But if it hit the surface obliquely, it is thrown off or reflected in a different direction.

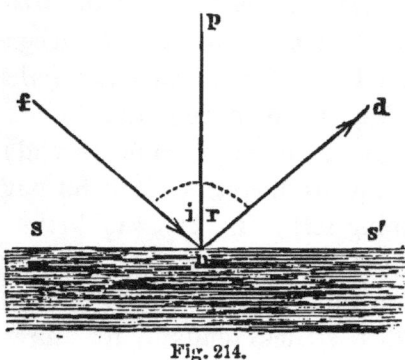

Thus a ball thrown from p upon the surface $s\,s'$ at the point n will rebound on the same line, $n\,p$; but if it be thrown from f, it will be reflected in the direction $n\,d$. The angle i between the lines $f\,n$ and $n\,p$ is called the angle of incidence, and the angle r between the lines $p\,n$ and $n\,d$ is called the angle of reflection: these two angles are always equal if the body projected be perfectly elastic. Waves of sound are reflected in accordance with the same law. The reflection of sound is the cause of *echoes*. In order that an echo be perfect, the sound must be reflected back to the ear from a plane surface of some size. Sometimes successive plane surfaces of rocks in a valley, or along a river, cause a series of echoes. Thus in Fig. 215

Fig. 215.

is represented a locality on the Rhine where a sound is reflected at successive places, 1, 2, 3, 4. The rolling of thunder, though sometimes caused by the different distances of parts of the same flash of lightning, is commonly owing to reflections of the sound among the clouds. From this cause the report of a cannon is more apt to be a rolling sound when there are clouds above than when the sky is clear. Sound is continually reflected in every variety of direction from obstacles with which it meets. Thus in a room it is reflected from the walls and from all the objects in the room; and the more varied are the surfaces, the more varied and confused are the reflections. You know that a voice has a very different sound in an empty room from that which it has when the room is filled with an audience. Indeed, a blind speaker can estimate very nearly the size of his audience by the sound of his own voice. The explanation is, that with a full audience the surfaces for reflection are vastly multiplied, and so deprive the sound of the sharp and ringing character which is given to it by reflection from comparatively few surfaces which are plane and firm. The effect produced by an audience upon the voice of the speaker is quite analogous to that of muffling upon the sound of a drum.

156. **Whispering-Galleries.**—The reflection of sound from curved surfaces gives us some interesting phenomena. The waves of sound in being reflected from a concave surface are gathered together at some point. If the surface be a perfectly spherical one, and the sound issue from the centre, the reflection will be from all points to the centre. But suppose the concave surface have the curve of an ellipse, as represented in Fig. 216. This, instead of having a centre, has two foci, c and g. Now if a sound proceed from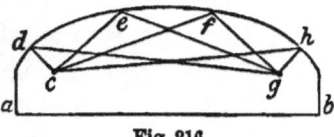

Fig. 216.

one focus, c, the waves of sound, as represented by the lines $c\,d$, $c\,e$, $c\,f$, $c\,h$, will all be reflected to the other focus, g; so that if a person speak in a very low tone or even whisper at *c*, he may be heard distinctly by another at *g*, though persons at other points may hear nothing. This phenomenon may occur with a curved wall extending even several hundred feet; and such structures are called whispering-galleries. If in one of these galleries a person standing in one focus speak softly, he will be heard by others at any point by the *direct* waves of sound; but the reflected sound will be added to the direct in the case of one standing at the other focus. The whispering-gallery in the dome of St. Paul's Cathedral, London, is a celebrated example well worth visiting.

157. Concentration of Sound.—It is by the reflection of sound that it can be concentrated in various ways. Thus in using a speaking-trumpet the waves of sound, instead of moving in all directions as soon as they escape from the

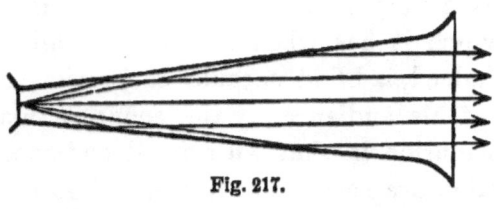

mouth, are reflected by the sides of the instrument towards a central line, as represented in Fig. 217. The waves or vibrations, being thus concentrated, have more intensity, and are thrown to a greater distance than if they issued directly from the mouth. In a similar manner a speaking-tube, confining the vibrations, carries the voice to distant parts of a building. For the same reason the voice can be heard much farther through a narrow street than in an open space. A speaker can be heard more distinctly in a hall than when addressing an audience of the same size in the open air. The "sounding-board," once so fashionable in churches, was really of considerable service in preventing the escape of the vibrations of the voice of the preacher upward, and directing them downward upon the audience. In the hearing-trumpet, Fig. 218, the vibrations are collected in the broad open end of the instrument and by reflection are thrown together

Fig. 217.

Fig. 218.

into a narrow compass before they enter the ear to strike
upon the drum. We often instinctively make the palm of
the hand act as an ear-trumpet when we do not hear distinctly. Many animals have the external ears movable, so
that they can direct their concave surface towards the
point from which they wish to hear. Such ears act like
movable ear-trumpets.

158. **Difference between a Musical Sound and a Noise.**—
The difference between a musical sound and a noise is very
analogous to the difference between a crystal and the same
substance destitute of the crystalline arrangement. In
both sound and noise there are vibrations, but in musical sound they recur at equal intervals of time; while in a
noise the vibrations are irregular, and there is confusion.
Indeed, so regular are the vibrations of musical sounds that
the rules and principles of music have all the rigid exact-
ness of mathematics.

Musical sounds differ among themselves in three par-
ticulars—loudness, pitch, and quality. The loudness, or intensity, depends, as already stated (§ 154), upon the width
of the vibrations of the sounding body. The pitch depends
upon the rapidity of these vibrations; the quicker the vibration, the higher is the note. Thus a short and small
string on a violin or in a piano gives a higher note than a
long and large string, because its vibrations are quicker.
The tension of the string also has an influence, the note
being raised by increasing the tension. In tuning a violin
the right pitch is given to each string by lessening or increasing the tension by means of the screws to which the
strings are attached. In playing upon it various notes are
made upon each string by shortening the vibrating portion
more or less by pressure of the finger.

In wind instruments the note depends on the length and
size of the column of air contained in them. This may be

L

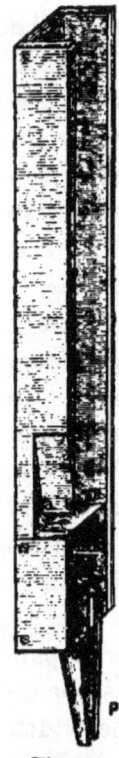

Fig. 219.

illustrated by an organ-pipe, Fig. 219. It is one of the pipes of what is called the flute-stop. It is constructed very much like a boy's willow whistle. The air from the bellows of the organ enters at P, and, passing through the narrow slit, *c d*, is projected against the edge of the mouth, *a b*, and causes a vibration of the whole column of air in the pipe. The pitch of the musical note depends in part on the length of the column of air set in vibration, and in part on the size of the pipe.

It is owing to the difference in rapidity of vibration that a large bell gives a graver note than a small one. When musical sounds are produced by passing the moistened fingers over the edges of glass vessels, the larger the vessel, the graver its note. A tumbler will give a graver note than a wine-glass.

The third peculiarity of musical sounds, called quality, is that which enables us to distinguish the notes of different instruments even when their pitch is the same. It also gives the distinctive character to the voices of different animals, and even of different persons. It depends upon the *form* of the vibratory motion, as illustrated by Fig. 220. The three waves, A B, C D, and E F, there represented have the same width and the same length, and consequently the sounds corresponding would have the same pitch and intensity; but, having different *forms*, the sounds

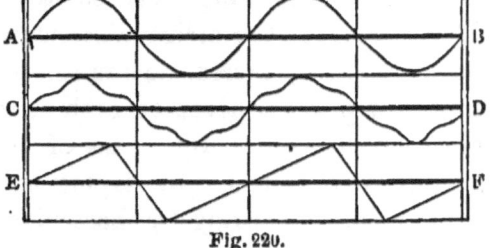

Fig. 220.

would be unlike in quality. This method of exhibiting the subject does not rest on theory alone, but can be demonstrated to the eye by a beautiful experiment of a delicate character which we cannot here detail.

159. **Human Voice.**—The principles developed in relation to musical instruments apply to the voice. The musical instrument of man, by which the voice is produced, is contained in a very small compass. It is that box at the top of the throat commonly called Adam's apple. Across this, from front to rear, stretch two sheets of membrane, leaving a space between their edges. In our ordinary breathing these membranes are relaxed, and the space between their edges is considerable, to allow the air to pass in and out freely. But when we speak or sing these membranes, or vocal chords, as they are termed, are put into a tense state by muscles pulling upon them, and the opening between them is lessened. The voice is produced by the air that is forced out from the lungs, which, striking on the chords, causes them to vibrate. The nearer their edges are together, and the more tense, the higher the note. The sounds are produced precisely like those of the Æolian harp, the air causing in the one case a vibration of strings, and in the other of edges of membranes.

160. **Harmony.**—When a number of notes, sounded at the same time, are agreeable to the ear, they are said to harmonize. Now this harmony depends on a certain relation between the vibrations. The more simple the relation, the greater is the harmony. For example, if we take the first note, termed the fundamental note, of what is called the *scale* in music, it harmonizes better with the octave than with any other of the eight notes, because for every vibration in it there are just two in the octave. Take in contrast with the octave the second note. Here to every eight vibrations of the first note we have nine of the second, and the consequence is a discord when they are sounded together. The difference between the two cases is this: In the first case the commencement of every vibration in the fundamental note coincides with the commencement of every second

vibration in the octave. But in the other case there is a coincidence at only every eighth vibration of the first note with every ninth of the second. Next to the octave, the most agreeable harmony with the fundamental note is that of the fifth note of the scale. Here we have three vibrations to every two of the first note, and so every second vibration in the first note coincides with every third vibration of the fifth. Next comes the harmony of the fourth, there being here a coincidence at every third vibration of the fundamental note. The more frequent, you see, are the coincidences between the vibrations, the greater is the harmony. In the three cases just stated the coincidence is in the first at the commencement of *every* vibration of the fundamental note; in the second case, at the commencement of every *second* vibration; and in the third, at the commencement of every *third* vibration.

161. **The Gamut, or Diatonic Scale.** — In order that you may see the relative numbers of the vibrations for each of the notes, we give them for the whole scale. They are as follows :

$$1 \quad \tfrac{9}{8} \quad \tfrac{5}{4} \quad \tfrac{4}{3} \quad \tfrac{3}{2} \quad \tfrac{5}{3} \quad \tfrac{15}{8} \quad 2$$
$$C \quad D \quad E \quad F \quad G \quad A \quad B \quad C$$

According to this, the note D has nine vibrations to every eight vibrations of C, E has five to every four of C, etc., the octave C having just twice the number of vibrations as the fundamental note C. By this means is expressed the *proportion* between the numbers of vibrations in the different notes. Suppose, then, that you know the number of vibrations in a second required for C, the fundamental note, you can readily calculate the number of vibrations of each of the other notes. It is done by multiplying the number which C has by the fractions placed over the other notes. Thus if the number of vibrations in a second in the fundamental note be 128, by this process we make the vibrations of all the notes to be thus :

C	D	E	F	G	A	B	C
128	144	160	170	192	213	240	256

There are really but seven notes in what is called the diatonic scale, the eighth note, C, being truly the first of seven other notes above, having relations to each other similar to those of the notes below, and constituting another octave. So we may have several octaves, one above another.

It is interesting to observe that the proportionate lengths of strings required to produce the eight notes of the scale have an exact numerical relation, but the *reverse* of that of the numbers of the vibrations. Thus if you have eight strings of the same size, their vibrating lengths required for the notes are as follows:

C	D	E	F	G	A	B	C
1	$\frac{8}{9}$	$\frac{4}{5}$	$\frac{3}{4}$	$\frac{2}{3}$	$\frac{3}{5}$	$\frac{8}{15}$	$\frac{1}{2}$

For the notes of the octave above, the lengths are:

C	D	E	F	G	A	B	C
$\frac{1}{2}$	$\frac{4}{9}$	$\frac{2}{5}$	$\frac{3}{8}$	$\frac{1}{3}$	$\frac{3}{10}$	$\frac{4}{15}$	$\frac{1}{4}$

162. **Unison.**—In tuning instruments so as to make them harmonize, the result is obtained when the corresponding parts of the instruments have the same number of vibrations. Thus the string in one violin that gives any particular note must vibrate just the same number of times in a second as the strings giving the same note in other violins, or it will not be in perfect unison with them. The same is true of other strings for other notes, and also of the corresponding parts of all kinds of instruments.which are to be played together. When, in tuning instruments together, it is said that a string of a violin, for example, is too *flat*, the difficulty is that it does not vibrate with sufficient rapidity, and it is therefore tightened to make its note *sharp* enough, as it is expressed, to be in unison with the note of the corresponding strings or parts of other instruments.

163. **Mysteries of Sound and Hearing.**—There are many things of a mysterious character in relation both to sound and the manner in which it causes the sensation of hearing. We will barely notice but two of these. The effect, or rather the chain of effects, resulting in hearing is wholly mechanical, until we come to the nerve of hearing, which branches out with minute fibrils in the halls of audience of the internal ear. It is merely a series of vibrations. Now, how the mere agitation of a fluid enclosed in hard bone can communicate through fine white fibres to the brain, and through that to the mind, the impression of all the various sounds produced is a great mys-

tery. All that we know is that the nerve is the medium of the communication, but of the manner in which it performs its office we know absolutely nothing. Again, while it is sufficiently mysterious that this information can thus be given to the mind when one sound after another communicates its vibration to the liquid in the ear, the mystery is greatly enhanced when various sounds come to the ear at one and the same time. To get a distinct idea of the very complex and wonderful character of the process of hearing in such a case, we will suppose that a full band of music is playing, and at the same time mingled with its sounds there are various other sounds heard, some of them perhaps discordant. What a diversity of vibrations we have here! We have the slow vibrations produced in the grave notes, and the quick vibrations of the higher ones, all travelling together through the air to the ear, and each preserving its distinctive character. And more than this, after they arrive at the ear they are communicated unaltered through the drum, the chain of bones, the second drum, and the liquid where the nerve is, so that a correct report of each of all the notes is given through the nerve to the mind. Then, too, if there be any discord, its vibration travels along with the rest, and so do the vibrations of other sounds, as the roaring of the wind, the report of cannon, and the noise of the people. And besides all this, in the multiplicity of the vibrations thus transmitted through so many different substances the mind gets a true report of the comparative loudness of the sounds, and even of their character, so that the sounds of drum, fife, trumpet, etc., are all accurately distinguished. In view of such wonders, how significant is the question, "He that planted the ear, shall he not hear?"

QUESTIONS.

147. What is the meaning of Acoustics? Define sound. Mention cases in which the vibration of sounding bodies is manifest to the sight and touch. What is said of wind instruments?—148. State the analogy of a sounding body to a pendulum. How does the air transmit sound? What is the connection between sound and rapidity of vibration?—149. Describe the process by which the sensation of sound is produced. Where does the vibration caused by the sounding body stop in the ear? What is transmitted thence to the brain? Give examples of the transmission of sound through various substances. — 150. State the experiment by which it is shown that sound is not transmitted through a vacuum. What is said of sound at great heights?—151. How far has the sound of a volcano been heard? If the same sound were made in space at that distance from the earth, why could not the inhabitants hear it? What is the cause of the noise of bodies passing through the air? Why do the heavenly bodies, moving so rapidly, produce no sound?—152. Cite examples showing the different velocities of sound in different media. What is said of the uniformity of the velocity of sound?—153. Show how we can measure distances by sound as compared with light in velocity.—154. Upon what does the loudness of sound depend? Illustrate this point. What is said of the diffusion of sound?—155. What of its reflection? What of echoes? What is said of multiplied and mingled reflections of sound?—156. Explain the operation of whispering-galleries by Fig. 217.—157. Explain the operation of the speaking-trumpet. Give other examples of the concentration of sonorous vibrations.—158. What is the difference between a musical sound and a noise? What is said of the exact regularity of musical vibrations? Name the three points in which musical sounds differ. Upon what does the pitch depend? How are different notes produced in stringed instruments? Upon what does the note depend in wind instruments? Explain the operation of the organ-pipe. What is said of the notes of bells and of musical glasses? What is meant by quality? Upon what does it depend? Illustrate this.—159. Explain the mechanism of the human voice.—160. What is harmony? Upon what does it depend? Between what two notes of the scale is there the greatest harmony? What note next to the octave harmonizes best with the fundamental note? And what note next? Show why the second note, in contrast with the octave, is so discordant with the fundamental note.—161. State the proportions between the numbers of the vibrations in the different notes. If you know the number of vibrations of the fundamental note in a second, how may you determine the number

of vibrations in the other notes? What is said of the number of notes in the diatonic scale? What of the proportionate lengths of strings for different notes?—162. What is said of tuning instruments? What is meant by saying that a note is too sharp or too flat?—163. State in full what is said about the mysteries of sound and hearing.

CHAPTER XV.

HEAT.

164. **Heat and Cold.**—In common language we speak of heat and cold as two distinct and opposite things. That this is not strictly correct may be shown by the following experiment: Take three vessels, and fill the first with ice-cold water, the second with hot water, and the third with tepid water. If you place your right hand in the first and the left in the second, and let them remain a little time, on taking them out and plunging them together into the third vessel, the water in it will feel warm to the right hand and cold to the left. Thus the air of a cellar seems warm to you in winter and cold in summer in contrast with the air outside. For the same reason water of a temperature that would ordinarily be refreshingly cool to us seems warm when drunk after eating ice-cream. It is manifest, then, that there is no fixed dividing-line between heat and cold; they are merely relative terms. There is, in fact, no such thing as cold. Substances are cold from being deprived of heat; and no substance ever has all its heat taken from it. Sir Humphry Davy proved that there is heat in ice by rubbing two pieces together in a very cold room until they were gradually melted. Now, this was not done by the air, for that was at a temperature below the freezing-point; the heat which melted the ice resided in the ice itself.

165. **Nature of Heat.**—We have just stated. that there is no such thing as cold, and we now assure you that there is no such thing as heat. That is to say, there is no substance to that which we call heat. A hot body weighs no more than the same body after it has cooled. When we heat a substance we add nothing to it, and when we cool it we take nothing ponderable from it. And yet the effects of heat are everywhere present: "we know hot iron, hot water, or hot air; but nature nowhere presents to us, nor has art succeeded in exhibiting to us, heat alone." The old theory of the nature of heat was that it is an imponderable, or unweighable, and consequently very subtile substance, pervading all matter, and tending to diffuse by the mutual repulsion of the particles. This view has given way to another supposition, now generally received, that heat is merely motion of a certain kind among the material particles of bodies. As with sound (§ 147), this motion is vibratory, and the width and velocity of the vibrations determine the temperature of the body, the hottest substances being those in which the particles vibrate with greatest rapidity. It is further assumed that the transfer of heat from one body to another is effected by means of an imponderable elastic and subtile fluid called *ether*, which fills all space, both celestial and intermolecular (§ 7). This ether transmits with immense velocity the vibratory motion of the particles; and its motion produces heat, just as the motion of aeriform bodies produces sound.

According to the old view, then, heat existed as a kind of matter, called "caloric." Under the new view, heat is a "mode of motion."

Many philosophers have contributed to the establishment of the latter theory, but the first reliable experiments in this connection were made in 1798 by our countryman Benjamin Thompson, better known as Count Rumford. Having entered the service of the Elector of Bavaria, he had,

among other duties, charge of the Munich arsenal. When engaged in boring cannon, he observed the enormous amount of heat generated, a phenomenon which in his opinion was insufficiently explained by the common
theory that the heat was furnished by the abrasion of the metal. He accordingly made a series of experiments which showed that enough heat
was generated in boring a metallic cylinder (by means of horse-power) to
raise water surrounding it to boiling. Reasoning upon these remarkable
results, he was led to the conclusion that the heat generated could not be a
substance or material, but was in all probability motion. Subsequently
this theory has received confirmation through the labors of many distinguished men, and it has been further proved that an exact relation exists
between the amount of heat generated and the amount of mechanical force
exerted in its production. Not only, however, is mechanical force capable
of being transformed into heat, but the latter can be converted into the
former; in other words, they are mutually convertible. Whenever motion
is arrested, some of the force is transformed into heat. Of this we have
innumerable examples: a blacksmith hammers a piece of cold iron until it
glows with a red heat; a bullet fired against an iron target flattens out and
becomes quite warm.

We have already stated that force, like matter, is indestructible (§ 9).
When it appears to be destroyed, as in the case of arrested motion, it is
really converted into some other manifestation of force. Such are a few
of the phenomena and reasonings which have aided in establishing the
present theory of the nature of heat. You will learn in another chapter
that heat and light, as well as electricity, are also mutually convertible,
phenomena on which is based the grand law of the "correlation and con-
servation of force."

166. Sources of Heat.—The principal source of heat on
our earth is the *sun*, though that body is ninety-two
millions of miles distant from us. As the heat, in travelling all this long journey, is becoming more and more diffused or scattered, we can have no conception of the intensity of the heat in the sun itself. We can, however,
form an approximate idea by observing the effects of heat
when some of its separated rays are gathered to a point
by a powerful lens, as represented in Fig. 221. A lens
which concentrated the heat ten thousand times melted

platinum, gold, quartz, etc., in
a few seconds. And since the
heat at the sun is supposed to
be vastly more intense than
this, none of the most solid
substances of our earth would
remain solid if present, but
many of them would become
liquid, and others even vapors.
The heat which the sun constantly radiates to the earth

Fig. 241.

pervades all substances, producing motion, and awakening
life everywhere; so that, in the expressive language of the
Bible, "There is nothing hid from the heat thereof."

Another source of heat is *within the earth itself*. It has
been found that as we descend into the earth the temperature constantly increases the farther we go. This
internal heat is attributed in part to subterranean fires
and various chemical actions. Here and there we see
external evidences of this in the eruptions of volcanoes,
the boiling springs, the jets of steam and sulphurous
vapors, etc. In very deep mines the temperature rises
as you descend, becoming positively uncomfortable below a depth of 1800 to 2000 feet. But that the heat in
our earth which comes from these subterranean sources is
small compared with that which comes from the sun is
evident from the fact that the rate of increase of heat at
great depths is much less than it is nearer the surface.
This would seem to show that although fires within the
earth may have considerable influence in heating its crust,
on which we live, it derives the most of its heat from the
sun, at least to a very great depth.

Another very common source of heat is *chemical action.*
We see it continually produced in chemical experiments.

Combustion, which is the development of heat and light
accompanying chemical combination, as will be shown to
you in the Second Part of this series, is the most common
of all the chemical sources of heat. Animal heat is also, for
the most part, a result of chemical action.

Mechanical action is a common source of heat. The rubbing of a match producing heat enough to occasion flame
is a familiar example. The spark produced by striking together flint and steel is an incandescent particle of steel ignited by the blow. The American Indians were accustomed
to procure fire by rubbing together two dry sticks until they
learned an easier way from civilized neighbors. An im-
proved method of obtaining fire by the friction of wood
against wood is shown in Fig. 222. The board B is pressed

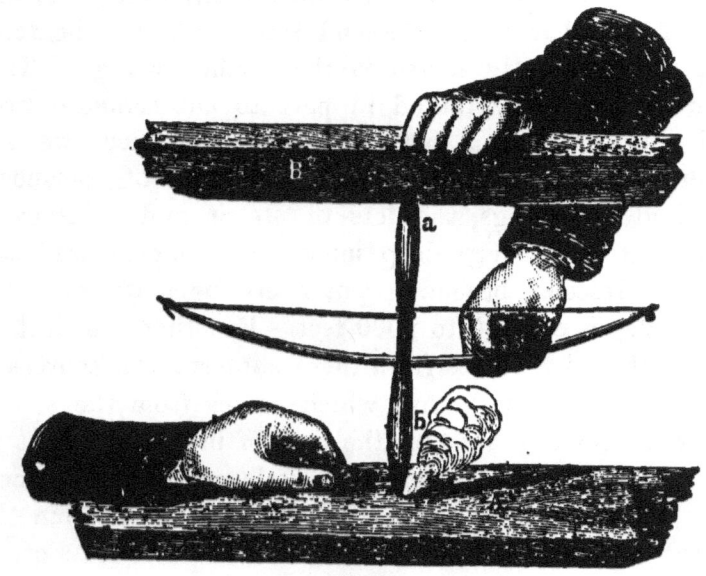

Fig. 222.

strongly against A, while the upright piece of hard wood
is rapidly revolved by means of the instrument known as
a " fiddle-bow." As soon as there are any indications of

fire, a second person approaches a piece of tinder. This affords a striking example of the conversion of motion into heat (§ 165).

The blacksmith, previous to the invention of phosphorus matches, often lighted his fire by touching a sulphur match to a nail made red-hot by rapid and continued hammering. Machinery has sometimes been set on fire by friction, and the water around a mass of metal has been so heated by boring as even to boil (§ 165). If you stretch a piece of India-rubber several times in quick succession, and then apply it to your lips, you will perceive that the motion has warmed it.

Heat is sometimes accompanied by light, and sometimes the latter force is absent: its presence depends upon the rapidity of the vibrations communicated to the ether surrounding the source of heat, heat waves being less rapid than those of light. To this we shall again refer in the chapter on Light (§ 211).

167. **Expansion of Solids.**—The principal effects of heat are expansion, liquefaction, and vaporization; each of these requires your attention. You have already learned in § 5 that heat acts in opposition to the attraction of cohesion, tending to separate the particles, and so produces an expansion of any substance. This may be exemplified in the experiment represented in Fig. 223, in which A B is an iron rod of such a size that at the ordinary temperature it will fit into the space C D in a bar of iron, and easily pass through the hole, E. If the rod be heated, it will be enlarged or expanded in all directions, so that it will neither fit into C D nor pass into the hole, E.

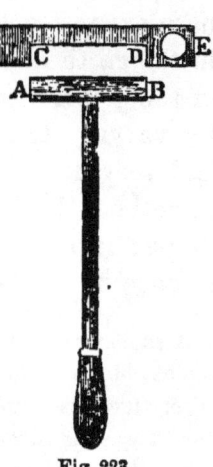

Fig. 223.

When the wheelwright puts a tire upon a wheel, he uses the expansive force of heat to make it fit tightly and firmly. The tire is purposely made a little too small to fit the wheel; but by being heated it is so expanded that it will readily go around the wheel, and then in contracting as it cools it compresses the fellies very tightly. Water is poured on to cool the iron quickly, and thus prevent it from burning the wood. Iron hoops are put on barrels in a similar manner, the compression caused by their contraction binding the staves together very strongly. In like manner the plates of boilers are fastened together; the rivets are put in red-hot, so that by their contraction they may bind the plates closely together. If an iron gate just shuts into its place in cold weather, expansion will prevent it shutting in warm weather. In order to avoid this difficulty, allowance must be made in fitting it for the expansion to which it will be subjected by heat. So in laying the rails of a railroad in cold weather care must be taken not to put the ends too near together. Nails often become loose after the lapse of years from the wear of the wood around them, occasioned by their alternate expansion and contraction. The leaking of gas-pipes in the earth is often undoubtedly caused by the loosening of the joints from contraction and expansion of the pipes by varying temperatures of the soil, especially when not laid very deep. If a stopper stick fast in a bottle, it can sometimes be loosened by applying to the neck a cloth dipped in hot water, because the neck becomes expanded at once by the heat.

A similar expedient was once very ingeniously made use of in repairing the machinery of the steamer *Persia* at sea, and was perhaps the means of saving the vessel and the lives of all on board. The accident which occurred was the breaking of the port crank-pin of the engine. The problem to be solved was the removal of this pin, which weighed nearly a ton, and the substitution of a sound one which they had on hand in its place. But it was found impossible to start the broken pin from its socket with all the force which could be brought to bear upon it by a sort of battering-ram constructed extemporaneously for the purpose. It was then determined to try the expansive force of heat. An iron platform was built under the socket, and a hot fire made upon it. The socket soon expanded, and the pin was then readily knocked out by the battering-ram, just as the stopper of a bottle is easily removed when the neck is heated.

The walls of a very large building in Paris, which had bulged out and were in danger of falling, were restored to their upright position by the expansion of iron. It was done in this way: Long rods of iron were run through the walls after the plan represented in Fig. 224, their ends being made with a screw-thread, with nuts fitted to them. Alternate rods were first heated, and as they lengthened the nuts were screwed up tight to the walls. On cooling,

Fig. 224.

their contraction would of course draw the walls together. The other bars were then heated and managed in the same way. The one set, you see, were made to hold on by their nuts to what had already been gained, while the other were expanding. By many repetitions of this process the walls were straightened and the building saved. The same mode has been adopted successfully in other cases of a similar character.

Different substances expand at different rates; copper

expands more than twice as much as glass, and zinc nearly twice as much as copper, with the same increase of temperature. Advantage is taken of the unequal expansion of metals to regulate the length of pendulums, upon which, as you learned in § 93, depends the correctness of time-keepers. Various arrangements are adopted, but that known as the gridiron pendulum is the best, in which ingenious use is made of the fact that heat expands brass nearly twice as much as it does steel. A simple form of this pendulum is given in Fig. 225. The middle rod is made of brass, and the side rods, b and c, of steel. Suppose that the brass rod expands or increases in length half an inch. The rod c would be drawn upward by it, and the rod b downward, each one quarter of an inch; but this effect is counteracted by the expansion of

Fig. 225.
each steel rod, which is half that of the brass —that is, one quarter of an inch. The ball, d, therefore, always retains the same distance from the point of suspension, e. Fig. 226 represents a gridiron pendulum of a more complex character, part of the bars being steel and a part brass.

168. **Expansion of Liquids.**—Liquids are expanded by heat to a greater degree than solids, but very unequally so. Thus water is expanded more than twice as much as mercury, and alcohol six times as much. We have a frequent example of the expansion of water by heat in our kitchens. If the tea-kettle be put over the fire filled to the brim, it will run over long before the water begins to boil. All liquids occupy more space in summer than in winter, and in the former case weigh less—that is, have less of real substance in them than in the latter. If, therefore, alcohol, or oil, or molasses be bought by the

Fig. 226.

gallon in winter and sold in summer, there will be a profit afforded by the expansion.

The influence of the expansion of heat upon the specific gravity of liquids may be very prettily shown by the following experiment: Throw some little bits of amber—a substance having nearly the same specific gravity as water—into water in a glass vessel, and heat the water by a spirit-lamp, as represented in Fig. 227. That portion of the water which is heated passes upward because it becomes specifically lighter, and colder water continually comes down to take its place. An upward current passes up in the middle, as indicated in the picture, the downward coming down at the sides. This will be made manifest by the little bits of amber. This experiment also illus-

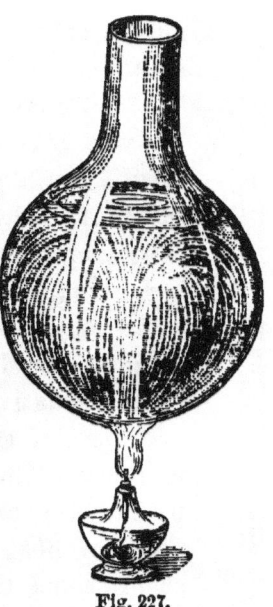

Fig. 227.

trates the manner in which heat is transmitted in liquids by *convection*, as will be more fully explained in § 183, Chapter XVI.

169. **Thermometers.**—The expansion of liquids by heat affords us a convenient means of measuring differences of temperature: the instrument commonly employed is called a thermometer, the word being derived from two Greek words signifying together "heat - measurer." The form and general construction of a thermometer are familiar to all; the liquids used for filling the bulb are alcohol and mercury. The latter answers well except in the extreme cold of the polar regions; for mercury becomes solid at about 39 degrees below zero, while alcohol cannot be frozen by any known degree of cold. The manner in which a thermometer indicates temperatures is very simple: heat expands the liquid in the bulb, and the only way in which it can occupy more space is by rising in the tube. The removal of heat, on the other hand, causes contraction

and of course a proportionate fall of the mercurial column.

170. **Fahrenheit's Thermometer.**—The thermometer was invented in the beginning of the seventeenth century, but it is not decided who was the inventor. There may have been in this case, as in others, more inventors than one, the same ideas having, perhaps, entered several inquiring minds at the same time. Various fluids were used by different persons. Sir Isaac Newton used linseed-oil. Fahrenheit, a native of Hamburg, who flourished in the first part of the last century, was the first to use mercury. Though various propositions were made by Newton and others in regard to the measurement of heat by thermometers, no thermometric scale seems to have met with general reception till that of Fahrenheit, which was introduced about 1720. His zero is the point at which the mercury stood in the coldest freezing mixture that he could make; and he supposed that this was the greatest possible degree of cold, as it was the greatest that he knew. He next found the point at which the mercury stood in melting ice. This he called the freezing-point, because the temperature is the same in water passing into the solid from the fluid state as in water passing into the fluid state from the solid. In other words, this point in the scale marks the transition line between the two states. From this point Fahrenheit marked off 32 equal spaces or degrees down to zero. He then found the point at which the mercury stands in boiling water, and called this the boiling-point. Marking off the space on the scale between this and the freezing-point in the same manner, there are 180 degrees—that is, the boiling-point is 212 degrees above zero. The degrees

Fig. 223.

above zero are commonly designated by the mark +, plus; and those below by the mark —, minus. Thus, +32° signifies 32 degrees above zero, and —32° signifies 32 degrees below.

171. **Thermometric Scales.**—Fahrenheit's thermometer is the one commonly used in this country. But there are several other thermometers on different scales, as the Centigrade, Réaumur's, and De Lisle's. Fig. 229 shows the scales of these thermometers placed side by side. In the Centigrade thermometer, which is in use in France, and indeed in a large part of Europe, the zero is placed at the freezing-point; and the space between this and the boiling-point is divided into 100 degrees, which gives it the name Centigrade. It is also called Celsius's, after its inventor. Réaumur's, which is in use in Germany, has the same zero-point, but has only 80. degrees from this to the boiling-point. De Lisle's which has gone entirely out of use, has its zero at the boiling-point. In the arrangement of Fahrenheit the zero is a mere arbitrary point, and the division of the scale into 212 parts is very inconvenient. The Centigrade thermometer, on the other hand, having two points easily determined and invariable—viz., that of the freezing and boiling of water—having also a

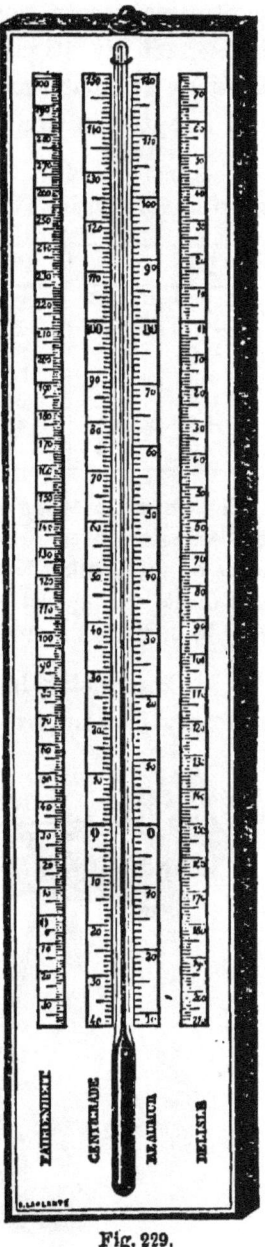

Fig. 229.

centesimal scale, possesses great advantages for scientific and exact investigations over every other style. It is now used by scientific men almost exclusively, and is the standard adopted in this series of works.

The following short table may be useful for comparing temperatures given in Fahrenheit and Centigrade degrees:

C.	F.	C.	F.	C.	F.	C.	F.
−20°	−4°	15°	59°	45°	113°	75°	167°
−15	+5	20	68	50	122	80	176
−10	+14	25	77	55	131	85	185
−5	+23	30	86	60	140	90	194
0	32	35	95	65	149	95	203
+5	41	40	104	70	158	100	212
10	50						

For temperatures not given in the above table the following rules may be used: to convert degrees on Fahrenheit's scale to corresponding degrees on the Centigrade scale subtract 32°, multiply the remainder by 5, and divide the product by 9; to convert Centigrade to Fahrenheit multiply by 9, divide the product by 5, and add 32.

172. Expansion in Aeriform Substances.—Heat produces a vastly greater expansive effect in air, the gases, and vapors than it does in liquids. The expansion of air by heat may be shown very prettily in this way: Take a glass tube having a bulb on one end, and, placing the other open end in water (as represented in Fig. 230), apply the palm of your hand to the bulb. The heat of the hand, being communicated to the bulb, will expand the air, and bubbles of air will escape through the water. On re-

Fig. 230.

moving the hand, and allowing the bulb to cool, the air in it will be condensed, and water will enter the tube in proportion to the amount of air which has escaped. A bladder partly filled with air will swell out to plumpness if heated sufficiently, and a full one may be so heated as to burst from the expansion of the air. Chestnut and other porous wood snap very much when burned, because the heat expands the air and moisture contained in their pores.

Balloons.—The first balloons used were filled with heated air. You have already seen, in § 129, why balloons rise. The hot-air balloon becomes lighter than the surrounding atmosphere, because the contained air is expanded by heat. Of course such a balloon is not so effective as the gas balloon, for the air within it loses its comparative lightness as it becomes cooled; while the coal gas used, being very much lighter than air at the same temperature, does not lose its lightness as the balloon ascends. You learned in § 131 that the atmosphere becomes thinner as we go upward. The gas balloon, therefore, rises until it arrives at that point where the air is of about the same specific gravity with the gas, and there it stops. It is made to descend by letting out some of the gas from a valve. Gas was not used for balloons till 1782. Hydrogen gas was employed at first, being over fourteen times lighter than air. Of late the common coal-gas, carburetted hydrogen, is generally used, because it can be so readily obtained from gas-works.

173. **Currents in the Air from Heat.**—Heat is the grand mover of the atmosphere. Any portion of it that becomes warmer than surrounding portions rises, or rather is pushed up, for the same reason that a hot-air balloon rises, the only difference between the two cases being that in the one the air is confined, and in the other is left free, and so becomes diffused. And it is this expansion that causes nearly all the movements witnessed in the air. We see this exemplified in various ways wherever there is a fire. The air heated by the fire is forced upward by the colder air, which, on the principle of specific gravity,

seeks to get below the warmer and lighter air. The hot air that comes through the registers of a furnace is pushed up by colder air below. For the same reason the heated air around a stove-pipe is constantly rising. This is very prettily shown by the toy represented in Fig. 231, which is a paper cut spirally, and suspended upon the point of a wire. The upward current makes the paper revolve rapidly around the wire. It is owing to the rising of heated air that the galleries of a church are warmer than the space below. In a common room the air is so disposed that its warmest portions are above and the colder below. For this reason our arrangements for producing or introducing heat are placed at as low a point as possible.

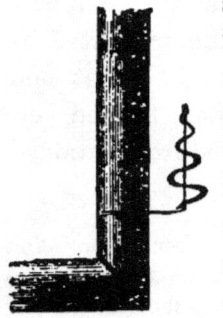

Fig. 231.

Chimneys.—We speak of the *draught* of a chimney, and we say of one which does not smoke that it *draws* well, as if the smoke were in some way actually drawn up. But the same principles apply here as those above developed. The smoke, which is a combination of heated air and gases, with some solid matters in a fine state, is *forced* up the chimney. When a chimney does not draw well, we open a door or a window for a little while until the fire is well started. This is in order that we may let denser air into the room, so that the smoke may be pushed up more forcibly. When the chimney becomes well heated there is ordinarily no difficulty, because then the smoke in it is not obliged to part with much of its heat to the walls of the chimney, and therefore is so much lighter than the air in the room that it is very easily forced upward. The principal reason that a stove-pipe generally draws better than a chimney is that there is much less heat expended in establishing and maintaining the upward current. Especially is this true if the chimney be a large one. In such a case both a great extent of brick and a large body of air must be heated to establish an upward current.*

---

* The author was once consulted in regard to a smoking stove. It was an open Franklin stove, the pipe of which went through a fire-board into

174. **Winds.**—If you open the door of a heated room, the flame of a candle held near the floor will be blown inward, while one held near the top of the door will have its flame blown towards the cold entry. (Fig. 232.) This is a good illustration of the man-ner in which winds are pro-duced. Wherever the wind blows it is caused by air push-ing out of the way other air that is warmer, in order that it may, in obedience to gravitation, get as near the earth as possible. Take, for example, the land and sea breezes, as they are called.

Fig. 232.

During a hot summer's day the sun heats the earth power-fully, while the ocean receives but little of its heat. The heated land heats the air above it; and as the air over the ocean is cooler, and therefore heavier, it pushes upward the air of the land, for the same reason that water pushes up oil; and as this goes on continuously, a regular current is es-tablished. The wind blows in upon the land, as represented in Fig. 233, while the warmer air passes upward into the higher regions of the atmosphere, and turns towards the sea. The arrows show the course of the currents. The re-semblance of all this to the effect upon the candle held near the open door is very obvious, the cold air from with-out blowing in below representing the breeze from the ocean, and the warm air of the room blowing out above

an enormous chimney. He recommended that a pipe with a knee should extend from the pipe of the stove a little way up the chimney. The expe-dient was successful, because but a small body of air, that in the pipe, needed to be heated to establish an upward current.

Fig. 233.

representing the passage of the warm air of the land out towards the ocean. At night this is apt to be reversed. The earth becomes cooled, and with it the air above. The result is that the cooled air of the land then pushes upward the warmer air of the sea, as shown in Fig. 234.

Fig. 234.

175. **Winds Affected by the Rotation of the Earth.**—The heat of the vertical sun in the tropics causes a rise of heated air into the upper regions, while there is a rush of colder air towards the equator from both north and south. This effect is represented in Fig. 235, E being the sun,

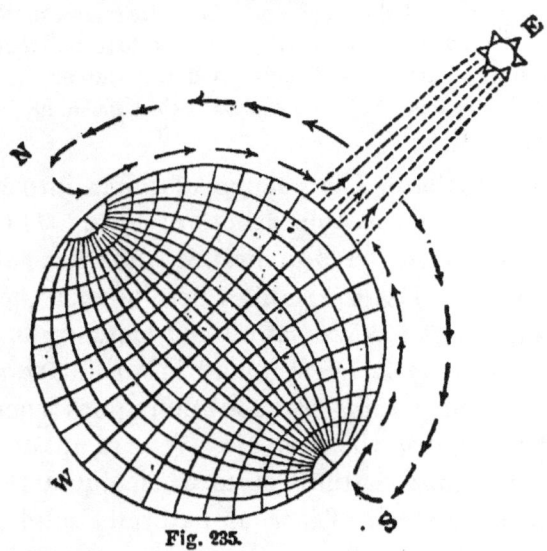

Fig. 235.

N the north pole, and S the south pole. An effect similar to that represented in Figs. 233 and 234 is produced here, but it is on a much larger scale. But the diagram does not present the matter in its true light in all respects. The prevailing winds in the equatorial regions are not north and south winds, as would appear from this diagram; but they are from the northeast and southeast. Fig. 236 will explain this. As the earth turns on its axis, it is plain that there is no part of the surface of the earth that moves so rapidly as the equator, E W, for that moves in the largest circle. And the nearer you go to either pole, N or S, the less is the rapidity of the revolution. Now, the atmosphere partakes of the motion of the earth; the air, therefore, at the equator is moving from west to east with the earth faster than any-

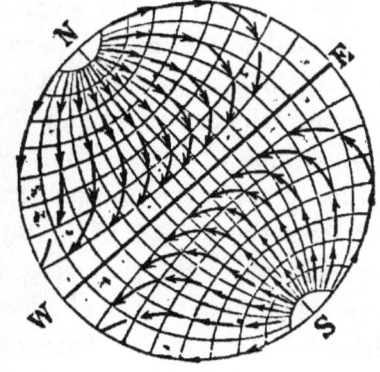

Fig. 236.

M

where else, and the nearer you go to either pole, the slower its motion. Hence any portion of air blowing from the north or the south towards the equator, and coming from a point where it was moving east slower than air at the equator, would from its lesser momentum lag behind the air of the equator; and the wind would be curved towards the west, as indicated by the arrows. The result would be that the northern wind would be converted into a northeaster, and the southern into a southeaster. All this can be made more clear with a globe, or, indeed, with any round object.

176. **Liquefaction.**—The change of solids into liquids is one of the most noticeable effects of heat. This change requires different degrees of heat in different substances. Thus while iron melts at the high heat of 1530°, lead melts at 330°, sulphur at 115°, ice at 0°, and mercury at 39.4° below zero. Mercury is never found in a solid state, but it sometimes becomes solid in the arctic regions when carried there and exposed in the open air. We are apt to think of water as in a more natural state when liquid than when solid, just as we think of iron as naturally solid and mercury liquid. But in all these cases the state of the substance depends on its temperature, and this is varied by circumstances. Water at the equator is always liquid, and the idea of ice there is exceedingly unnatural; while near the poles it is the reverse, ice and snow reigning everywhere throughout the whole year.

177. **Evaporation.**—There are two ways in which the change of a liquid into a vapor occurs. One is a rapid change when heat is so applied as to raise the liquid to its boiling-point. This is commonly termed vaporization. The other mode is the ordinary gradual evaporation which goes on from the *surface* of the liquid. This process is going on continuously, not requiring any particular degree of heat, but occurring under all degrees of the temperature of a liquid. Its rapidity, however, is in proportion to the degree of heat, as may be seen by the rise of vapor from

heated water long before it begins to boil. The same thing can also be seen on a bright summer's morning, when the heat of the sun causes the moisture gathered from rain or dew to rise so abundantly from fences, boards, and roofs as to be visible like smoke.

178. **Moisture in the Atmosphere.**—Evaporation is constantly going on from every wet surface, except when the air is so loaded with moisture that it can take up no more. The vapor is not ordinarily visible, the particles of water passing quietly upward among those of the air, being mingled with the air just as some liquids mix with water. It becomes visible only when so much of it rises that the mixture of water and air is not readily effected. The readiness with which this takes place depends much upon the temperature of the atmosphere. Some very common phenomena illustrate this. In a very cold day the breath of animals, as it comes out of the mouth, seems to be loaded with moisture. Why? It is not because it contains more moisture than in warm weather, but because cold air condenses the aeriform water and renders it visible in minute drops. The same explanation applies to the smoking of wet fences and roofs in the sun of a summer's morning. The moisture is heated by the sun, but the air, not having become very warm as yet, cannot readily convert into vapor all the moisture that rises. The phenomenon is not apt to occur when the hot sun shines after a shower at midday or in the afternoon, because then the air is warm enough to take up all the moisture present.

- 179. **Clouds.**—The water which rises in the air by evaporation is variously disposed of. Some of it is deposited as dew or frost. Some of it forms fog. Some of it also mounts far upward and forms the clouds, which are really collections of fog high up in the air. In fog and in clouds the water which in its evaporation is invisible becomes visible. Let us see how this is. The atmosphere always contains more or less water, but the particles are so minutely divided and so thoroughly mingled with the particles of the air that they cannot be seen. But in a fog or cloud the particles of water are gathered together in little clusters, as we may express it. And it is supposed, some think ascertained, that each of these clusters of particles is globular and hollow. If so, then we may regard every cloud as a vast collection of minute bubbles or balloons careering through the air.

Shapes of Clouds.—Clouds assume a very great variety of shapes, the causes of which are for the most part not understood. They are generally divided into four classes: *cirrus, cumulus, stratus,* and *nimbus.* Besides these there are several intermediate forms known as *cirro-cumulus, cirro-stratus,* and *cumulo-stratus.* All these forms except the *nimbus* are shown in the full-page engraving Fig. 237; the *cirrus* being marked by one bird, the *cirro-cumulus* by two, the *cirro-stratus* by three, the *cumulus* by four, the *cumulo-stratus* by five, and the *stratus* by six.

The *cirrus* is a light, fleecy cloud, having graceful curves like curls, and hence its name, which is the Latin word for curl. Such clouds are commonly very high up in the air. It is the first cloud to appear after a period of fine weather, its delicate, waving, and thread-like forms stretching across the blue sky like pencilled lines of white. Although this cloud appears so light and airy, it is probably composed of minute masses of ice, for at the enormous distance at which it floats along the earth the temperature is very low even in summer.

The *cumulus* appears as heaps rounded upward, often looking like mountains of snow when they are illuminated by the sun. The name is derived from the Latin for heap. In Fig. 237 this form is marked by four birds. It is a cloud

Fig. 237.

of dense structure, and forms and floats in the lower regions of the atmosphere. It has been called the "cloud of day," being produced in the daytime by the currents of moist warm air rising from the heated earth.

. The *stratus*, from the Latin for a layer, is marked by six birds. Clouds of this form lie low in the horizon, stretched along like a sheet. It has been called the "cloud of night," because it usually forms towards nightfall, grows denser during the night, and is dissipated shortly after sunrise. It is caused by the vapors that rise during the day, and descend at evening with the falling temperature.

The *nimbus*, or rain-cloud, is represented in Fig. 238; it

Fig. 238.

has a uniform gray or dark color. This form of cloud is also called the *cumulo-cirro-stratus*, a name suggesting the way in which it is formed, being a combination of the three named.

The remaining complex forms we need only briefly notice : the *cirro-cumulus* (marked by two birds, Fig. 237) is commonly called the "mackerel-sky," and is regarded as a quite sure indication of approaching rain. The *cumulo-stratus* (five birds) is formed of small fleecy clouds surrounding the cumulus, and often precedes a storm ; this form is sometimes called "thunderheads." The *cirro-stratus* (three birds) is sufficiently explained by the engraving.

Water is gathered into clouds undoubtedly, in part at least, through the influence of attraction. But what circumstances cause these various shapes is not known. Whatever they are, their influence is sometimes very extensive, giving a similar shape to all the clouds covering the whole arch of the heavens; and at other times they operate variously in different localities, producing different shapes, sometimes even quite near each other. Sometimes the edge of a cloud is irregular, or curved, or feathery; and at others it is a well-defined line, stretching along over a large portion of the horizon. In all these cases we have only divers arrangements of the same thing—a collection of vesicles of water containing air, which is made lighter than the air outside of the cloud by means which we shall explain in the next chapter.

180. **Rain, Snow, and Hail.**—When it rains, the vesicles or minute bubbles of which the clouds are composed are broken up, and each drop of rain contains the water from a multitude of these vesicles. But let us see exactly how this result is produced. Rain results from the contraction of the clouds by cold. A cold current of air coming in contact with a cloud will condense its bubbles into drops, and these, of course, will fall. The same result occurs if a cloud passes into a cold stratum of air. The first effect of cold upon the bubbles may be made clear by Fig. 239. If a bubble be contracted by the influence of cold, the water of its wall being made thicker, gravitation will cause a gathering at the lower part, as represented by the dotted line. You often see a similar effect in the soap-bubble; it rises filled with warm air from your lungs, and as it ascends is contracted by the colder air around it. This contraction makes the water hang from the bottom. And as the soap-bubble at length bursts in the air from the weight of this water, so it is with the vesicles in the cloud. And many of these, united together by attrac-

Fig. 239.

tion, form a drop. When the cold is sufficiently severe, it makes the water of the ruptured vesicles of a cloud arrange itself in snow-crystals instead of drops. And when cold acts with great rapidity upon a cloud, it presses the particles of water together so suddenly that there is no time for them to assume a crystalline arrangement, and hail is formed.

181. **Vaporization.**—The production of vapor by boiling differs in some respects from quiet evaporation. Here the liquid is raised in temperature to its boiling-point, and the formation of vapor is not confined to the surface. The liquid *boils* when the small bubbles of aeriform matter, forming at the bottom of the vessel, rise to the surface and thereby keep the liquid in a state of agitation. The temperature at which liquids boil varies for each; thus the boiling-point of water is 100° Centigrade (or 212° Fahrenheit), that of alcohol 78.4°, that of ether 35°, that of oil of turpentine 165°, and that of mercury 360° Centigrade.

The temperature at which a given liquid boils depends upon the amount of heat which it requires to overcome both the natural attraction of its particles and the compressing force of the atmosphere. Pressure restrains the production of vapor, whether it be formed by evaporation or vaporization. We know by experiments with the air-pump that the less the pressure of air upon the surface of a liquid, the more rapidly will evaporation go on. We have already mentioned the influence of pressure upon the boiling of liquids in § 142; we will give here a few additional illustrations. Ether boils when heated to 35°, about one and a half degrees below the heat of the blood in our bodies. If we place some of it in a vessel under the receiver of an air-pump, by exhausting the air we can so lower the pressure that the ether will boil at the ordinary temperature of the air in a room.

The restraint of pressure upon boiling is very strikingly shown in the *digester*, Fig. 240. This is a strong boiler, partly filled with water. A ther-mometer to indicate the temperature of the water may be inserted into mer-cury contained in the nar-row cup, *a*. Let, now, the boiler be heated till the wa-ter boils, the air being left to escape by the stop-cock. If the stop-cock be shut and we continue to apply the heat, we can raise the water to a very high tem-perature without its boil-ing, because of the press-ure of the condensed steam

Fig. 240.

upon its surface. To guard against the danger of explo-sion a safety-valve is provided, having a weight upon it which will keep it shut until a certain amount of pressure accumulates, and then it is forced open, letting out some of the steam. An apparatus somewhat after this plan, called *Papin's digester*, has sometimes been used in cooking. The great heat to which water can thus be raised causes it to extract the nutritious matter from bones and carti-lages, affording material for soup from that which is com-monly thrown away.

Distillation is a process whereby a liquid converted into vapor is again condensed into a liquid by cooling it in a suitable apparatus. The manner of conducting the opera-tion, and its application to the purification of substances and the separation of liquids possessing different boiling-points, will be explained in Part II. of this series—Chemistry.

M 2

182. Steam.—The cloud of steam, so called, which you often see escaping from a locomotive is not really steam. Steam is transparent and invisible. This may be rendered evident by watching the spout of a tea-kettle whence steam is issuing. For the space of an inch or so from the end of the spout nothing is visible; but, at a greater distance, the steam coming in contact with the cooler surrounding air is condensed to water, and this mixture of water-drops and steam is plainly seen.

The Steam-Engine.—It has been shown in § 138 that compressed air, in an air-gun for instance, possesses great power by virtue of its elasticity. Compressed steam in like manner exerts enormous force, constituting, in fact, the motive power of the steam-engine. This machine, complex as it appears to the casual observer, is not so difficult of comprehension as supposed by many; and, Dr. Arnott justly remarks, any one " who can understand a common pump may understand a steam-engine." It is, in fact, only a pump in which the fluid drives the piston instead of the piston impelling the fluid; in other words, the fluid passing through the cylinder acts as the *power* in the steam-engine and as the *resistance* in the pump. We cannot, in this elementary work, enter upon an elaborate description of the steam-engine; but we wish to show you the source of the power in this wonderful invention, and for this purpose shall consider only the simplest form.

The steam is generated in a boiler, having, like the boiler of Papin's digester, a valve with a weight attached to it. This valve is called a safety-valve, because when the steam has reached a certain degree of condensation it lifts the valve, and, as some of the steam escapes, such an increase of pressure as would occasion an explosion is prevented. The expansive force of steam in a boiler is estimated in pounds on the weight of the valve, and hence the common expression that there are so many "pounds of steam" on. But the boiler is only the generator of steam, and it remains to show how the steam is used in moving machinery. This is done by allowing the steam to pass from the boiler into a cylinder, in which it moves a piston back and forth by its expansive force. The manner in which this is done may be made clear by the diagram, Fig. 241. Let e be a piston in a cylinder, f, which has four openings, a, b, c, and d. Each of these is provided with a valve, not shown in the diagram. The steam is supplied from the boiler to the cylinder through a and c, and makes its escape from b and d. Suppose, now, the piston be near the bottom of the cylinder, as represented. The valve at a is opened that steam may enter to push up the piston, and the valve at b shuts that the steam may not escape. At the same time, in order that the pressure may be removed from the upper surface of the piston, the valve d opens that the steam may escape, and c shuts that none may enter. When the piston is to be forced downward, all this is reversed—c opens to admit the steam, d shuts to prevent its escaping; and below, b is opened to let the steam escape, and a is shut to prevent any from entering. This is the plan of what is called the high-pressure engine. The low-pressure engine differs from it in causing the steam, as it escapes from the cylinder, to pass into water to be condensed. The latter requires less pressure of steam to work it, and is therefore the safest. The manner in which the motion of the piston is made to drive various kinds of machinery cannot be here explained.

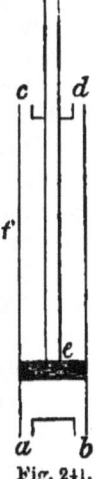

Fig. 241.

QUESTIONS.

164. Describe the experiment with the three vessels of water, and the inference from it. What other facts sustain this inference? How did Sir Humphry Davy prove that ice contains heat?—165. What are the two theories of heat? What is said of Count Rumford and his experiment? Give examples of the conversion of motion into heat.—166. What is the chief source of heat for the earth? What is said of the heat of the sun itself? What is said of the universal influence of the heat of the sun on the earth? What of the heat supplied from within the earth itself? What is said of chemical action as a source of heat? Give examples of the production of heat by mechanical action.—167. Show the expansive influence of heat by describing the experiment with a bar of iron. Give familiar examples of this expansion. How can you loosen a stopper stuck fast in a bottle? Give the anecdote about the *Persia*. Give the statement about the building in Paris. Explain how the unequal expansion of metals regulates the length of pendulums. What is a gridiron pendulum?—168. What is said of the expansion of liquids by heat? How may the influence of this expansion upon specific gravity be shown?—169. What is said of thermometers?—170. What of the invention of the thermometer? Explain the graduation of Fahrenheit's thermometer.—171. Give the plans of other thermometers. Why is Celsius's thermometer, on the whole, the best? Show how to convert degrees of one scale into those of another.—172. What is said of the expansion of gases by heat? Describe experiments in illustration. What is said of balloons?—173. What of the influence of heat on the atmosphere? Give examples of this influence. In heating apartments, why is heat introduced at as low a place as possible? Explain the draught of a chimney. Why does a stove-pipe generally draw better than a chimney?—174. Describe the experiment with the candle and the door. What is the explanation of the occurrence of wind? Explain the land breeze. Explain the sea breeze.—175. How are winds affected by the rotation of the earth? Show why the prevailing winds at the equator are northeast and southeast.—176. Mention the melting-points of various substances. What is said about the natural state of water and other substances?—177. What are the two modes of changing a liquid into vapor? What is said of the rapidity of evaporation?—178. What is said of moisture in the atmosphere? What influence has heat upon the moisture of the air? What phenomena illustrate this?—179. What becomes of the water that rises in the air? What is said of the formation of fog and of clouds? Mention the different shapes of clouds and their names. What

is said of the influences that give shape to clouds?—180. State how rain is produced, and explain Fig. 239. How are snow and hail formed?—181. What is said of vaporization? What influence has pressure upon the formation of vapor? Describe the experiment with ether in illustration. Describe the apparatus represented in Fig. 240. What is said of Papin's digester? What is distillation?—182. What is said of steam? In what consists the power of the steam-engine? How is the expansive force of the steam in the boiler estimated? Describe the manner in which steam moves a piston within a cylinder. What is the difference between high and low pressure engines?

CHAPTER XVI.

HEAT (CONTINUED).

183. Communication of Heat.—Heat has a constant tendency to an equilibrium. If, therefore, any warm substance be in the neighborhood of a cooler one, heat passes from the former to the latter. This communication of heat occurs in three different ways, called convection, conduction, and radiation. We will speak of each of these separately.

Convection.—This mode of diffusion of heat operates in those substances that are mobile—viz., liquids and aeriform substances. We have already alluded to examples of this mode in speaking of the movements which heat causes in these substances. The heat accompanies the particles which are moved, or is *conveyed* along with them, and hence the term convection. In this movement the heated particles always ascend, for the reason given in § 168. Of the multitude of examples of convection we will present but a few.

The upward current about a stove-pipe furnishes an example of convection, the heat generated being carried upward by the particles of this current. This being so, the heat of a stove has no effect upon the air *below* it by convection, though it does by radiation, as you will soon learn. Any hot fluid becomes cool chiefly by convection. The air coming in contact with it, taking some of its heat, rises, and other air becomes heated in turn, and so on till the fluid becomes of the same temperature as the air, and then the currents of air cease. The liquid cools more rapidly by stirring it, because the air is brought into contact with a greater extent of surface, and so the heat is conveyed away more rapidly. The result is the same whether we disturb the surface by stirring it or by blowing upon it. In the latter case, however, the effect is increased by causing the air to come more rapidly upon the disturbed surface. Thus in fanning, it is the rapidity with which the air is brought in contact with the surface of the body that causes the more rapid convection of heat from it. Every one must have observed the fact that a buckwheat cake cools much more quickly than a flour or rice cake. It is because it contains so many pores and little projections, and presents a much larger surface to the heat-conveying air than the smoother and more solid cakes. Viscid fluids, such as molasses, oil, chocolate with milk, etc., when heated do not cool so readily as water, because their particles are not so mobile, and therefore heat is not conveyed so rapidly upward.

184. **Conduction.** — In this mode of diffusion the heat passes through or among the particles of substances. For example, if one end of a bar of iron be held in the fire, it travels through or along the particles to the other end. The gradual progress of the heat may be seen by the following simple experiment: Take a rod of iron and attach to it some little balls of wood by means of wax. By heating one end with a lamp the balls will drop one after another, as the heat passing along melts the wax which holds them. By making this experiment with two rods of different metals, as shown in Fig. 242, you will observe that heat does not travel through them at the same rate. (§ 185.)

Fig. 242.

185. Conductors and Non-Conductors.—Heat is conducted more rapidly through some substances than through others: in this respect there is great variety. That this is the case, even among those which are reckoned good conductors, is shown by the experiment represented in Fig. 243. Cones of the same size made of seven different substances —copper, iron, zinc, tin, lead, marble, and brick—each tipped with a little wax, are placed on a stove. The wax will melt on the copper cone first, showing that this is the best conductor of all; and on the brick one last, showing that this is the poorest conductor. The conducting powers of the rest are according to the order in which we have mentioned them.

Fig. 243.

Another way of making this comparison is to substitute small pieces of phosphorus for the wax, each piece taking fire, and burning with a brilliant flame and white smoke as it becomes sufficiently heated by the conducting power of the metallic cone. Care should be taken in experimenting with phosphorus lest it be ignited by the warmth of the fingers, and inflict very serious burns. The best plan is to cut it under water, and to use very small pieces.

Those substances which allow heat to pass through them very slowly are called non-conductors. The term, though convenient, is not strictly correct, for there are no substances which do not conduct heat in some degree. Wood is one of these poor conductors, and hence wooden handles are put upon various instruments and vessels used about fires, such as the soldering-irons of the tin-man, the metallic teapot, etc. Since cloth is a non-conductor, a cloth holder is used in taking off the tea-kettle, and in using the flat-iron. Glass is so poor a conductor that if you hold a glass rod or tube in the flame of a spirit-lamp or gas-burner, and heat it even to redness, you can place your fingers comparatively near to the heated portion with impunity. We had occasion recently to bend a small glass tube in this way, and we observed some water in it quite near the heated part, which remained undisturbed through the process. It is the non-conducting quality of glass that makes thick pieces so liable to break when exposed to any sudden change of temperature. For example, if hot water be poured into a thick glass vessel, the inner surface is quickly expanded; but the outer surface does not expand so rapidly because the heat is not readily conducted through, and this irregularity in expansion causes a fracture. It is for this reason that the flasks, retorts, etc., used by chemists are made very thin, especially where heat is to be applied.

186. **Davy's Safety-Lamp.** — One of the most beautiful applications of the conduction of heat is found in the safety-lamp of Sir Humphry Davy, an invention which has been the means of saving the lives of multitudes of miners. It is represented in Fig. 244, which presents a sectional and an external view. The bottom part contains the oil in which the wick lies coiled; when lighted, the

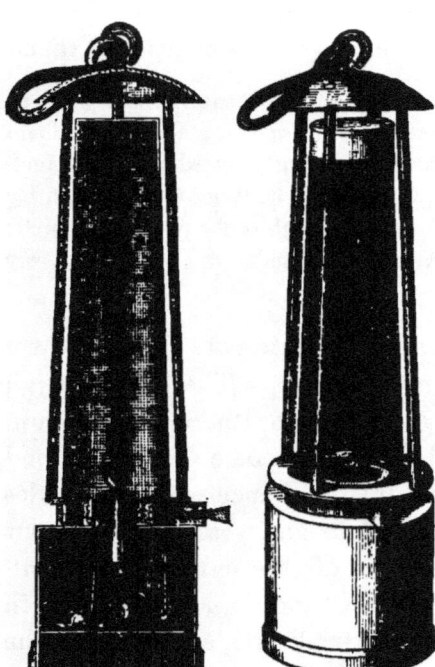

Fig. 244.

flame is entirely surrounded by wire gauze, as shown in the engravings. With this lamp one can safely go into deep mines containing the most explosive gases. All that prevents the flame within from setting on fire the gases without is the cylinder of wire gauze. This, being a good conductor, carries off the heat of the flame within so rapidly that it cannot go through the openings *as flame*, and so does not set fire to the gas without. The facts upon which the construction of this lamp was based were discovered by trying many experiments. Among them were the following : A piece of wire gauze was held over a candle so that its flame struck against it. The smoke issued above, but no flame. Then a stream of gas was allowed to pass through the gauze, as shown in Fig. 245, and was set on fire above. It burned without inflaming the gas below. This proved that flame cannot pass through wire gauze, provided the meshes are not too large. The safety-lamp burns well in pure air; but, when dangerous gases accumulate in the mines, the flame is extinguished, and this serves as a warning to

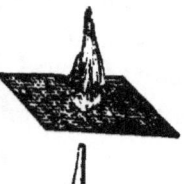

Fig. 245.

the miner. The terrific explosions constantly occurring in coal-mines in England and in Pennsylvania are usually the result of gross carelessness on the part of the miners; they open the lamp to light a pipe, or strike a match for some purpose, instantly igniting the explosive gases. Of the nature of these gases you will learn fully in Part II. of this series of works.*

---

* As in the case of many other inventions, the same idea was originated and put to practical use by more minds than one. George Stephenson, who from being a common engine-wright in a colliery rose step by step till he invented the locomotive, constructed a lamp which illustrates in another way the same principle—in other words, he invented another safety-lamp. But

187. Relation of Density to Conduction.—Generally, the more dense a substance, the better its conduction of heat. Thus metals are better conductors than wood, marble than brick, solids than liquids, and liquids than aeriform substances. A good illustration of the difference in conducting power of stone and brick is seen in the melting of snow on sidewalks. If a light snow fall in the spring, after the earth has become somewhat warm, it will melt on the stone walks much sooner than on the brick ones. This will be the case especially if the snow be melted chiefly by the warmth of the earth without the agency of the sun. The explanation is obvious. The stone is a better conductor than the brick, and therefore the heat of the earth comes up through the former more rapidly than through the latter.

You must not suppose, however, that the conducting power of the metals is directly proportioned to their density; platinum is much denser than silver, but has only one twelfth its power of conducting heat. The relative position of the metals in this respect is shown in the following table, the figures giving the approximate conductivity compared with silver taken as a standard:

Silver	100.0	Iron	11.9
Copper	77.6	Steel	11.6
Gold	53.2	Lead	8.5
Brass	23.6	Platinum	8.4
Zinc	19.0	Palladium	6.3
Tin	14.5	Bismuth	1.8

this does not in the least detract from the glory which the invention has given to the name of Davy, for each acted independently. In Davy's case, it is to be remarked, there was a long course of scientific reasoning and investigation which led him at length to the invention, the record of which is exceedingly interesting. No invention or discovery is made without thought, though accident may suggest the thought; but here is an invention which, without any suggestion by accident, was evolved by laborious and long-continued thought, proceeding step by step to its conclusion.

188. Conduction in Liquids.—That liquids are poor conductors of heat may be shown by a simple experiment. Place a few pieces of ice at the bottom of a test-tube (a thin glass tube closed at one end), and add water until nearly full. By heating the upper portion of the tube carefully, the water may be boiled without a particle of the ice melting in the bottom. The stand for holding the tube in this experiment is shown in Fig. 246.

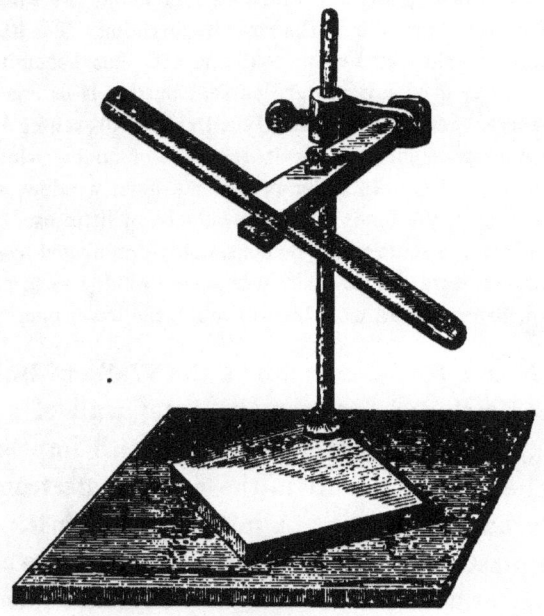

Fig. 246.

If the heat were applied at the lower part of the tube, the ice would melt, and then the heat would be diffused by convection.

189. Air as a Non-Conductor.—Heat is rapidly diffused in air by convection; but this takes place only when the air is free. When the air is confined in spaces or pores, or among fibres, heat makes its way through it very slowly, for then it can be diffused through it only by conduction. The variety of ways in which air is of service to us as a non-conductor is almost endless. We will notice some of them.

Double Windows.—The efficacy of double windows depends upon the air imprisoned between them. In the case of the single window a great deal of the heat inside is lost in this way: The warm air of the room, coming in contact with the window, imparts to it some of its heat, and, being thus cooled and therefore condensed, passes downward. As this process goes on continually, this downward current along the window is constant. The current outside is in the opposite direction. The heat imparted to the window is taken up by the cold air, and thus becoming warmer, it passes upward. And this upward current outside is as constant as the downward current inside. Now nearly all this is prevented by the non-conducting quality of confined air in the case of double windows. If a pane were removed from the upper part of the inner window, and another from its lower part, the inner window would be of little use, for then the heat of the air in the room would be continually diminished by convection, as if the window were single. The warm air would pass in at the upper opening, and, being cooled, pass down through the lower one.*

190. **Air as a Non-Conductor in the Walls of Buildings.**— The spaces included between the outer wall of a building and the plastering inside, being filled with imprisoned air, prevent the heat of the air in the apartments from passing readily through the wall. A house built of brick or stone, with the plastering placed directly upon the wall, would be kept warm with greater difficulty in winter, because the solid wall would more readily carry off the heat to the external air. For a similar reason, such a house would be very warm in summer, because the heat of the sun and of

---

* The author once adopted a contrivance for a small conservatory, which he wished to keep warm from the heat of an adjoining room. In each space of the window-frames were inserted two panes of glass, leaving nearly half an inch of space between them. In this way nearly all the benefit of double windows was secured, with less expense and a less cumbrous arrangement.

the **external air** would be so rapidly communicated to the air of the house. In this connection we may mention a contrivance to prevent the spreading of fires in blocks of buildings, which, though very effectual, is seldom used, partly because it occasions some trouble and expense, and partly because it takes up a little room : A small space is left in the division wall between two adjoining houses extending from top to bottom, and containing, of course, a body of imprisoned air. With such an arrangement the interior of one house may be entirely consumed without transmitting sufficient heat through the imprisoned air to fire the other.

191. **Fur, Hair, and Feathers.**—Animals living in cold climates are provided with suitable coverings for their protection. Quadrupeds, for example, are covered with fur, and birds have an abundance of downy feathers. These coverings have no warmth in themselves, though in common language we speak of them as being warm. They are simply non-conductors, and prevent the heat generated in the body of the animal from escaping as fast as it otherwise would. But why are they non-conductors? It is partly because the substance of which they are made is a non-conductor, and partly because that great non-conductor, air, is imprisoned among their numberless fibres. But if the fur or down were compressed into a thin hard plate upon the animal, it would prove of far less service as a protection against cold. Down is much more abundant on the birds of cold climates than on those in warmer regions, because more air can be confined among the fibres of down than among those of common feathers. Quadrupeds that are natives of warm climates generally have hair instead of fur. When, therefore, the horse is taken to a cold climate, he requires in winter the protection of a blanket; and the ox needs, under the same circumstances, to be better housed. The elephant be-

ing a native of a very warm climate, has scanty and coarse hairs. Formerly there were elephants in the cold regions of Siberia, as has been ascertained by remains found there. But the elephant of Siberia had under its hair, close to the skin, a fine wool to serve as a protection against the cold. Animals living in cold climates are provided with coverings finer in fibre in the cold season of the year, to give them the additional protection which they then need. And when animals with a furry covering are carried into a warm climate, their fur becomes coarse, and approximates the condition of hair.

192. **Clothing.** — Man has no covering to guard him against cold, because he is capable of contriving clothing suitable to the various degrees of temperature to which he may be exposed. The object of clothing is not to make the body warm, but to keep it so. The heat of the body is continually generated within itself, and under all circumstances this heat is maintained quite uniformly at 98° Fahrenheit (36.6° Centigrade). This is a much higher degree than that ordinarily possessed by the atmosphere. We are all the time, then, giving off heat to the air around us, except when the air gets up to 98°. We are comfortable only when we are giving off heat to a considerable amount, for the point of temperature which is most agreeable when we are at rest is 70° Fahrenheit (21° Centigrade), or a little less—that is, 28° Fahrenheit (or 15.6° Centigrade) lower—than the temperature of our bodies. When the temperature is below this we need extra clothing. In making choice of clothing for various degrees of temperature we practically apply the principles just developed. Those articles of clothing which can confine or entangle, as we may say, the largest quantity of air among their fibres are the best non-conductors, or, in common language, are the warmest. And loose clothing is warmer than tight, on account

of the amount of air between the clothing and the body. Thus a loose glove is much warmer than a tight one. The same general fact is illustrated by the coverings of straw placed around tender trees and shrubs in winter. It is the air confined in the tubes of the straw which makes these coverings so effective a protection.

Cocoons.—Many insects pass through their pupa or transition state in cocoons. When this is done during warm weather, as in the case of the silk-worm, the cocoon is simple. But when the pupa state lasts through the winter special provisions are made in the arrangements of the cocoon to guard the insect against the cold. We will cite as an example the cocoon of one of our largest moths, the Cecropia. This cocoon, fastened to some shrub, keeps its inmate secure from the rigors of the winter by a very beautiful arrangement. The real cocoon is similar to that of the silk-worm; but it has a very dense air-tight outer covering, and the space between these two coverings of the pupa is filled with a loose substance, which acts the part of a blanket for the insect.

193. **Buds of Plants in Winter.**—In the latter part of summer buds are formed on trees and shrubs, and these contain the germs of the branches, leaves, and flowers which are to come out the next year. These, of course, must be guarded against the cold of winter, and it is done very much as the pupa is guarded in the cocoon. Each bud has an air-tight covering of scales inside of which is a soft downy substance, the blanketing of the bud. In these coverings, which have been called by some one the "winter-cradles" of the buds, the infant vegetation of another year rocks back and forth in the wintry winds secure from the cold, till the warm sun of spring wakes its hidden life into activity.

Snow a Protection to Plants.—Snow acts as a good blanket to the earth, keeping its warmth from escaping into the cold air. This is because it contains a quantity of air mingled with its feathery crystals. If snow fall early, before the ground and the plants in it have become frozen, it will keep them from freezing through the winter, provided it remain during all that time. It is curious to observe the peculiar arrangement of the snow in the arctic regions for the preservation of vegetation. First in the autumn come soft light snows covering up the grasses and heaths and willows. Then, as winter advances, on top of these are laid the denser snows, making a compact, stout roof over the lighter snows in which the scanty but precious vegetation of those regions is imbedded. On top of this roof are deposited the snows of spring. As these melt the water runs off from the icy roof down the slopes, leaving untouched the plants underneath, which lie there alike secure from the rush of waters and from the nightly frosts until the season is sufficiently advanced to bring them with safety out from their concealment. Then the icy roof melts, and with it the light snows that have so long encircled the plants, and the sun wakes them from their long sleep to a new life.

194. Influence of the Conduction of Heat on Sensation.—
If you place your hand upon fur hanging at the door of a fur-store, it does not feel so cold as the wood from which it hangs, and the wood does not feel so cold as the iron bar of the shutter close by. Why is this, when these substances are exposed to the same atmosphere, and really have the same temperature? It is because the iron conducts the heat from your hand more readily than the wood, and the wood more readily than the fur. For a similar reason the iron handle of a wooden pump feels colder than the pump, and the pump colder than the snow around it; and the rug or the carpet in a cold room will not feel so cold as the poker and the hearth. If water has stood long enough in a room to acquire the same temperature as the air, your hand will feel colder in the water than in the air, because the water is the better conductor. So much for the sensation of *cold*. On the other hand, heated substances which are good conductors convey the sensation of heat, while non-conductors do not. As they receive heat readily, they also readily impart it. For this reason, with a brisk fire the hearth-stone feels very hot, while the rug before the fire does not.

195. Radiation of Heat.—Every substance constantly sends heat into space in straight lines in every direction. These lines are radii, and hence the term radiation is applied to heat diffused in this way. As explained in § 165, the transfer of heat through space is effected by the imponderable fluid called ether. That the sun radiates heat in all directions is very obvious. The same can be perceived in the case of a heated iron ball. In whatever direction you hold your hand, above, below, or laterally, you feel the heat. And it makes no difference whether the ball be red-hot or not. That is, heat is radiated either with or without light. When a room is warmed by a furnace, it is warmed altogether by convection; but when it is warmed by a fire, either in a fireplace or a stove, we have both convection and radiation. The heat which we receive from the sun comes altogether by radiation; it travels earthward with about the same velocity as light. When radiant heat passes through air, through glass lenses, or any other medium, they are not warmed by it. The intensity of radiant heat decreases as the square of the distance from the source—a law which we have several times explained, first with respect to attraction, and secondly as to sound. Now, this law holds good also for light; and this fact, together with the others just mentioned, has led philosophers to the conclusion that heat and light are merely different manifestations of the same force. In fact, both are effects of motion (§ 165). All surfaces that radiate will absorb equally well the heat radiated upon them. All rough and dark surfaces both absorb and radiate freely; but all light-colored and polished surfaces do both slowly. For this reason the black, rough tea-kettle is well fitted to heat water in; but it is not fitted to retain the heat in the water. On the other hand, the bright, polished teapot absorbs heat poorly, but retains it well.

N

196. Reflection of Heat.—Radiant heat is reflected from polished surfaces, and, as in the case of motion and of sound, § 155, the angles of incidence and reflection are equal. Some interesting experiments in relation to the reflection of heat can be tried with concave metallic mirrors. Thus, if we take two such mirrors, Fig. 247, and place in the focus of one a thermometer and in the focus of the other a small flask of hot water, or a heated iron ball, the mercury in the thermometer will rise, although the mirrors may be many feet apart. Observe how the effect is produced. Rays of heat pass from the flask directly towards the thermometer, as represented by the lines in the figure; but that the effect is not produced by these can be proved by removing the mirrors, leaving the flask and thermometer in the same positions. When the experiment is tried in this way, no effect is produced on the thermometer, because it is too far from the source of heat, the flask, to receive any perceptible influence in this way. The effect comes from the rays of heat which pass to the mirror near the flask, and are reflected to the other mirror, and then are reflected upon the thermometer, as represented by the dotted lines. There is another way, besides that already mentioned, of showing that it is not the *direct* rays that produce the effect. After arranging the apparatus, put a screen between the thermometer and the mirror near it, and the effect will be prevented because the reflection is cut off. If a piece of ice be substituted for the flask of hot water, the thermometer will fall—an effect opposite to that produced in the previous experiment. This would seem to show that cold is radiated; but since there is really no such thing as cold (§ 164), the effect must be attributed to the radiation of heat from the thermometer to the ice. If a hot ball be placed in the focus of one mirror and a piece of phosphorus in that of the other, as represented in

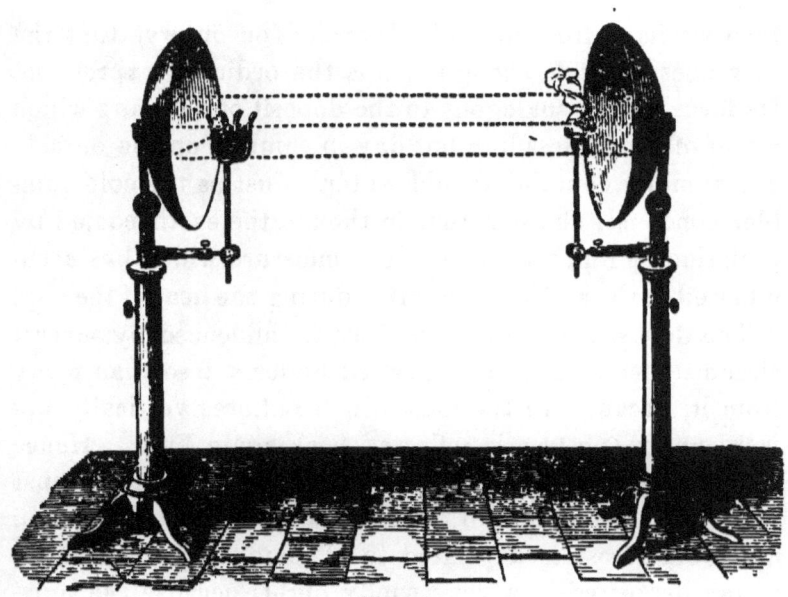

Fig. 247.

Fig. 247, the phosphorus will be ignited even when the mirrors are twenty or more feet apart.

The reflection of heat may be exhibited very prettily by a much simpler apparatus. A sheet of bright gilt paper is rolled up in the shape of a funnel, with the metallic side inward. Holding the larger end towards a fire, the rays of heat coming from the fire into the funnel are reflected towards a central line, and so pass out of the smaller end of the funnel. A bit of phosphorus or a lucifer-match held a little distance from this end of the funnel will be set on fire.

197. **Formation of Dew.**—Dew is formed by the radiation of heat from the surface of the earth. The earth as well as the sun is constantly radiating heat into space. In the daytime the earth receives a great deal more than it gives out; but at night this is reversed, and the earth is cooled. The cooled earth condenses the atmospheric moisture, and this moisture is deposited in the form of little drops of water. If the weather be very cold, this is frozen, and

then we have frost instead of dew. You observe that the dew does not *fall*, though this is the ordinary expression. Its formation is analogous to the deposit of moisture which we so often witness in a hot day in summer on the outside of a tumbler containing cold water. Just as the cold tumbler condenses the moisture in the air, the earth, cooled by radiation at night, condenses the moisture which has accumulated in the air by evaporation during the heat of the day.

The deposition of dew and frost is influenced by several circumstances. Less is deposited under a tree than away from it, because all the heat which radiates vertically upward under the tree is reflected back again by it. Hence the efficacy of a covering over plants as a defence against frost. Clouds operate in the same way, and for this reason no dew or frost is deposited in a cloudy night. Neither is any deposited in a very windy night, because the moving air promotes evaporation, and thus prevents the accumulation of moisture.

Dew is deposited in different amounts on different substances. This is owing to a difference in radiation. Grass and leaves radiate heat better than earth, and earth better than stone; and, therefore, while stones and gravel-walks may be dry or nearly so, the loose earth may be moist and the grass and leaves thoroughly wet. So you see that not even the dew, plentiful as it is, is wasted by the Creator, but is deposited just where it is wanted to refresh the parched earth and its vegetation.

Gideon's Fleece.—If you spread a fleece of wool upon the ground, it is so poor a radiator of heat that no dew will be deposited upon it, although the dew may be abundant on the grass and leaves in its neighborhood. But this was reversed in the case of Gideon's fleece. The laws of nature were set aside, and the fleece was wet with dew, while all around was dry.

198. **Dew-Point.**—What is called the dew-point of the air is that degree of temperature to which any substance must be reduced in order that dew may be deposited upon it. This depends upon the amount of water present in the atmosphere, the dew-point rising in proportion to the

increase of moisture. When water condenses on a cold tumbler in a hot day, there is much more water in the air, and the dew-point is higher, than when no moisture is condensed upon the tumbler. Thus after a very hot clear day the earth need not be much cooled to produce a deposit of dew, because the air has become so highly charged with moisture through the evaporation of the earth under the hot sun. We can at any time very readily ascertain the dew-point. Take a glass of water, and, placing a thermometer in it, drop into it some pieces of ice, and watch the outside of the glass. As soon as it begins to be dimmed with moisture, read the thermometer and note the dew-point.

On clear still nights water is sometimes frozen by radiation, even when the temperature of the air is considerably above the freezing-point. Advantage is taken of this in the tropical climate of India to procure ice: large flat and shallow pans containing water are placed on straw or other non-conducting material and sunk slightly into the earth; the water freezes even when the temperature of the air is 10° Centigrade.

199. **Latent Heat.** — As shown by the experiments in § 164, our sensations do not inform us accurately of the amount of heat in any substance. The same is also true of the thermometer, which indicates only the *sensible* or free heat. A great deal of heat may be locked up, as we may say, in the substance, that can be brought out or made free by a change of state in that substance. This heat thus locked up is called *latent* heat.

Whether a substance assumes the form of a solid, liquid, or gas depends upon the amount of heat latent in it. If you take a piece of ice and melt it in a vessel, the ice and the water resulting both remain at 0° Centigrade until the ice is all melted. Yet all this time heat is being communicated to the ice and water. What becomes of it? It is all taken up by the ice as it changes from its solid to its fluid state, and becomes latent in it. In fact, *every particle of ice must absorb a definite amount of heat in order to become fluid.* If water be heated to the boiling - point (100° Centigrade), and be kept boiling, the water will remain at that point till it is all vaporized. All this time the water is receiving heat, which, instead of raising its temperature, is becoming latent in the particles as they change from the liquid to the gaseous state. As in the change from the solid to the liquid state, so in this case, *every particle of the liquid must absorb a definite amount of heat in order to become aeriform.* Whenever, therefore, any solid substance becomes liquid, or liquid becomes aeriform, heat is absorbed and becomes latent. On the other hand, whenever any aeriform substance becomes liquid, or liquid becomes solid, latent heat is given out, and becomes free and sensible. The freezing of water, then, is a source of warmth to the air in its neighborhood —a fact which is practically made use of when tubs or pails of water are placed in conservatories to keep plants from freezing; and the thawing of snow and ice is a source of cold, as is exemplified by the chilliness of the air occasioned by this process.

200. **Recent Theory of Latent Heat.** — The expression *latent* heat was introduced into the science of physics at a time when the prevailing doctrine concerning the nature of heat admitted its existence as a material substance; when heat disappeared or was rendered *latent*, it was supposed to enter the spaces between the molecules of matter; and when it was rendered sensible again, it was supposed to be squeezed out, as it were, from the infinitely small recesses of the body. Now, however, since heat is recognized to be a mode of motion, the explanation is different; when a solid becomes liquefied, a great

deal of heat is rendered *latent;* that is, the heat communicated to the solid is consumed in accomplishing the separation of the molecules and in overcoming the cohesion necessary to convert it into a liquid. It is evident that if the mobility of molecules be so much increased as occurs in the passage of solids to liquids, the heat which effects this change cannot simultaneously do the work of free heat or render its presence appreciable by a thermometer. The expression *latent* heat is, then, liable to mislead students; yet it has taken so firm a hold upon the language of heat-science, or pyronomics, as it is sometimes called, that we cannot well dispense with it. You should remember, however, that heat is motion, and latent heat simply motion diverted to do other work.

Further Illustrations.—This subject may be further illustrated by some experiments, easily made, the results of which are apparently quite paradoxical. When equal quantities of hot and cold water are mingled, the whole becomes lukewarm, each degree lost by the hot water becoming a degree gained by the cold. Mix, for example, 300 cubic centimetres of water at 21° Centigrade, and the same amount of water at 54° Centigrade; the temperature of the mixture will be the mean of the two—viz., 37.5° Centigrade, the hot water losing 16.5° and the cold water gaining 16.5°. In like manner, a mixture of equal weights of water having the temperature of 0° and 78° respectively would yield water at 39°. Now, suppose we repeat the last experiment, using *ice* at 0° instead of water, and water of 78°; you might expect to obtain a liquid having the temperature 39° as before; but this is not the result, the mixture after the ice has melted will have only the temperature of 0° Centigrade. What has become of the heat? It has accomplished the work of converting the solid ice into the mobile liquid, water, and is not now capable of affecting the thermometer.

The refreshing beverage iced tea is drunk by many persons in warm weather. Those making it are often surprised to find that boiling-hot tea poured into a tumbler of ice is very quickly cooled to the temperature of the ice itself; whereas should ice-cold water instead of ice be added, a large quantity would be needed to cool the tea, consequently spoiling it as a beverage. The reason of this, however, is quite plain from the explanations just given of latent heat.

201. Capacity for Heat.—The more heat a substance can absorb and render latent, the greater its *capacity for heat*, as it is expressed. Thus water has a much greater capacity for heat than mercury. This can be proved by various experiments. Take two vessels just alike, and place a certain quantity of water in one, and the same quantity of mercury in the other; if you then expose them to the same degree of heat, it will take much longer to raise the water to any specified temperature than the mercury. Why is this, when they are both receiving the same amount of heat? It is because the water renders a much larger portion of the heat latent than the mercury does. We can reverse this experiment. Take these same vessels with their contents raised to the same temperature, as indicated by the thermometer, and allow them to cool in the air side by side. The mercury will cool faster than the water, because it has much less latent heat to part with. The difference in capacity for heat between water, oil, and mercury may be shown by the experiment represented in Fig. 248. Put one hundred grammes of water

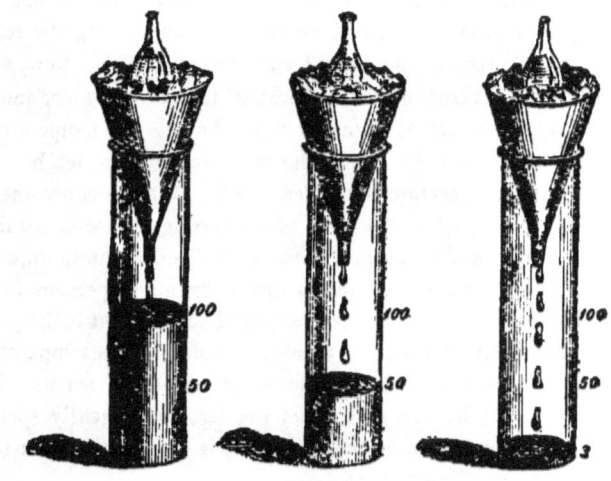

Fig. 248.

into one Florence flask, one hundred grammes of olive-oil into another, and the same amount of mercury into a third. Heat the contents of each flask to 100° Centigrade, and then place them in funnels filled with pounded ice, the funnels resting in glass jars of the same size. Now, in cooling these fluids down to a certain point, say 0° Centigrade, different amounts of the ice will be melted, in the proportions of 100 and 50 and 3. This shows the proportions of latent heat which become sensible or free as their temperatures are lowered.

202. **Relation of Latent Heat to Density.** — The more dense a substance becomes, the less its capacity for heat.

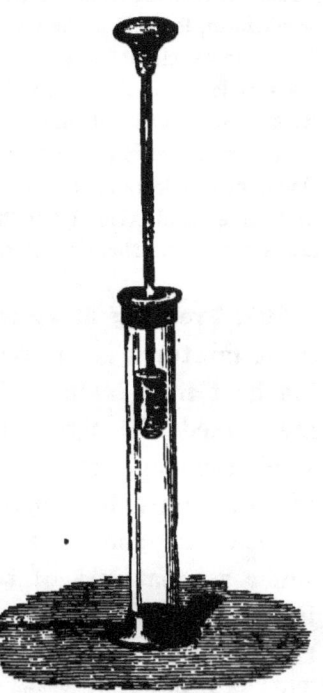

The heat produced by hammering iron is the latent heat rendered free by condensation, this lessening the capacity of the iron for heat. The same thing can be better illustrated in the condensation of a very compressible substance, as air. Fig. 249 represents a glass syringe closed at one end. If a piece of tinder, or a little bit of cotton wool moistened with ether, be placed in this end, and the piston be forced downward very quickly, the tinder or the ether will be set on fire. This is because the compression of the air lessens its capacity for heat so much that a great deal of its latent heat is made sensible or free.

Fig. 249.

You learned in § 131 that the atmosphere is rarer the farther you go from the earth. It is very

N 2

rare, therefore, on the summits of high mountains. This is the chief reason why it is so cold there; for the rarer the air, the greater its capacity for heat, and the more free heat, therefore, can it render latent.

Clouds and Latent Heat.—The water of which clouds are composed is heavier than air. Why, then, does it remain suspended? Why is it necessary that it should be collected into drops to cause its descent? This question can be answered by looking at the manner in which clouds are formed. A cloud, as stated in § 179, is made up of minute vesicles, or bubbles, containing air. Now, the air in these bubbles is lighter than the air surrounding the cloud, because it is warmer. But how does it receive its heat? In order to understand this, observe from what the bubble is made. It is made from the water which was in the air in a state of vapor, or aeriform, for this is the state of water in the atmosphere. But when it forms the vesicle, it leaves this state and becomes a liquid, for the wall of the vesicle is liquid. Now, in passing from the aeriform to the liquid state some latent heat must be made sensible. This sensible heat heats the air in the vesicle, and so makes it like a heated air-balloon. Thus all clouds are collections of innumerable heated air-balloons, and the reason that some clouds rise higher than others is perhaps that their vesicles contain warmer and therefore lighter air.

203. Freezing Mixtures.—The intense cold produced by these mixtures is the result of the change of free or sensible heat into latent. For example, when salt and snow are mixed, the two quickly produce a liquid. In this sudden change of a solid into a fluid a great quantity of heat must be rendered latent, and therefore objects with which the freezing mixture comes in contact will suffer a great loss of sensible heat. The process in this instance is the opposite of solidification. A portion of the snow, after melting with the salt, becomes solid ice. This is because it gives up its sensible or free heat to portions of the melting snow, which are causing heat to become latent.

The low temperatures obtained by certain freezing mixtures are remarkable: for example, pulverized Glauber's-salt, moistened with hydrochloric acid, lowers the temperature from 10° to —17° Centigrade—a fall of 27 degrees; a mixture of one part by weight of snow and one of dilute sulphuric acid lowers the temperature from —7° to —51° Centigrade. Still greater degrees of cold can be obtained by other similar means, as given in the table at the end of this chapter.

204. **Cold Produced by Evaporation.**—If you pour a little ether into the palm of your hand, it will rapidly disappear in vapor, producing a sensation of great cold. This sensation is due to the fact that, in the passage of the liquid to the aeriform state, some of the sensible heat of your hand is abstracted to become latent in the vapor. The evaporation of water also produces cold, though not so decidedly as ether, because its change into vapor is not so rapid at ordinary temperatures. We make a practical use of the evaporation of water in many different ways. Thus we sprinkle water in a hot day upon the floors of piazzas, steps, etc., that much of the sensible heat about our houses may be rendered latent by the evaporation. For the same purpose, in hot climates, apartments are often separated from each other by mere curtains, which are occasionally sprinkled with water. In like manner, the inhabitants of such climates often cool their beverages by wrapping a wet cloth around the vessels containing them.

Evaporation is an important remedy in many cases of disease. For example, if the head be hot, a steady application of a wet cloth to the forehead, though a simple remedy, is often effectual, and sometimes is very important. Most people make the application in a wrong manner. They put on several thicknesses of cloth, when a single thickness is the best, because it will best secure the evaporation, which is the cause of the relief afforded.

205. **Freezing in the Midst of Boiling.**—It is owing to

the quantity of heat rendered latent by evaporation that water can be frozen in the midst of boiling ether; and, paradoxical as it may seem, the boiling of the ether causes the freezing. The experiment is performed in this way: Place a test-tube or a little thin vial with water in it in the midst of some ether contained in a shallow vessel under the receiver of an air-pump. On exhausting the air the ether will boil, evaporation taking place rapidly because the pressure of the air is removed. As the ether passes into vapor it extracts so much free heat from the water that it is cooled down to the freezing-point, and becomes solid. Water can be frozen even by its own evaporation. It is done in this way: Place in a shallow vessel, *b*, Fig. 250, a little water, and in the vessel *c* oil of vitriol or

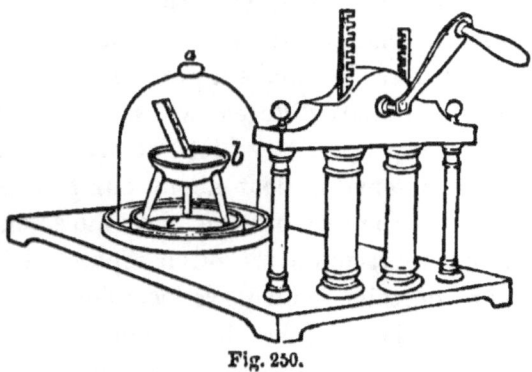

Fig. 250.

sulphuric acid; and cover the whole with the receiver, *a*. When the air is exhausted by working the air-pump, the pressure of the air is removed from the water, and vapor rises from it freely. Since the sulphuric acid has a great attraction for water, it absorbs this vapor, and thus vapor continually rises from the water; this proceeds the more rapidly because the vapor formed is absorbed, instead of remaining to cause pressure on the water. By this rapid formation of vapor, requiring a great quantity of heat to become latent, so much heat is at length abstracted from the water remaining that it freezes.

20G. Degree of Heat Endurable by Man.—It was formerly believed that the human body could not endure with impunity, even for a short time, a much higher temperature than that met with in hot climates. But in the year 1760 it was accidentally discovered that a much higher temperature than this could be endured. At that time an insect was destroying the grain gathered in some parts of France, and it was found that if the grain were subjected to a high temperature, the insect was killed, and yet the grain was not injured. In trying some experiments in regard to this matter, the experimenters wished to know the point at which the thermometer stood in a large oven. A girl attending the oven offered to go in and mark the thermometer. She did so, remaining two or three minutes, while the thermometer was at 127° Centigrade—that is, 27° Centigrade above the boiling-point of water. As she experienced no great inconvenience from the heat, she remained ten minutes longer, when the thermometer rose to 142° Centigrade. These facts were published, and prompted scientific men to try other experiments. In England, Dr. Fordyce, Sir Charles Blagden, and others, went into rooms heated even to 115° and 127° Centigrade, and remained long enough to cook eggs and steaks, and yet themselves suffered little inconvenience. The pulse was quickened, the perspiration was very profuse, but the heat of the body, as ascertained by putting the thermometer under the tongue the moment they came out, was scarcely raised at all. The air in which they remained roasted eggs quite hard in twenty minutes, and, applied by a pair of bellows to a steak, cooked it in thirteen minutes. The question arises, why is it that this high degree of heat did not produce a more injurious effect upon the body? One reason is that the heat of the air in the immediate neighborhood of the body was continually reduced by the evaporation of the free perspiration, sensible heat being thus converted into latent. Another reason is that air is not a good conductor, and therefore did not communicate its heat readily to the body. Dr. Fordyce and his friends found that they could not safely touch any good conductor, such as metals, and they were obliged to wear upon their feet some non-conducting substance.

207. Formation of Ice.—Before dismissing the subject of heat, we must notice the grand exception to some of the operations of heat in the formation of ice. Heat generally produces expansion. But in the case of water this law of expansion is set aside, and the reverse is established. This is the case, however, only within a small range of temperature—viz., from the freezing-point up the scale about four degrees. In all degrees above that the usual expansion by heat takes place. The exception occurs at this part of the scale for a special purpose—viz., *in order that water, in distinction from other substances, shall become more bulky, and therefore lighter, when it takes the solid form.*

In order to make the process of freezing clear to you, we will describe it as it ordinarily occurs—that is, from the action of cold air upon the surface of water. The uppermost layer of the water imparts some of its heat to the air in contact with it. This air rises and colder air takes its place, which, being warmed, rises in its turn to make way for more cold air. A constant current of warmed air rises, therefore, from the water. In the meantime a current of a different character forms in the water—a downward one. As fast as the water at the surface parts with heat to the air it falls, other warmer water taking its place, to cool in its turn and descend. This descent of the cooled water goes on regularly until a portion becomes cooled down to +4° Centigrade—that is, 4° above the freezing-point. This layer does not sink, but remains at the surface, for it is lighter than the warmer water below. This is because the law that heat expands matter is here reversed. Below this temperature the colder the water, the lighter it is. As the cooling now proceeds as before, the cooled water at the surface continually increases. At first it is merely a single layer of particles, but after a while quite a body of cold water rests on the warmer water below. At length some of it is cooled down to 0°, the freezing-point, and a thin film of ice then forms. The state of things just at this stage of the process may be explained by the aid of a simple diagram, Fig. 251. Let the line *a* represent the film of ice. The space between *a* and *b* is the portion of water cooled down below 4°. The space below *b* is occupied by the water of a higher temperature. In the space between *a* and *b* the cooler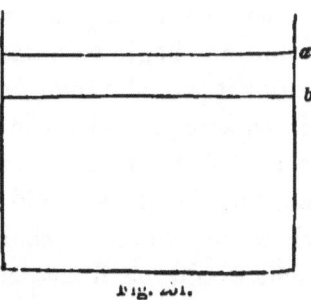

Fig. 251.

water is nearer the surface. That is, from the line *b*, where the water is exactly at 4°, the water lessens in temperature as you ascend, it being successively 3°, 2°, 1°, etc., till, just in contact with the film of ice, *a*, it is at 0°. The ice thickens gradually by additions below. But it is to be remembered that ice is a good non-conductor, so that the very first layer of ice makes the cooling of the water proceed more slowly than before. And the thicker the ice becomes, the slower the cooling. This prevents too great a formation of ice.

208. **Why the Above Exception to Expansion by Heat Exists.**—That we may understand the reasons in part for the grand exception to the general law of expansion by heat above illustrated, let us examine some of the results if the exception did not exist. In that case the process of freezing would be as follows: The water would communicate its heat from the surface to the air, as before described, and there would be a constant downward current of the cooled water. When any portion of the water became cooled by the air down to 0°, it would become ice, and would sink to the bottom. And after the process of freezing had once begun, there would be a continual accumulation of ice at the bottom so long as the air remained cold enough to cool the water with which it comes in contact down to 0°.

The result may be stated in general thus: Freezing would not begin so quickly as it now does; but when once begun it would prove very destructive. It would not begin so soon, because the whole of any body of water must be cooled down to a temperature very close to 0° before it could begin. This would not take long in ponds and streams where the water is shallow; all shallow bodies of water, then, would be frozen up quite early in the winter; and since water is a poor conductor, and thawing would proceed from above downward, some of them would not be thawed out again fully till quite into the next summer. And where the water is quite deep, ice would at length begin to form, and when formed it would be exceedingly slow in thawing. In some cases it would never thaw with such a body of non-conducting water to guard it against the warmth above. It is easy to see that the heat of spring and summer would not thaw out so large a quantity of ice as it now does. The reign of ice and snow on our earth would therefore be vastly more extensive than now, and, what is worse, it would extend more and more every year. Under such circumstances great destruction of both animal and vegetable life would result. We will mention, however, but a single item, for it would occupy too much space to go into this subject more fully. In the water under the ice, which is always above 4°, except that which is close to the ice during its formation, there is a vast amount of busy life which would be destroyed if ice were formed at the bottom, chilling all the water above.

209. **Force of Expansion in Ice.**—Since ice occupies one seventh more room than the water from which it is formed, it exerts in its formation an expansive force which, under various circumstances produces varied and often remarkable results. Of the numerous experiments which have been tried to show the force of this expansion we will mention,

but one made many years ago in Quebec. A bomb-shell was filled with water and closed with an iron plug which was driven in with great force. When the water froze, the plug was thrown a distance of more than 450 feet (150 metres) by the expansion (Fig. 252). This expansion is some-

Fig. 252.

times an inconvenience to us, as in bursting water-pipes; but besides the great service which it does in the earth, already noticed, it is of advantage also in loosening the soil, and in supplying it with requisite ingredients from the rocks by breaking them up and pulverizing them in small quantities from year to year.

210. **Scale of Temperature.**—The phenomena effected by heat are exceedingly varied, and it is interesting to examine the degrees in the general scale of temperature at which certain changes ensue. In the following table, taken from Dr. Arnott's Elements of Physics, only a few facts have been selected by way of comparison. The figures are partly from actual observations and partly calculated; they are given in both Centigrade and Fahrenheit degrees. The extremely low and high temperatures are only approximations.

TABLE OF HIGH AND LOW TEMPERATURES.

	Degrees Centigrade.	Degrees Fahrenheit.
Estimated " absolute zero " (Ganot)	−273	−460
Greatest artificial cold produced by nitrous oxide and carbon disulphide in vacuo (Natterer)	−140	−220
Greatest cold from a bath of carbonic acid and ether in vacuo (Faraday)	−110	−166
Liquefied nitrous oxide freezes	−101	−150
Liquefied sulphurous anhydride freezes	−77.2	−105
Greatest natural cold observed (in Siberia by Erman)	−57.7	−72
Liquefied carbonic acid freezes	−57.2	−71
Estimated temperature of planetary space (Fourier)	−50.	−58
Mercury freezes	−39.4	−39
Mixture of equal parts of sal ammoniac and ice (Fahrenheit's zero)	−17.7	0
Air on the summit of Mont Blanc, February, 1876, 3 P.M.	−12.2	+10
Ice melts (zero of Celsius and Réaumur)	0	+32
Animal heat (the " blood-heat of the human body ")	+36.6	+98
Highest natural temperature observed in India	+60.	+140
Steamship engine-room, West Indies	67.7	154
Alcohol boils	78.8	174
Water boils	100	212
Tin melts	227.7	442
Bismuth melts	260.	500
Lead melts	322.2	612
Mercury boils	343.3	650
Black heat	371.1	700
Zinc melts	411.5	773
Antimony melts	482.2	900
Red heat visible in the dark	537.7	1000
" " " daylight	593.3	1100
Heat of a common fire	616.1	1141
Bright-red heat	648.8	1200
Silver melts	1022.7	1873
Gold melts	1249.9	2282
French wrought iron melts	1500	2732
Hydrogen burned in air	1503.8	2739
Cast iron melts	1530	2786
English wrought iron melts	1600	2912
Wind-furnace, white heat	1804.4	3280
Combustion of hydrogen in oxygen	3025.5	5478

QUESTIONS.

183. What is said of the communication of heat? How many and what are the modes of communication? What is the mode called convection? Give examples of convection.—184. What is the conduction of heat? Describe the experiment showing the gradual progress of heat in an iron rod.—185. Describe the experiments showing the different rates of conductivity of different substances. What is said of non-conductors of heat? Give the examples cited.—186. Explain Davy's safety-lamp. What is said of explosions in mines? Give what is stated in the note about Stephenson and Davy.—187. What is said of the influence of density on the conduction of heat? Give the illustration about melting snow.—188. State the experiments which show that liquids are poor conductors of heat.—189. What is said of air as a non-conductor of heat? What is said of double windows?—190. What is said of arrangements of the walls of buildings? What of an arrangement for preventing the spreading of fires in blocks? —191. How are animals in very cold regions protected from the cold? What is it in their coverings that affords the protection? What is said of the coverings of quadrupeds that are natives of warm climates? What of the elephants whose remains are found in Siberia? What changes take place in the coverings of animals carried from a cold to a warm climate, and the reverse?—192. Why has man no covering against the cold? Explain the object of clothing. What is said of articles of clothing? What of loose clothing? What of straw coverings on trees? What of bricks compared with stones? What is said of cocoons?—193. What of buds of plants in winter? What of snow as a protection of plants? State the arrangement of snow observed in the arctic regions.—194. State in full what is said of the influence of the conduction of heat upon sensation.— 195. What is meant by the radiation of heat? Give examples of it. What is said of the relations of heat and light? What is said of the relation between absorption and radiation?—196. What of the reflection of heat? State the experiment with the mirrors and the thermometer and flask. Explain the experiment with the ice. Give the experiment with phosphorus. Describe the experiment with a cone of gilt paper.—197. Explain the formation of dew. State the analogy of the tumbler. What is said of the circumstances that influence the deposition of dew and frost? What is said of different substances in regard to the deposition of dew? What about Gideon's fleece?—198. What is the dew-point? How can you ascertain it? What is said of the freezing of water in India?—199. What is meant by sensible heat, and what by latent? Upon what does the form of

a substance depend? State in full what is said of the melting of ice and vaporization of water. What is said of the expression latent heat? Explain its meaning. Illustrate by reference to mixtures of cold water with hot, and of ice with hot water. What is said of iced tea?—200. What is said of capacity for heat? Describe and explain the experiment with water, oil, and mercury of the same temperature.—201. What is the relation of heat to density? Give the illustrations. What is the reason that the air is so cold on great heights?—202. State in full what is said of latent heat in reference to clouds.—203. Explain the operation of freezing mixtures. —204. State the examples of the production of cold by evaporation.—205. Describe and explain the experiment with ether and an air-pump; also that with water and sulphuric acid in vacuo.—206. Give the facts stated in regard to the degree of heat which man can endure. Give the reasons why the heat did not produce a greater effect in these cases.—207. What effect does heat produce upon the bulk of substances? What is said of water as an exception? Describe the process of freezing as illustrated by the diagram. —208. What would be the process if the exception did not exist? State what would be the results.—209. What is said of the force of expansion in ice? State some of the benefits which come from this expansion?—210. Quote from the table some of the remarkable temperatures at which certain phenomena occur.

CHAPTER XVII.

LIGHT.

211. **Nature of Light.**—The exact nature of light is not known. There are two suppositions in regard to it. One is that of Sir Isaac Newton, called the theory of *emission*. According to this, light is a material substance, but so subtile as to possess no weight and to be impalpable; and, being thrown off in all directions from the sun, it passes with inconceivable rapidity through the atmosphere, and even through substances of great density. The minute particles of this substance striking the eye produce the sensation of light, just as particles thrown off by an odorous body affect the organs of smell.

The other supposition, first definitely advanced by Huyghens, is known as the *undulatory* theory. The advocates of this, which is now quite generally received, believe light to consist of undulations, waves, or vibrations in an imponderable fluid or ether which is supposed to exist everywhere, pervading all space and every substance. You perceive here an analogy to sound, but the vibrations of sound are in a palpable medium, and those of light (as well as of heat) are in an imponderable ether.

The vibrations of light, moreover, differ essentially from those of sound in the manner of their transmission. You have seen in § 148 that the pulses or vibrations of air causing sound are in the direction of the line of transmission, whereas in light the vibrations are at right angles to this direction. This kind of wave motion is shown in Fig. 253,

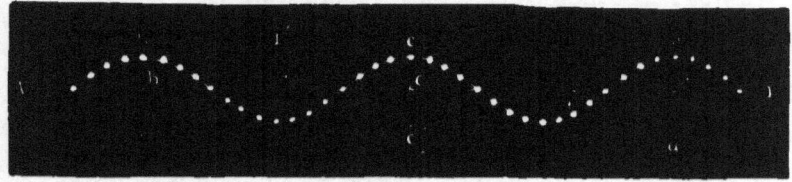

Fig. 253.

in which the white dots represent particles of ether, and the light is supposed to pass in the direction A B. Each particle in succession makes a to-and-fro motion in the direction $b'\,b''$, $f'\,f''$, etc., coming finally to rest on the line A B. This peculiar wave motion may be illustrated by shaking a stiff cord or rope from one end, when waves will appear to run along the rope, though, of course, there is no actual transfer of the particles of which the rope is made. Referring again to Fig. 253, the distance $b'\,c'$ is called a *wave-length*, and the distance $b'\,b''$ or $c'\,c''$ the *amplitude* of the wave. These distances differ in different kinds of light as will be explained in the latter part of this chapter.

The two theories just named correspond to the two theories of the nature of heat as explained in § 165: heat is supposed to be due to the vibrations of the universal ether; and light being ascribed to the same, it only remains to distinguish the two. The difference between heat and light is believed to depend on the *velocity* of the vibrations of the ethereal substance, those of light being far more rapid than those of heat. In both cases the vibrations of the ether are believed to be excited by the motions of the molecules of the substances generating the heat or light. This theory assists in explaining what is known as the convertibility of heat into light. Any solid substance heated to a sufficiently high temperature emits light. Of this we have numerous examples: a blacksmith hammers a piece of iron until it becomes at first warm, then hot, and eventually *red-hot;* in other words, until it emits light. The frequent association of light with heat in the phenomenon of combustion is another example. In these cases the particles or molecules of the substances are set in motion, and we observe heat with or without light according to the rapidity of their vibrations: when circumstances quicken immensely those vibrations which produce the effect we call heat, they then impart the sensation of light.

212. **Sources of Light.**—Any substance capable of communicating light-vibrations to the surrounding ether is said to be *luminous.* Our chief source of light is the sun. What particular conditions exist in the sun enabling it to send out such prodigious quantities of heat and light during such enormous periods of time is not well understood. The light is probably the result of intense heat, but whether this heat arises from combustion similar to that on the earth or from some other causes has not been determined. Other sources of light are the stars which, like the sun, are self-luminous. The moon, however, as you know, shines by

reflected light, that of the sun. Chemical action is a very important source of light, all combustion is merely intense chemical action accompanied by heat and light, as will be shown in Part II. of this series—Chemistry. Hydrogen gas burns with a very feeble non-illuminating flame; but if you introduce a platinum wire, it will glow or become incandescent on account of the intense heat. The illuminating power of ordinary kerosene or oil lamps is due to the incandescent solid particles floating in the flames.

As given in the table on page 310, the temperature at which red heat becomes visible in the dark is about 538° Centigrade (=1000° Fahrenheit), and in daylight 593° Centigrade (=1100° Fahrenheit); but these figures are only approximate.

Electricity is a source of light; this is seen in the lightning flash, the spark of a frictional machine, and the effects produced by a galvanic battery (Chapters XVIII. and XIX.).

There are many other inferior sources of light: the phosphorescence of decaying animal and vegetable matter, the so-called phosphorescence of certain bodies which appear luminous after exposure to the sun's rays, and the phosphorescence of certain living animals, as fire-flies, glow-worms, and animalcula in the sea.

213. **Opaque and Transparent Bodies.**—Bodies which permit the free passage of light through them are said to be transparent; those which obstruct the rays of light are called opaque. Intermediate between these are certain bodies which allow light to pass through them dimly: these are said to be translucent. Transparent and opaque are relative terms, some substances usually called opaque becoming translucent when made excessively thin. Of this we have an example in gold, which transmits a greenish light when beaten out into very thin leaf.

Light, like heat and sound, radiates in straight lines in
all directions from its source. We can see this to be true
by admitting rays of light into a darkened room through
small openings in the shutters, the rays making straight
lines across the darkness, as shown by the motes flying
in the air. The fact is recognized by the marksman in
taking aim, and by the engineer in making his levels.
The carpenter acts upon it when he tests the smooth-
ness of any surface by letting the light pass along over
it to his eye.

The manner in which opaque bodies cast shadows also illustrates this fact, as shown in Fig. 254; light radiating from

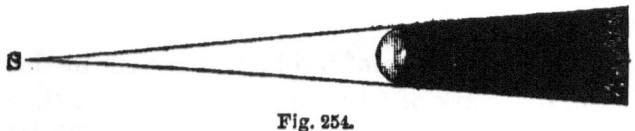

Fig. 254.

the point S and striking the opaque body forms a conical
shadow. The shape of the shadow is determined by drawing straight lines from the point S to either edge of the intervening substance, and continuing these lines indefinitely.

214. **Intensity of Light.**—As light passes in all directions
from any body or point, the farther we go from its source,
the less bright will the light be. The farther we trace any
two rays of light from their source, the farther are they
separated from each other, and that which is true of any
two rays is true of all the rays. It follows that the farther
any surface is removed from a source of light, the less light
will fall upon it. This decrease of light in proportion to
distance is perfectly regular, being as the square of the distance; or, in other words, the intensity of light is inversely
as the square of the distance (§ 28). To illustrate this experimentally, place a screen, a candle, and a square piece of
pasteboard between them at one foot from each, as shown

in Fig. 255. The shadow on
the screen covers a space four
times as large as the pasteboard; that is, the light that
shines on the pasteboard, if allowed to pass on to the screen,
would be diffused over four
times the space, and therefore
would have only one quarter

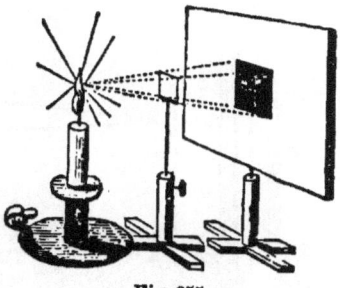

Fig. 255.

of the intensity. So if the screen be placed at twice the
distance from the pasteboard (Fig. 256), the shadow will

cover a space nine times
as large, and therefore
the light there would
have one ninth of the
intensity. Again, it is
seen by Fig. 257 that if
the screen be placed at
the distance of three feet,
the intensity of the light

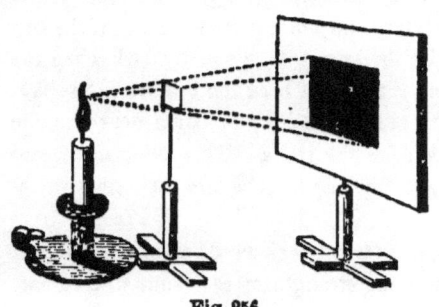

Fig. 256.

is one sixteenth of that which it is at the pasteboard.
While the dis-
tances, therefore,
are 1, 2, 3, 4, etc.,
the intensity of the
light is *inversely* as
the numbers 1, 4, 9,
16, etc.; that is, *in-
versely as the square
of the distance.*

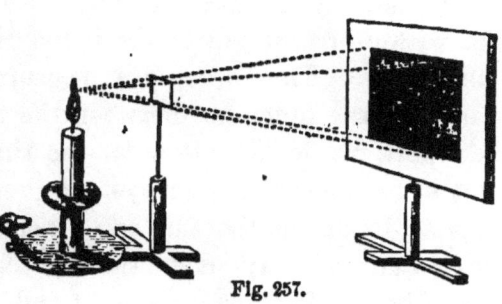

Fig. 257.

Advantage is taken of this regularity to compare the intensities of lights
from different sources. Fig. 258 represents a very simple method of testing the comparative strength of two lights. C D is a white surface in front
of which the small rod, *s*, is fastened. When a light is placed at *l* and an-

O

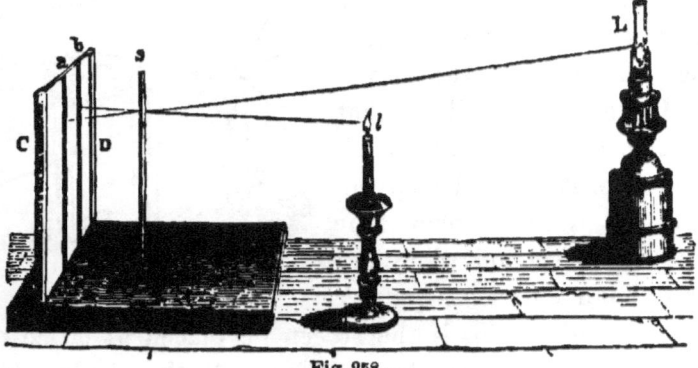

Fig. 253.

other at L, two shadows are formed upon the white surface. Now, if the two lights be of equal intensity, both the shadows will be equally dark when the lights are at the same distance; but if the light L be the brighter, the shadow *a* will be darker than *b*; and to obtain shadows of equal darkness, the stronger light must be moved back from the rod *s*. This being done, and the distance of each light measured, 't is easy to calculate their relative intensities.

Instruments called photometers, from two Greek words signifying "light-measurer," are employed to examine the strength of the flame of coal gas. These instruments operate on a principle similar to that just described, and the gas light is compared with that of candles of standard size and weight.

215. Velocity of Light.—The velocity of light is so great that within ordinary distances it may be considered as instantaneous. Thus when we measure the distance of a cannon by the difference between the time of its flash and the report, we do not allow for the time consumed by the light in its passage to the eye. But when we look at objects as distant as the sun and other heavenly bodies, we allow in our calculations for the time consumed by the passage of light. It takes light eight and one quarter minutes to travel from the sun to us, a distance of ninety-two millions of miles. With the telescope stars have been seen which are ascertained to be at such a distance that it requires over ten years for their light to reach the earth.

Others have been seen which are much farther off, but their distances have not been absolutely ascertained; some are supposed to be so remote that the light occupies many centuries in its passage to the eye of the astronomer!

Roemer's Observations.—The velocity of light was first determined by Roemer, a Danish astronomer, in 1676. It was done by calculations and observations of the eclipse of one of Jupiter's moons. After making the calculation of the time it would take for the satellite to pass through the shadow of the planet, he observed its passage, and found that it did not emerge from the shadow so soon as his calculation required by fifteen seconds. What was the difficulty? If the earth had remained in one spot from the beginning to the end of the passage of the satellite, the observation would have agreed exactly with the calculation. But the earth had moved in its orbit during this time (about forty-two hours and a half) the immense distance of 2,880,000 miles. The light of the emerging satellite therefore had to travel over this additional distance to overtake the earth, and it took fifteen seconds to do it. If we divide, then, this distance by 15, we get the distance which light travels in a second, which is 192,000 miles. All this can be made clear by the diagram Fig. 259. Let S be the sun,

J Jupiter, and C one of its moons emerging from its shadow. Let A be the earth as it is when the eclipse of Jupiter's moon begins. When it emerges, the earth has passed to B, and the light from the satellite has to travel as

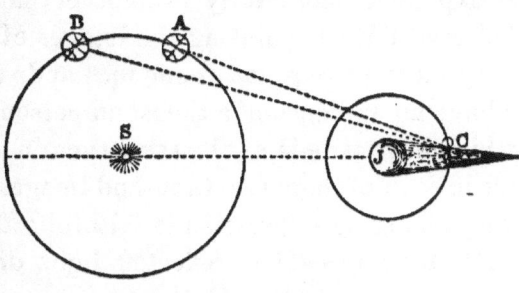

Fig. 259.

much farther to reach it now as B C is longer than A C. Roemer made other observations with the earth at some other parts of her orbit with similar results.

216. **Reflection of Light.**—Light, like sound and heat, is reflected in straight lines when it strikes upon any resisting substance. This is evident when it strikes any smooth and plane surface. And it is true of light, as it is of heat,

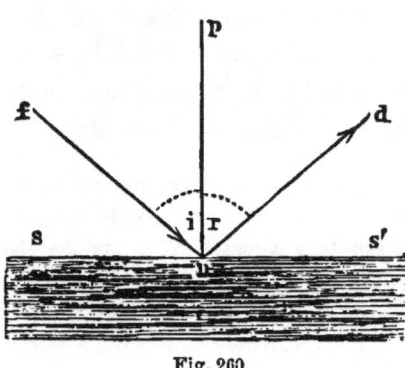

Fig. 260.

that the angles of incidence and reflection are equal. Thus if *s s'* (Fig. 260) be a reflecting surface, and *n p* a line perpendicular to it, then a ray of light, *f n*, will be reflected in the line *n d*, and the angle of incidence, *f n p*, will be equal to the angle of reflection, *p n d*.

The various objects around us are made visible by the light reflected from them. Every point of each surface we see reflects rays or vibrations of light to our eyes. Thus when we see a person, rays of light are reflected into our eyes from every part of him. These rays form an image of him in the back part of each eye, and it is by this image that we see him, as will be explained more fully in another part of this chapter. Reflected light is painting the images of objects in the eye every moment in great abundance and variety. If a speaker have an audience of a thousand persons, each one looking at him, his image is at the same time in two thousand eyes, and in each of these two thousand images every motion and every changing expression is faithfully depicted.

217. **Mirrors.**—That reflected light does thus form images of objects is seen in the common mirror. The image of any object formed is produced by the light reflected from that object to the glass. Then in seeing the image light is reflected from it into the eye, there to form a similar image, though of much less size. By using two or more mirrors the reflections of the image can be multiplied, and by certain arrangements to a very great extent. When you look at the reflection of any object in a mirror, the image appears to be at the same distance beyond the surface

as the object before it: this is owing to the fact that the
reflected rays come from the glass at the same angle that
the incident rays strike upon it (§ 215). This may be shown
from Fig. 261. Suppose $m\,m'$ is a looking-glass, and the
arrow, A B, is placed before it. Rays of light pass from it
at all points to the glass. We
will consider the path of only
two of these rays at each end of
the arrow. The ray A g will be
reflected to the eye at the same
angle in the ray $g\,o$, and the ray
A f will be reflected in the ray
f E. And the reflected rays
will have the same rate of divergence as the incident rays. The
same can be shown in regard to
rays from B or any other point
on the arrow. Now, if the lines

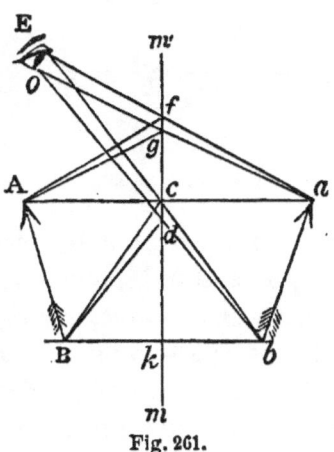

Fig. 261.

$o\,g$ and E f be prolonged, they will meet at the point a,
which lies at the same distance behind the mirror as A is
before it. The same thing can be shown of the rays from
B or any other point. Therefore the image of the arrow
will appear to the eye to have the same relative position
behind the glass that the arrow itself has before it.

218. **The Kaleidoscope.**—We have already noticed the
multiplication of the images of objects by using two or
more mirrors. In the scientific toy called a *kaleidoscope*,
by a particular arrangement of mirrors the images are
multiplied, and by changes in the position of the objects
the relative positions of the images are infinitely varied.
Fig. 262 will serve to explain the operation of the instrument. Let A B and B C be two plane mirrors placed
at right angles to each other, and a an object before
them. Let I be the position of the eye looking at the

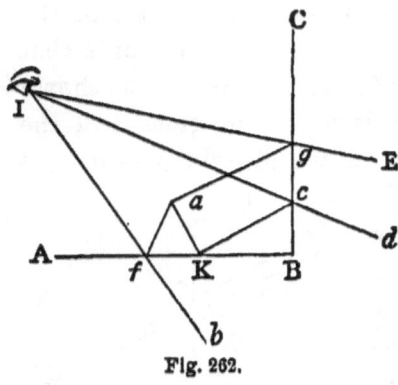

Fig. 262.

mirrors. The rays $a f$ and $a g$ will be reflected to I as represented, and the eye will see two images, which appear to be at b and E. But the ray a K will be reflected to c, and then to I, so that a third image will be seen at d. In this case but a single second reflection, or reflection of an image is formed; but by placing the mirrors at an angle of $60°, 45°,$ and $30°$ the images may be increased in number to five, seven, and nine, having a circular arrangement. In the kaleidoscope two mirrors are placed in a tube at an angle of $30°$; and variously colored pieces of glass in the farther end of the instrument, changing their relative position with every movement, give an endless variety of images symmetrically arranged.

219. **Effect of Curved Mirrors.**—Curved mirrors are chiefly of two kinds—*concave*, or hollowed out, and *convex*, or bulging out. The manner in which light is reflected from a concave mirror may be illustrated by Fig. 263. If parallel rays strike upon the mirror, they will be made to *converge*, or come together, at the point a, called the *focus*. If, however, the source of light be placed at the focus itself, the rays striking the mirror, as shown, will *diverge*, and the reflected rays will be parallel. If the light or object

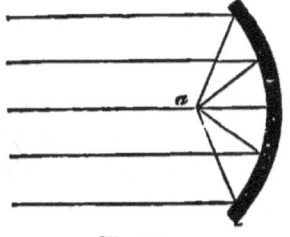

Fig. 263.

be nearer to the mirror than the focus, and the rays, of course, diverge more, then the effect of the mirror will be to lessen the divergency when the rays are reflected.

You see that the tendency is to make the rays converge. And hence concave reflectors are much used when it is desired to throw a great amount of light in one direction. The effect of the concave mirror upon the apparent size and position of objects placed before it varies with the relation of their position to the focus.

The action of a convex mirror upon light is the opposite of that of the concave. Its tendency is to make the rays diverge. Thus, if parallel rays strike upon a convex mirror (Fig. 264), they diverge, as if they came from a focus behind the mirror at *b*, indicated by the dotted lines.

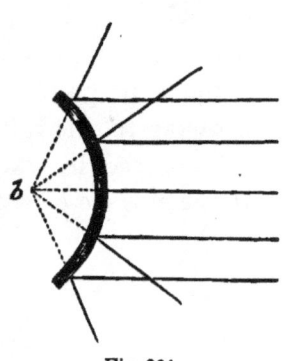

Fig. 264.

220. **Refraction of Light.**—The substance through which light (or any other agent) moves in passing from one point to another is called a medium; thus air, glass, water, etc., are media. Now, light moves in straight lines so long as it passes through a medium of uniform density; but if it pass from a denser into a rarer, or from a rarer into a denser medium, it is bent from its course. This bending, or *refraction,* as it is called, takes place only when the ray of light enters or leaves the medium in an oblique direction. Thus, if a ray of light passing from air into the denser medium water strike the latter perpendicularly to the surface, *p'p*, in Fig. 265, no refraction or alteration in its course takes place; but if it strike the

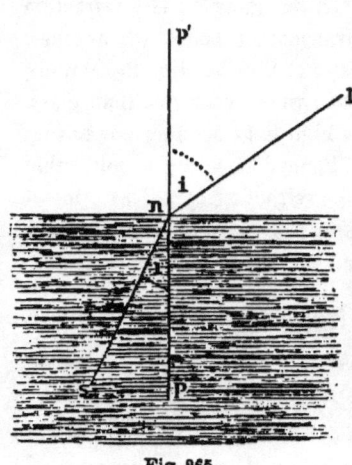

Fig. 265.

water obliquely, as indicated by the line *l n*, it is bent in the direction *n s*. When light passes into a denser medium, it is refracted *towards* the perpendicular let fall upon the point of contact, the angle *r* being smaller than the angle *i*; when, however, light passes into a rarer medium, the reverse takes place: it is bent *from* the perpendicular let fall upon the point at which the emergence occurs.

It is owing to this refraction of light that a straight stick partly immersed in water appears to the eye to be bent just at the surface of the water.

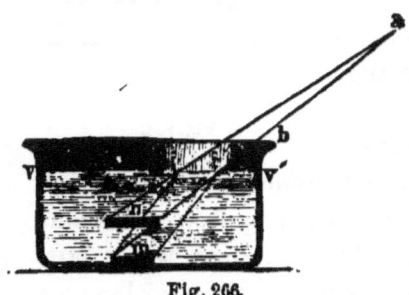

Fig. 266 also furnishes another illustration. Let the vessel shown be empty, and place a coin at *m*. Then place your eye at such a point, *a*, that the rim of the vessel just conceals the coin from view. If you then pour in water up to a certain level, say *v v'*, the coin will come in view. This is because light reflected from the

Fig. 266.

coin to *i i* is bent in the direction *i a*, and therefore appears to the eye to be at *n*. In this case the refraction is *from* the perpendicular.

We have taken as examples water and air, but more or less refraction occurs whenever light passes from one transparent body into another. Different substances differ much in their power of thus bending light; thus, glass possesses a higher refractive power than water, rock-salt than glass, and the diamond most of all. It is to this high light-bending power that the diamond owes in part its sparkling brilliancy. Bearing in mind that light is vibrations of ether, we may regard the refraction by a denser medium as a retardation of these vibrations, and the refraction which takes place when light enters a rarer medium as an acceleration of the same.

221. **Dawn and Twilight.**—The light of the sun, in passing from space into our atmosphere, is refracted. Otherwise we should have no daylight preceding the rise of the sun, nor twilight after its setting; but light would burst upon the darkness of night as soon as the sun appeared .

above the horizon, and darkness would suddenly succeed
to the light of day at sunset. As it is, in the morning the
light bends towards us in passing through the atmosphere
long before we see the sun, and after the sun has disappeared from view at evening its light bends towards us in
the same manner. And, further, we really see the sun in
the morning before it gets above the horizon, and in the
evening after it has gone below it. This may be made clear
by Fig. 267. Let the central ball represent the earth.

Now, since the atmosphere is
most dense near the earth,
and is rarer as you go outward from the earth, it is represented in the figure as having different layers in order
that the operation of the refraction may be more easily

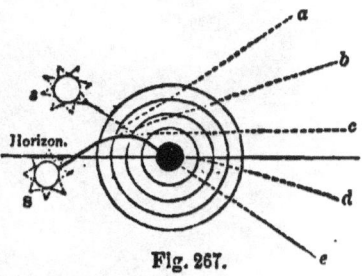

Fig. 267.

explained. The light issuing from the sun, S, below the horizon, on reaching the first layer of air, instead of passing on
straight to *a*, as indicated by the dotted line, bends towards the earth. Then in entering the second layer, instead
of passing on to *b*, it is bent or refracted still more; and so
on through all the layers, being refracted in each more
than in the previous one. The result is, that the sun,
though really below the horizon, appears to be above it.
The path of light from the sun, as it passes through the air,
is a curved line. This is because the air, instead of being
of uniform density, lessens in density as we go from the
earth. If it were of uniform density, the light would be
refracted in straight lines, as in the experiments in § 220.

222. **Mirages.**—Sometimes inequalities occur in the density of the lower portions of the atmosphere, causing, of
course, unequal refraction, and producing strange appearances, termed *mirages*. For example, at Ramsgate, on the

coast of England, on a certain occasion, a ship was seen at such a distance that only her topsails were visible; and above in the air there were two complete images of the ship, the uppermost being erect and the under one inverted (Fig. 268). Captain Scoresby, in a voyage to Green-

Fig. 268.

land, saw an inverted image of a ship so well defined that he decided that it was the image of his father's ship, the *Fame*, which was afterwards verified. The ships were at that time separated a distance of thirty miles.

An incident in the early history of the author's place of residence may be cited as an example of mirage. A ship left New Haven for England freighted with a valuable cargo, and having on board a large number of the best citizens of the colony. Some time after there was immense excitement in New Haven, because the inhabitants saw, with great distinctness, what they supposed to be this vessel, at only a little distance, apparently sailing against the wind. But it soon disappeared from view, little by little, until the whole was gone. The ship itself was never heard from, and it was supposed at the time that this appearance was a manifestation of Providence for the purpose of informing the colonists what had become of their friends. But that which was seen was undoubtedly the reflected image of this or some other ship. Such appearances as these have given rise to the stories sometimes told of phantom ships. Mirages are very common in the extensive deserts in hot climates, exhibiting to the eye of the traveller various deceptive appearances, as islands, lakes, etc. In Bonaparte's campaign in Egypt such an appearance on one occasion caused whole battalions of thirsty soldiers to rush forward, supposing at the moment that a plentiful supply of water was at hand.

223. **Visual Angle.**—In order that you may understand the operation of lenses in relation to vision, we must first explain what is meant by the visual angle. In Fig. 269

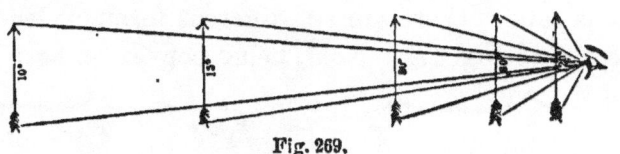

Fig. 269.

are represented arrows of the same size at different distances from the eye. From the ends of each of the arrows are drawn lines to the eye. The angle which these lines make in each case as they meet at the eye is termed the visual angle. Now, the apparent size of an object depends upon the size of this angle. The degrees of the angles are marked upon the arrows. Thus the visual angle of the nearest arrow is 120 degrees, and that of the second is 60, only half as large. The first arrow therefore appears twice as large as the second. For the same reason, it appears four times as large as the third, eight times as large

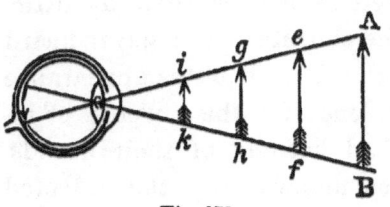

Fig. 270.

as the fourth, and twelve times as large as the fifth. The same thing is illustrated in another way in Fig. 270. Here the arrows $e\ f$, $g\ h$, and $i\ k$ appear to the eye as large as A B, because they have the same visual angle, and for this reason make an image of the same size in the eye, as indicated in the figure. It is hardly necessary to say that what is true of objects as a whole is also true of any part of them. Each part, however small, has its visual angle, and this governs its apparent size.

224. Lenses. — Lenses are transparent bodies having curved surfaces and possessing the power of increasing or diminishing the convergence or divergence of the rays of light which pass through them. Lenses are usually made of glass, but they can be made of any transparent solid, as quartz, ice, etc. There are six different forms of lenses, as represented in Fig. 271. No. 1, being convex on both sides,

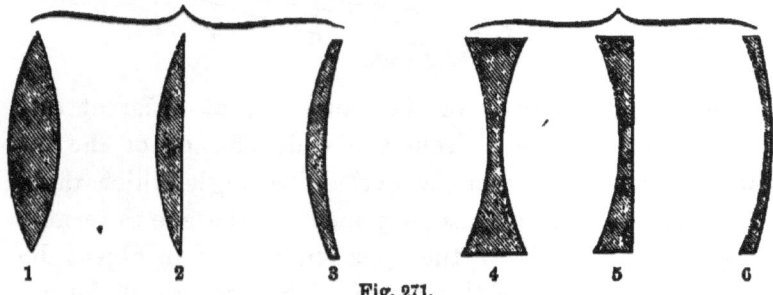

Fig. 271.

is called a double convex lens; 2 is a plano-convex lens; 3 is concavo-convex; 4 double concave; 5 plano-concave; and 6 concavo-convex. 3 and 6 are also called meniscuses. The lenses in most common use are the double convex (1) and the double concave (4). The explanation of the manner in which these act upon light will sufficiently illustrate the operation of the others. They act by refraction, the convex collecting the rays, or bringing them nearer together, and the concave spreading them farther apart. You can at once see that a convex lens, by causing the rays coming from an object to converge more, increases the visual angle, and therefore makes the object to appear larger than it otherwise would. This effect is illustrated by Fig. 272. The rays of light reflected from the arrow are made by the lens to converge so as to meet at *a*, instead of at *b*; that is, by passing through the lens they have a larger visual angle, and therefore the object is magnified. The distance between *c* and *d* shows the size which the arrow would appear to have to the eye placed at *a*. In a similar manner the double concave lens causes the rays to diverge, and, consequently, objects seen through it appear to be smaller.

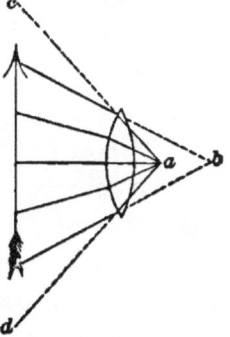

Fig. 272.

The magnifying power of any lens depends on the degree of its convexity, or the bulging of its surfaces; for the less it bulges, the nearer it is to a plane surface; and the more it is curved outward, the more obliquely will the rays fall upon its surface, and the greater therefore will be the refraction which brings them to a focus. A minute sphere forms a very powerful magnifying-glass.

225. **Microscopes and Telescopes.**—What has been said of the action of the convex lens upon the visual angle explains also the operation of the microscope. Microscopes may be simple or compound. Simple microscopes contain only one lens, the action of which has been described. Compound microscopes used to magnify and render visible exceedingly minute objects contain combinations of two or

more lenses. The operation of a simple form of this instrument is shown in Fig. 273. Rays from the object, E F, passing through the first lens, or object-glass, as it is called, form a magnified inverted image, G H, which is still more magnified by the eyeglass, C D.

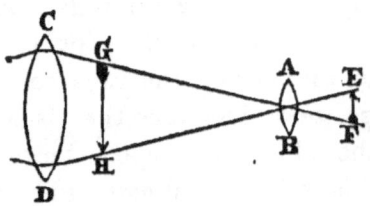

Fig. 273.

In the telescope, convex lenses are also employed, but the arrangement is different, since the objects to be magnified are very distant. Fig. 274 shows the path of the rays of light through the two convex lenses, forming

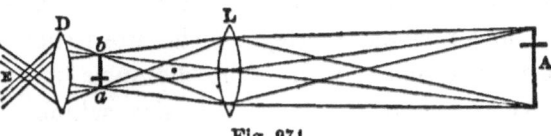

Fig. 274.

a simple astronomical telescope. Light proceeding from the object, A, passes through the object-glass, L, to form an image at *b a*, which image is seen through the eye-piece, D, the rays entering the eye at E. Of course the image is seen in an inverted position, but this is of no consequence in viewing heavenly bodies, which are spherical.

226. **Magic Lantern.**—This is an instrument by which greatly enlarged transparent pictures are made visible by being thrown upon a screen. The pictures are either painted with transparent colors upon glass, or directly photographed upon the same material. The apparatus, in its simplest form, consists of a metallic lantern, A A, Fig. 275, having a concave reflector, *p q*, and two convex lenses, *m* and *n*. At *c d* is a space between the lenses into which the pictures are introduced. L is a strong light, which is in the focus both of the mirror and the lens *m*. The picture is therefore illuminated strongly by the rays reflected from the mirror and passed through the lens. The lens *n*, which is movable, is so ad-

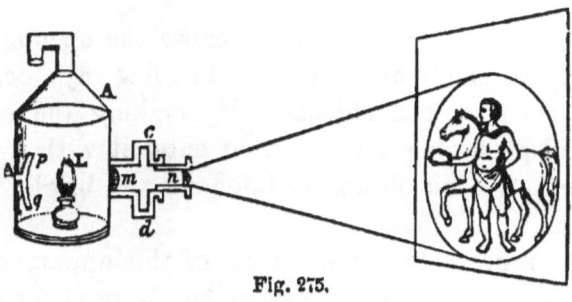

Fig. 275.

justed as to throw a highly magnified image of the picture
upon the screen. Since the image is an inverted one, the
pictures must be inserted upside down. In a more modern
form of this instrument, to which the name stereopticon has
been given, an oxyhydrogen light is substituted for the common lamp. The solar microscope is essentially similar, the
sun serving as the illuminator and three lenses being used.

227. **Camera Obscura.**—This instrument differs from the
magic lantern in giving us diminished images of objects.
An instrument of this kind can be arranged extemporaneously by any one. Thus, if into a darkened chamber light
be admitted through a small opening, inverted images of
any objects in front of the opening will be formed upon a
white screen in the opposite part of the chamber. Such an
arrangement is represented in Fig. 276. The images in

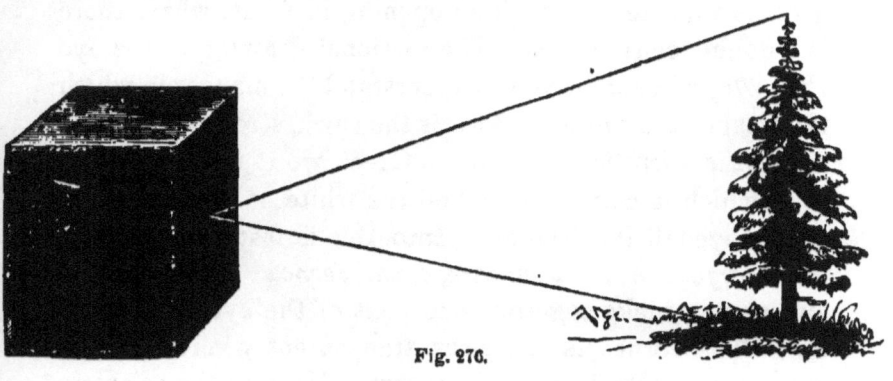

Fig. 276.

such a case, however, are faint, because the opening must necessarily be small, and therefore but few rays, comparatively, come from the objects. By making the opening larger, and gathering the rays that enter it with a double convex lens, we can obtain well-defined and bright images of objects.

Fig. 277 represents another form of this apparatus, called a *camera lucida*, used for sketching single objects and landscapes. In this the rays of light reflected from objects strike upon a mirror, A B, and are reflected through a convex lens, C D, upon white paper on the bottom, E F, of the box, where the outlines of the images are traced by the sketcher. The light can enter only at the opening above, for on the side of the box which is open there hangs down a curtain on the back of the artist while he sketches.

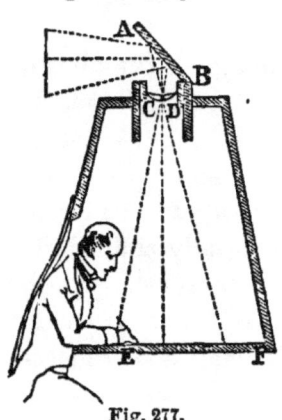

Fig. 277.

228. **The Eye.**—The eye is essentially a camera obscura. It is a dark chamber in which images are formed upon a screen in its back part, and the light which comes from objects.is admitted through an opening in front, where there is a double convex lens. The sectional drawing of the eye (Fig. 278) will enable you to understand the manner in which the images are formed. At *a* is the thick, strong white coat called the *sclerotic* coat, from a Greek word meaning hard. This, which is commonly called the white of the eye, gives to the eyeball its firmness. Into this is fastened in front, like a crystal in a watch-case, *e*, the *cornea*. The sclerotic and cornea make together one coat of the eye—the outer one. The cornea is the clear, transparent window of the eye through which the light enters. Next to the sclerotic

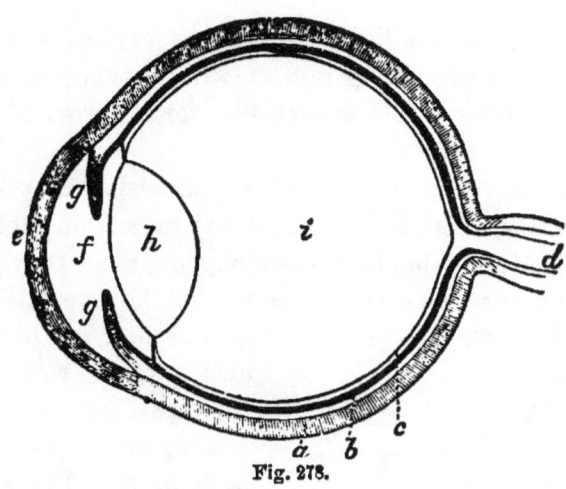

Fig. 278.

coat comes the *choroid* coat, *b*, which is dark, to prevent too much reflection back and forth in the eye. The very thin membrane, *c*, called the *retina*, serves as a screen on which the images are formed. This is composed chiefly of the fine fibres of the *optic nerve, d*. To return to the front of the eye where the light enters — behind the cornea is the *iris, g g*, on which depends the color of the eye; this is immersed in a watery fluid, *f*, called the *aqueous humor*. The light passing through the cornea and the aqueous humor, reaches the *crystalline lens, h*, which is a double convex lens. Passing through this and through a jelly-like substance called the vitreous humor, which fills all that large space, *i*, it strikes upon the retina, *c*, where it forms the images of the objects from which it came. These impressions are then transmitted to the brain by means of the optic nerve, *d*.

You now see how the eye resembles a camera obscura: like the latter, it has the dark chamber with its screen; the opening through the iris, the pupil, for the admission of the light; and just behind this opening the lens for gathering or concentrating the light before it falls upon the retina.

The refraction of the light is not, however, wholly due to this lens. The projecting cornea, with its contained aqueous humor, refracts it considerably, for it forms a convex lens.

229. **Distinct Vision.**—In order that vision may be perfectly distinct, it is necessary that the rays from each point of the object seen should, on converging, meet together, or be brought to a focus on the screen of the eye, the retina. Thus, in Fig. 279, the rays from *a*, one end of the arrow,

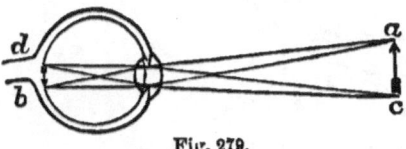

meet on the retina at *b*, and those from *c*, the other end, are brought to a focus at *d*. The muscles of the eye have the pow-

Fig. 279.

er of adjusting the eye to objects at different distances, so as in most cases to bring the rays together exactly at the·retina. They fail to do it with objects that are very near. You can perceive that this is the case if you bring any object, as your finger, nearer and nearer to the eye. You will find that at length you cannot see it distinctly. The reason is, that the rays from it diverge so much that the cornea and lens cannot make them converge enough to meet at the retina. This divergence of rays at different distances is illustrated in Fig. 280. When you look at a very minute object, the nearer you bring it to the eye, the better you can see it, up to a certain point. There the rays are so divergent, as is evident in the figure, that the lenses of the eye cannot make them converge suf-

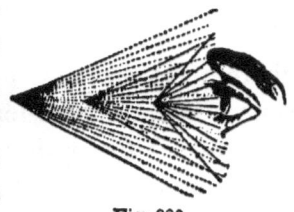

Fig. 280.

ficiently for distinct vision. The microscope assists the eye by causing these divergent rays to come nearer together before they enter the window of the eye, the cornea.

Near-sighted and Far-sighted Eyes.—Some persons have eyes so shaped that they cannot fully adjust them to objects at different distances. Thus, the near-sighted can see with distinctness only those objects comparatively close to the eye. This is because the rays converge too much, and are brought to a focus before they fall on the retina, as represented in Fig. 281. The images, therefore, of distant objects are indistinct. If the retina could in any way be brought forward a little, the difficulty would be obviated. But since this cannot be done, concave glasses are resorted to, which counteract the effect of the too great refractive power of the eye. In

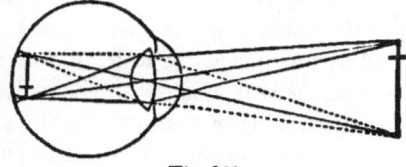

Fig. 281.

the far-sighted the difficulty is of an opposite character. The refractive power is so feeble that when near objects are viewed, the rays are not brought to a focus soon enough, and fall at a point behind the retina, as shown in Fig. 282.

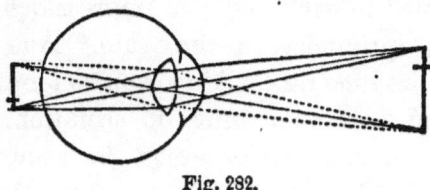

Fig. 282.

Convex glasses are used in this case, making the divergent rays of near objects less divergent before they enter the cornea.

230. Images in the Eye Inverted.—The images formed on the retina are inverted. This can be proved by taking the eye of an ox and carefully paring off the back of it, leaving little else than the retina itself. Holding now a candle before the eye, its image may be seen inverted upon the rear part. The question arises why is it that we see objects erect when their images on the retina are inverted? On this point we will quote from Hooker's Human Physiology :

"It has been supposed by some that we really see everything reversed, and that our experience with the sense of touch, in connection with that of vision, sets us right in this particular. And this, it is supposed, is the more readily done from the fact that our own limbs and bodies are reversed as pictured on the retina, as well as objects that are around us, so that everything is *relatively* right in position. But if this be the true explanation, those who have their sight restored after having been blind from birth should at first see everything wrong side up, and should be conscious of rectifying the error by looking at their own limbs and bodies. But this is not the case. The above explanation of erect vision, and other explanations of a similar character, are based upon a wrong idea of the office which the nerve performs in the process of vision. It is not the image formed upon the retina which is transmitted to the brain, but an impression produced by that image. The mind does not look in upon the eye and see the image, but it receives an impression from it through the nerve; and this impression is so managed that the mind gets the right idea of the relative position of objects. Of the way in which this is done we know as little as we know of the nature of the impression itself."

231. **Single Vision.**—Whenever we see any object with both eyes, an image is formed in each eye, and impressions pass from both eyes by the optic nerve to the brain. And yet these two impressions produce no double vision so long as the two eyes correspond with each other in situation. This is because the image in one eye occupies the same place on the retina as that in the other. The correspondence is ordinarily perfect, the two eyes turning always together in the same way, upward, downward, or laterally, without the least variation. You can observe the effect of a want of this correspondence by pressing one of the eyes with the finger while the other is left free to move in obedience to the muscles. When this is done, every object appears double, because its image occupies in one eye a different part of the retina from that which it does in the other, and hence two different impressions are carried to the brain. The same thing occurs in squinting, in which the action of the muscles of the two eyes does not agree. Ordinarily in squinting there is no double vision, because the mind has the habit of disregarding the impressions received from the defective eye; but when squinting occurs suddenly from disease, double vision ensues, for it takes a little time to form the habit referred to.

232. **Stereoscope.**— The images of objects in the two

eyes, though always similar, are generally not precisely alike. They are so only when the object presents a simple plane surface, as in the case of pictures. When the object presents two or more surfaces to the sight, the images are more or less unlike. This can be illustrated in a very simple way. Hold a book up straight before your eyes with its back towards you. You see the back and both sides. Now, if you shut your right eye, you will see with the left the back of the book and the left side; that is, these two parts of the book are imaged on the retina of the left eye. By shutting the left eye it will appear that the image in the right is different, for you see now with the back the right side of the book. These simple phenomena underlie the principles on which the instrument known as a *stereoscope* is constructed. In the right side of this instrument is placed the picture of the object as the object itself would appear to the right eye, and in the left side is placed the picture as it would appear to the left eye. Thus, if a book in the position alluded to above were the object, in the right picture there should be represented the back together with the right side of the cover, and in the left the back with the left side of the cover. The two impressions, carried to the brain by the optic nerves, give together the impression of a solid book. The same principles apply to the representation of all solids in the stereoscope.

233. **Thaumatrope.**—Each impression made upon the retina by light lasts about one-sixth part of a second; according to some, it is only one eighth or one eleventh of a second. No distinct impressions can be made, therefore, upon the retina unless they succeed each other with less rapidity than this. If, for example, in the revolution of a wheel, eight or more spokes pass by one point in a second, they cannot be seen as distinct spokes, but appear confused, producing one continuous impression. If a light be re-

volved so as to describe a circle in an eighth part of a second, it will appear to the eye as one unbroken circle of light. It is this persistence of impressions on the retina that makes small objects seen from a swiftly moving railroad-car appear to run in long lines along with us. This principle is made use of in the construction of a toy called the thaumatrope. Pictures are drawn on each side of a circular card, and when it is whirled around very rapidly by means of two strings fastened to it, the two pictures

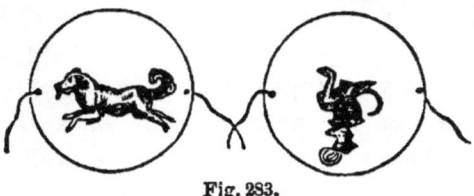

Fig. 283.

appear as one. Thus, in Fig. 283 are represented the two sides of such a card, one side having the picture of a dog, and the other that of a monkey. When made to revolve rapidly, the monkey will appear to be sitting on the back of the dog. Several other pieces of apparatus have been devised to illustrate the persistence of vision, such as the zoetrope, phenakistoscope, kaleidophone, etc., the popular expression "an optical illusion" being commonly employed to explain this whole class of scientific toys.

234. **Decomposition of White Light.** — The manner in which light is refracted in passing from one medium to another of greater or less density has been explained in § 220, but we must again return to the subject. Let A B, Fig. 284, represent the section of a piece of glass having parallel surfaces. When a ray of light passes through at right angles to its surface, as c d, it suffers no refraction; when, however, it strikes the glass obliquely, as e f, it is refracted both at f, where it

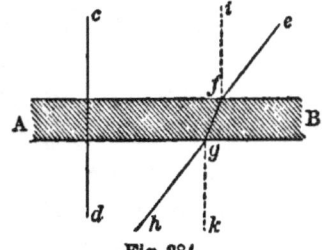

Fig. 284.

enters the glass, and at *g*, where it leaves it; its course on emerging, *g h*, being parallel to its original path, *e f.*

A piece of glass the section of which forms a triangle is called a *prism*. Fig. 285 represents such a prism. Somewhat similar pieces of glass were formerly used for decorating

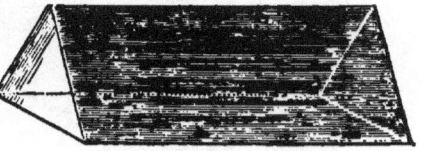

Fig. 285.

chandeliers. When a ray of light passes through a prism, it is twice refracted, just as in passing through a plate of glass with parallel surfaces, but, instead of issuing in a path parallel to that by which

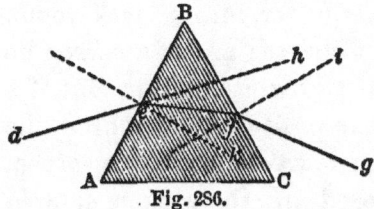

Fig. 286.

it entered, it is bent entirely out of its course; thus, the ray of light, *d e*, passing through the prism, A B C, Fig. 286, takes the course *d e f g.*

The explanation of this depends upon the fact already mentioned (§ 220), that a ray of light passing from a rare into a denser medium is bent *towards* the perpendicular, and from a dense into a rarer medium it is bent *from* the perpendicular let fall upon the surface. This will be easily understood by reference to Fig. 284, where the ray *e f* is bent *towards* the perpendicular *i f*, and *from* the perpendicular *k g* as it respectively enters and leaves the dense medium. In like manner the ray *d e*, entering the prism, A B C, Fig. 286, is bent towards the perpendicular, *k e*, and on emerging at *f* is bent from the perpendicular, *i f*, thus pursuing a very crooked path.

When we make the experiment of passing a ray of light from a small orifice through a glass prism, we observe, however, something more than the mere bending of the ray: we obtain a beautiful image, having all the colors of the rainbow, called a *spectrum*.

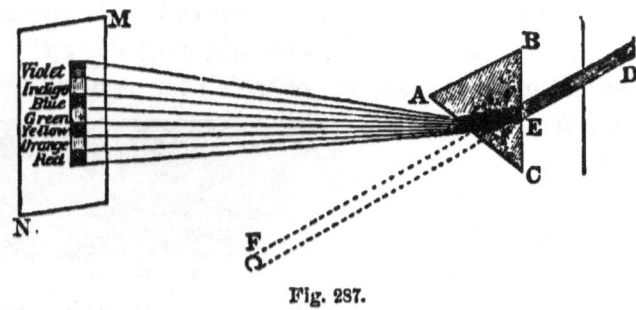

Fig. 287.

Fig. 287 illustrates this *dispersion* of white light, as the phenomenon is called. Let the beam of sunlight, D E, pass through a small opening in a shutter into a dark room. The rays will pursue a straight course; and if a screen be placed at F, they will form a spot of white light. But if a glass prism, A B C, be held in the position represented, the rays will be refracted, and when received upon the screen, M N, the light will be decomposed into the various colored rays of which it is constituted. The whole beam is refracted for reasons just given, and the separation or dispersion takes place because the rays of different colors are unequally refracted. If they were equally refracted, the light upon the screen would be white, as before the dispersion. The violet rays are most refracted, the indigo next, the blue next, etc., the red being the least refracted of all.

The light of the sun passed through a prism yields a *solar* spectrum, familiarly seen in the rainbow. When lights from other sources are similarly examined, different spectra are obtained characteristic of the sources. This fact gives rise to the science of spectrum analysis, which will be fully explained in the twenty-second chapter of Part II.

234. **Recomposition of White Light.** — The seven colors, violet, indigo, blue, green, orange, yellow, and red, are commonly spoken of as the primary colors. It is more correct, however, to accept the number of colors as infinite, for

they pass from one end of the spectrum to the other through imperceptible shades.

White, then, is, strictly speaking, not a color, but a combination of all colors: this is easily shown by another experiment. After decomposing light by passing it through a prism, we can combine the separated colors and form white light again. The manner in which this is done is represented in Fig. 288. The beam of light, after passing through the prism S A A', instead of proceeding in the direction indicated by the dotted lines to form the spectrum, is made to pass through the prism S' B B', placed in a reversed position, and its

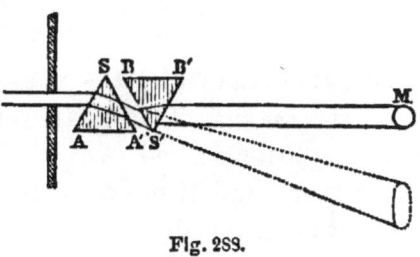

Fig. 288.

rays are refracted so as to assume their original relation, making a white beam, M. Here the second prism counteracts the effect of the first, because its position is exactly the reverse. Similar results are obtained by substituting a converging lens for the second prism.

Sir Isaac Newton, who first experimented on this subject, considered the decomposition and the recomposition of light as affording proof that white light contains seven colors. He tried various experiments to prove the same thing. Thus, he mingled together intimately seven powders having the seven prismatic colors, and found that the mixture had a grayish-white aspect. He also painted a circular board with these colors, and found that on whirling it so rapidly that the colors could not be distinguished the whole board appeared to be white. In order to succeed perfectly in this experiment a certain proportion between the colors must be observed, as indicated in Fig. 289. A very pretty way

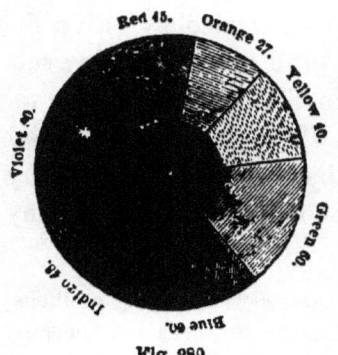

Fig. 289.

P

of illustrating the composition of light is to spin a top painted in this way. When the top whirls rapidly it is white, but as it slackens its motion the seven colors appear.

The names of these seven so-called elementary colors can be easily remembered by observing that the initial letters spell the meaningless but pronounceable word VIBGYOR; thus:

V	I	B	G	Y	O	R
i	n	l	r	e	r	e
o	d	u	e	l	a	d
l	i	e	e	l	n	
e	g		n	o	g	
t	o			w	e	

235. **Colors of Objects.**—The color of any object depends upon the manner in which it reflects light. Thus, if it be red, it reflects the red rays of the spectrum, absorbing the other rays; and if it be green, it reflects the green rays, etc. If it reflect all the colors together, it is white; and if it reflect none, or almost none, of the light, it is black.

You can readily see why the color of an object varies with the kind of light that falls upon it. If an object which is red in sunlight be exposed to a yellow light, such as a yellow flame, or to sunlight that has passed through a yellow-colored glass or curtain, it loses its red color, for there are no red rays to be reflected to our eyes. A person exposed to such a light has a deathlike paleness, the lips and skin losing entirely their red color. This effect can be witnessed at any time by mixing alcohol with a little salt on a plate and setting fire to it, or by throwing salt into a coal fire: the other lights in the room should be removed to obtain the full effect. This explains also why the colors of goods examined in the evening, especially by candle-light, often differ somewhat from those which they have in the day.

In some substances the colors are changeable with varying positions, though the light be the same. This is often seen in shells and minerals, as well as in some fabrics, as changeable silk. This is owing to the arrangement of the particles, by which variety in reflection results from changes of position.

236. **The Rainbow.**—There is no more gorgeous display of colors than that sometimes seen in the clouds at morning or evening. These colors are occasioned simply by refractions and reflections in the minute vesicles (§ 179) of which the clouds are composed. When, however, the moisture is condensed into drops, a still more magnificent spectacle results—the rainbow.

The rainbow is produced by the action of light on the drops of falling rain; both reflection and refraction are concerned in its formation. Consider, for example, that which takes place in a single drop, represented in Fig. 290. The sunbeam, S, entering the drop at A, is refracted, and passes to B, at the farther side of the drop. Here a portion of it is lost by its proceeding on in the line B C. The remainder is reflected to D, and passes to E, being again refracted as

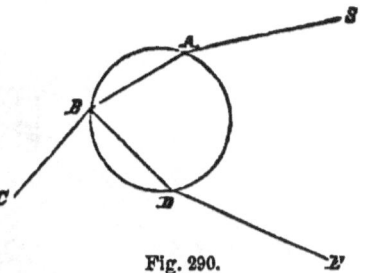

Fig. 290.

it passes out into a rarer medium, the air, thus producing a single reflection and two refractions. But in the second bow, which is sometimes formed, there are two reflections as well as two refractions, as represented in Fig. 291. The beam of light, S, from the sun enters the drop at A, is refracted, and passes to B. Here a portion proceeds on in the direction B C; the other portion is reflected to D. Some light is again lost by proceeding in the line D E; that which remains is reflected to F. This shows why

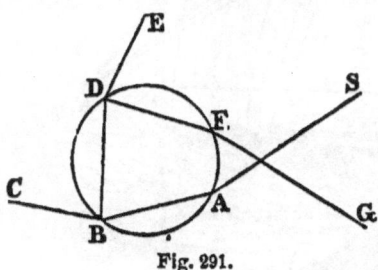

Fig. 291.

the second bow is not so bright as the primary one: in the latter there is but one reflection in each drop, and therefore loss of light occurs at but one point; while in the former there are two reflections, and therefore loss of light at two points.

237. **Circumstances under which Rainbows are Seen.**—A rainbow is seen when the spectator stands between the sun and falling rain. This can rarely be done except in the latter part of the day. It sometimes, though very rarely, happens that a shower passes from the east to the west in the morning, and then a rainbow can be seen in the west. Fig. 292 is intended to show under what circumstances a

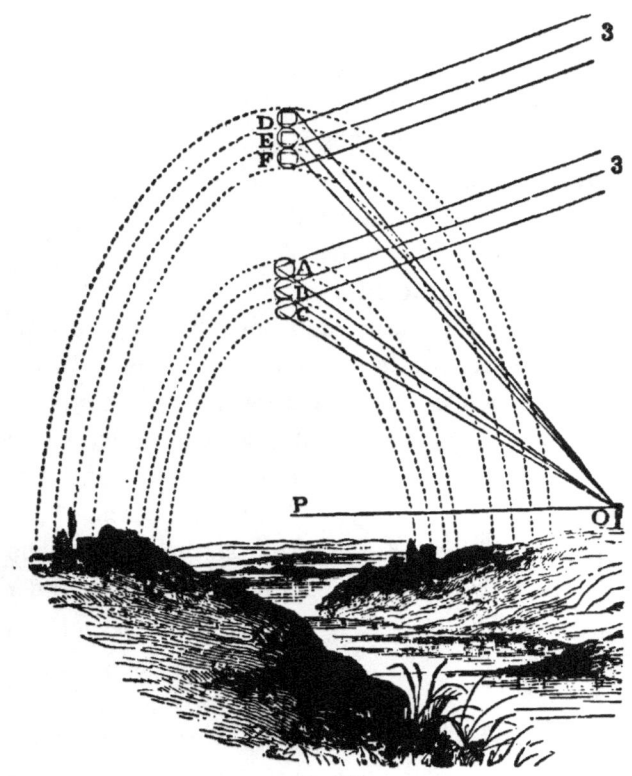

Fig. 292.

rainbow is seen. Let a horizontal line be drawn from O, the observer, to P, a point directly under the middle point of the arch. If this line were prolonged behind the observer, it would extend precisely in the direction of the sun. That is, the sun is directly opposite the middle of the bow. Now, if the drop at A reflect a red ray to the eye of the spectator, all other drops similarly situated in the arch will reflect red rays. If B reflect a green ray, all other drops similarly situated will do the same. And so of C, reflecting the violet ray. For the sake of clearness only three reflections are represented, but the same is true of all the seven colors. In the secondary bow the arrangement of the colors is reversed, the red being at the inner part of the bow and the violet at the outer part. The double reflections are manifest in the drops D, E, and F. That which we have described as taking place in a few drops takes place in countless multitudes of them in forming the bow. Since the exact position of the rainbow depends not only upon the direction of the rays of the sun, but also upon the position of the spectator, it is clear that no two spectators see precisely the same bow, for the drops that form it for the one are not the same as those forming it for the other. This is very obvious if the two be quite distant from each other; but it is equally true if they are very near together, although in this case the bow seen by one spectator would be very nearly coincident with that seen by the other. It is also true that the rainbow of one moment is not the rainbow of the next; for since the drops that reflect it are falling drops, there must be a constant succession of them in any part of the bow.

Colors in Dew-drops and Ice Crystals.—We often see something very analogous to the rainbow in the dew. If we look at the dew-drops when standing with our backs to the rising sun, we see all the colors of the rainbow glistening everywhere before us, as if the grass were filled with gems of every hue. This is occasioned by refraction and reflection in drops of water, and the resemblance fails only in the regularity of arrangement which the rainbow presents. We see the same thing also if the ground is strewed with bits of ice which have fallen from the branches of the trees, and the sun shines aslant upon them.

At the foot of Niagara Falls, when the sun shines favorably upon the dense mists, perfectly circular rainbows are seen, the spectator viewing the mist below him as well as that above him. The colorless *halos* seen around the moon are the result of reflections of light from the external surfaces of drops of water, without refractions.

238. **Heat, Light, and Chemical Rays.**—According to the undulatory theory (§ 211), light consists of vibrations of an all-pervading imponderable ether; and these vibrations have a wave-like character. It has been found that the rays of different colored light vary, both as regards the length of the waves and their velocity. Thus, the red waves are long and the violet waves short. Expressed in figures, the lengths are so small as to be entirely beyond our conception; and yet by delicate operations, which we cannot explain in this elementary work, these marvellously small distances have been determined experimentally. The number of vibrations per second is so enormous, on the other hand, as to be equally beyond our comprehension. The length of the waves constituting red light is 620 millionths of a millimetre, and the ether producing this light makes 514,000,000,000,000 vibrations in one second. Passing from red to violet through the spectrum, the waves of each successive color are shorter and shorter, and the number of vibrations of the ether greater and greater, until we reach the violet, for which the following figures have been determined: violet light has waves 423 millionths of a millimetre in length, and makes 752,000,000,000,000 vibrations per second.

In this respect there is an analogy with both sound and heat. As mentioned in § 148, the vibrations which produce sound are from 16 to 38,000 per second, the pitch depending on the velocity. The temperature also depends upon the rapidity of the vibrations of the molecules of the bodies and of the surrounding ether; and now we learn that the colors depend upon similar causes. The vibrations of the ether which produce light are, however, not the most rapid known: other rays exist which yield neither heat nor light, and the motions of which are far more rapid. The existence of these invisible rays is proved by the chemical effects they are capable of producing; they are consequently usually called *chemical rays.* The term *actinism* is also used to express this form of force.

When the combined rays of heat, light, and actinism are passed through a prism, they are dispersed in the manner shown in Fig. 293, the short rapid waves bending more

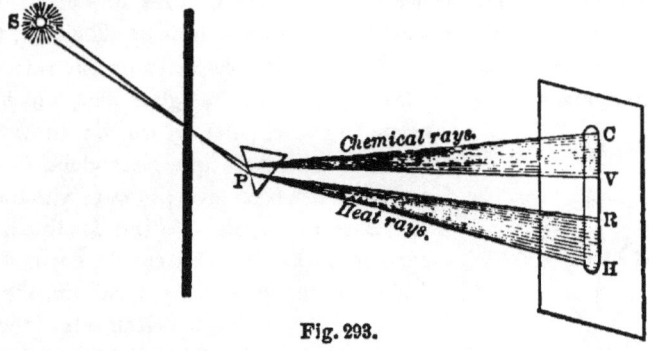

Fig. 293.

than the long slow ones. Between H and R on the screen are invisible heat rays; between R and V the colors in their order; and between V and C the chemical rays, which are powerless to affect our eyes, but affect strongly certain chemical substances. It is these rays which do the work in photography. The chemical effects of light will be described in Part II. of this series — Chemistry. The heat, light, and chemical rays overlap each other, and are not terminated by sharp lines, as you might judge from

Fig. 293.   The manner in which they are distributed is
shown in Fig. 294, in which the height of the curved lines
represents the intensity of the rays at each point; the
greatest heating effect being outside the red end of the
visible spectrum, and the greatest chemical effect residing
beyond the violet at the opposite end.

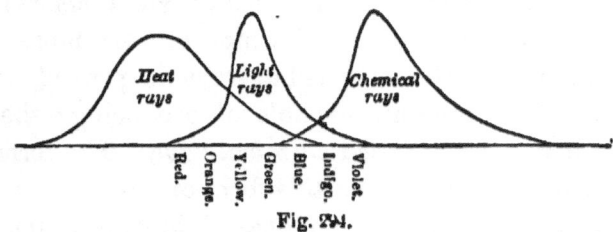

Fig. 294.

210. Crookes's Radiometer. — The intimate connection between
heat and light is manifested in the operation of an exceedingly curious little instrument recently invented by William Crookes, of London, which
he calls a *radiometer.* It consists of two cross-arms of aluminum, which
carry at their ends little thin plates of mica blackened on one side.   The
cross-arms are attached by melting to a vertical glass axis, which rests

below on a steel point, on which it turns freely.
The whole is placed in a glass globe, in which
a vacuum has been made, so as to withdraw the
resistance of the air.   In this condition, when
exposed to the light of a candle, daylight, etc.,
falling on one side of the apparatus, the arms
commence to revolve, the clean side of the mica
moving towards the light, while the velocity of
rotation is more rapid in proportion as the light
is stronger.   Thus, for instance, when placed
in the sunshine, the rapid rotation will at once
be retarded when a cloud passes over the sun;
hence it is that Crookes called the instrument
a "radiometer," attributing its motion to the
reaction of the reflection of the rays of light.
This explanation was not adopted by the ma-

Fig. 295.

jority of savans, who did not see any foundation for it, but found, by care-
ful experiment and consideration, that the motion was due to the action of
heated rays in a rarefied gas.   If, indeed, it were due to the reaction of

reflection, the polished side should move from the light, and the blackened side (which absorbs light) should move towards it; but the contrary is the case, and the motion is evidently due to the heat developed in the blackened side of the mica in the thin layer of rarefied air there, which by its expansion presses against this side, and, thus acting as a dilating spring, pushes it forward.

Others add to this that a perfectly dry vacuum is unattainable, and that therefore the watery vapor always condensed on the black pulverulent surface of the mica is evaporated by the least heat; and when emitted, by its reaction pushes the black surface forward, the same as the air in the former explanation.

QUESTIONS.

211. What was Newton's theory of the nature of light? What is the undulatory theory? Show the difference between sound vibrations and those of light. Explain the terms wave-length and amplitude. Show the analogy between heat and light.—212. When is a body luminous? What are the sources of light? What is said of chemical action as a source of light? —213. Define opaque and transparent. How may you see that light moves in straight lines? State various familiar recognitions of this fact.—214. Illustrate the fact that the intensity of light is inversely as the square of the distance. How are lights of different intensities compared? What is a photometer?—215. What is said of the velocity of light in regard to ordinary distances? How long is light coming from the sun to the earth? What is said of the light coming to us from certain stars? Give the observation of Roemer represented in Fig. 259.—216. What is said of the reflection of light? What of its reflection in relation to seeing?—217. What of the images formed in mirrors? Show by Fig. 261 why the image in a mirror seems to be at the same distance behind it that the object is before it.—218. Explain the operation of the kaleidoscope.—219. What are the chief kinds of curved mirrors? Explain the operation of a concave mirror. Explain that of a convex mirror.—220. What is meant by the refraction of light? Illustrate its refraction in passing from a denser into a rarer medium. Then from a rarer into a denser. How is the refraction in regard to a perpendicular in the two cases? What is said of the refractive power of different substances?—221. Explain dawn and twilight. Explain what is represented in Fig. 267.—222. What are mirages? Describe the mirage which occurred at Ramsgate. Describe that seen by Captain Scoresby. Relate the incident which occurred at New Haven. What is

P 2

said of mirages in deserts?—223. Explain what is meant by the visual angle. Explain Fig. 269.—224. What are lenses? What are the different kinds? What is the difference of effect in convex and concave lenses? Explain the effect of a convex lens on the visual angle. Upon what does the magnifying power of a lens depend?—225. What is said of microscopes and telescopes?—226. Describe and explain the magic lantern.—227. Describe and explain the camera obscura. Describe the arrangement of a camera for sketching.—228. How is the eye like a camera? Describe the arrangement of the parts of the eye. Show more fully the analogy between the eye and a camera. What is said of the influence of the cornea on the light?—229. Show what is required for distinct vision. Show why it is that objects brought very near the eye are not seen distinctly. What is said of the microscope? Explain the difficulty in the near-sighted. In the far-sighted.—230. How can you show that the images of objects in the retina are inverted? Give in full what is said of explanations of the fact that we see objects erect, notwithstanding this inversion.—231. Explain single vision. By what simple experiment can you show the explanation of single vision to be correct? What is said of squinting?—232. Explain the stereoscope. What is said of distinct impressions on the retina? — 233. Explain the thaumatrope. Name other contrivances for illustrating the persistence of vision. — 234. Explain the path of a ray of light through a piece of glass having parallel surfaces. What is a prism? What course does light take in passing through a prism? Why? What is a spectrum? Explain the experiment showing the dispersion of white light.—235. What is said of the recomposition of decomposed light? Give the illustration of the powders. The circular board. The top.—236. What is said of the colors of substances? What of the variations of these colors in different lights? What of variations with varying positions?—237. What of the colors of clouds? Explain the formation of the first rainbow by Fig. 290. Explain the formation of the second bow by Fig. 291.—238. What is said of the circumstances under which rainbows are seen? Explain in full the formation of the two bows as illustrated by Fig. 292. What is said of the bow as seen by different persons, and at different moments by the same person? What of rainbow hues in dew-drops and ice crystals?—239. In what respect do rays of different colors differ? What is said of the analogy of sound, heat, and light? What are chemical rays? At which end of the solar spectrum are they situated? Show how heat, light, and chemical rays are distributed in the solar spectrum. — 240. Describe Crookes's radiometer. What theories have been advanced to account for its motion?

CHAPTER XVIII.

ELECTRICITY.

241. The Effects of Electricity.—The ancient Greeks, so long as 2500 years ago, observed that when amber was rubbed with woollen cloth it acquired the singular property of attracting light substances—such as bits of paper, shreds of cloth, etc. The Greek word for amber being *electron*, the power thus excited has been called *electricity*. Little else was known of electricity until about 280 years ago, when Dr. Gilbert, physician to Queen Elizabeth, showed that other substances besides amber—such as glass, wax, sulphur, etc—possessed similar properties. Since then many philosophers have studied the remarkable phenomena exhibited by electricity; and, within comparatively recent times, it has become an important branch of natural philosophy.

The precise nature of electricity has not been definitely determined. Various theories have been offered to explain its existence and effects, of which we shall have more to say later on. Meanwhile let us examine, experimentally, some of the facts for the explanation of which the theories have been propounded.

The apparatus required to illustrate the fundamental facts of electricity is very simple and inexpensive. Any one provided with (1) a silk handkerchief or a woollen cloth, (2) a piece of sealing-wax, (3) a glass tube, (4) some round pieces of cork or of elder pith hanging from any support by means of silk threads, (5) a common fire-poker, and

(6) a few wine-glasses, can easily make himself acquainted with a large number of the effects of electricity.

If you rub the stick of sealing-wax with the silk handkerchief, and then hold it quite close to some bits of paper or shreds of cotton cloth, the latter will be attracted to the sealing-wax, springing a short distance towards it, and then adhering feebly to its surface. Amber and many other substances act similarly. As already mentioned, this experiment is more than 2500 years old.

Again, if you rub the well-dried glass tube or rod with the silk handkerchief, as shown in Fig. 296, and then hold

Fig. 296.

it near light articles, such as feathers, lint, bits of paper, etc., the glass, also, will attract them. Under favorable circumstances pith balls of some size will spring to the glass tube, as represented in Fig. 297.

The force thus developed by friction is called electricity, and the tube, ball, or other object in which the electrical action is excited is said to be _electrified._

Fig. 297.

We learn from the experiments described that one of the most common effects of electricity is *attraction;* another simple experiment will show that sometimes, however, *repulsion* is produced by electricity. Suspend the pith ball by a silk thread, and attach it to a glass rod inserted into a bottle (filled with sand to make it stand firm); and then, having rubbed the glass tube with the silk handkerchief, bring it near the ball, as represented in Fig. 298. The pith ball will be drawn towards the tube and follow it when the latter is moved about; if, however, the ball

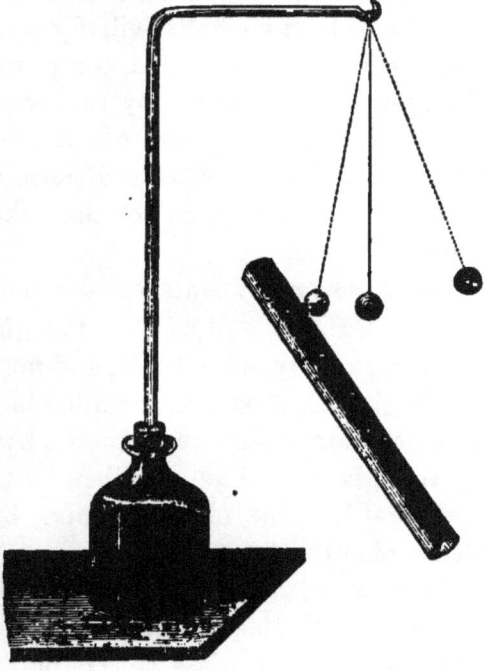

Fig. 298.

be allowed to touch the tube, then in an instant the ball will be repelled by it, and when the tube is moved the ball will try to escape touching it, and run away from it as if endowed with life.

After the ball has touched the electrified tube, it becomes itself electrified, and will attract any light object (as another ball) brought near it. So soon, however, as the two balls have touched each other, they will repel mutually, just as the glass tube and first ball do after contact. If either electrified ball be touched with the finger, it loses its electricity.

The following experiments will illustrate another point. Support the iron poker on two wine-glasses, and attach two pith balls by linen threads, or a fine wire to one end of the poker; then touch the other end with the excited glass tube, and the two balls will fly apart, showing that the metallic rod has permitted the passage of the electricity throughout its length. By this we learn that some substances act as *conductors* of electricity, while others are *non-conductors.* The wine-glasses, being *non-conductors,* form a suitable support for the poker and prevent the escape of the electricity.

All these experiments can be made with the silk handkerchief itself as well as with the glass tube. The effects, however, are far more feeble, and much care is necessary to obtain them. This singular difference, moreover, is noticeable: objects which are attracted by the glass are repelled by the silk, and those repelled by the glass are attracted by the silk. One of the rubbing bodies appears to gain that which the other loses.

Suppose, again, you rub the stick of sealing-wax with the silk or woollen cloth and hold it near a pith ball previously electrified by the glass tube; the ball will be attracted; and if you bring a glass tube, rubbed as before, near a ball electrified by the sealing-wax, this ball will also be attracted. Lastly, if you electrify one pith ball by means of seal-

ing-wax and the other by means of glass, the two balls will
attract each other. There seem, then, to be two kinds or
states of electricity having attraction for each other, but
each repelling itself.

242. **Theories of Electricity.**—To account for these and
other phenomena, several theories have been proposed. According to one, electricity is an imponderable fluid existing in all bodies to a greater or less degree, according to
their capacity for electricity. While a body is in its usual
state there is no manifestation of electricity. The fluid is
in a quiescent condition, because its particles are prevented
from repelling each other by the attraction which exists
between them and the particles of the substance. But this
quiescence can be disturbed by friction and other causes.
Thus, if a glass rod be rubbed with a piece of silk, the natural equilibrium is disturbed, the glass having an excess
and the cloth a deficiency of electricity. The glass is therefore said to be *positively* and the cloth *negatively* electrified. The equilibrium can be restored in the case of a positively electrified body by having its excess drawn off, and
in the case of a negatively electrified body by having its
deficiency made up by receiving electricity from other bodies. This was Franklin's view, and is known as the *single-
fluid* theory; opposed to this is the *two-fluid* theory, which
supposes the existence of two electric fluids which, like an
acid and an alkali, neutralize each other when present in
equal quantity, but which have a strong attraction or affin-
ity when separated; while, on the other hand, the particles
of either fluid are repellent to each other. When friction
or other cause excites electrical action, the two fluids are
separated and become evident to the senses. This theory
seems to explain the fact that sealing-wax and glass, or the
glass and a silk handkerchief, develop respectively electricities of a different character; and hence the electricity ex-

cited by friction of sealing-wax received the name *resinous* electricity, and that of glass *vitreous* electricity. The names positive and negative are also applied to vitreous and resinous electricity, these appellations being borrowed from the single-fluid theory of Franklin.

As we have already seen in the case of both heat (§ 165) and light (§ 211), the theories based upon the existence of *imponderable fluids* have little by little been discarded, and eventually gave way to the dynamical theory, or theory of motion.

In the same way the *fluid theories* of electricity have been abandoned, and it is now regarded as a mode of force operating on ordinary matter, the molecules of which it arranges in a definite direction. Moreover, it is convertible into the other modes of force—heat, light, magnetism, and chemical attraction.

Many of the terms commonly used in explaining the facts of electricity are, however, derived from the fluid theory, and still maintain their hold on the language of electricians. Thus the expressions "currents," "charged with the electric fluid," are remains of the abandoned theories.

243. Positive and Negative Electricity.—The terms positive and negative are retained as convenient for expressing the character of the electricity developed. Whether one or the other is excited depends upon the nature of the rubber. Thus, smooth glass rubbed with woollen cloth or silk will be positively electrified; while if it be rubbed upon the back of a cat, it will exhibit negative, or resinous, electricity. If a resin, as gumlac or sealing-wax, be rubbed with silk or woollen cloth, it will be charged with negative, or resinous, electricity; but if it be rubbed with sulphur, it will be charged with vitreous, or positive, electricity. The terms vitreous and resinous are therefore incorrect, for they are based upon the idea that one kind of electricity is always excited on glass, with whatever substance the friction may be made, and that the other kind is always excited on resins. The most decided illustration of the incorrectness of these terms we have in the fact that while smooth glass rubbed with silk or woollen cloth becomes charged with positive (vitreous) electricity, roughened glass rubbed with the same gives us negative (resinous) electricity. Below is a list of substances any one of which develops positive electricity when rubbed with any substance below it, and negative when rubbed with any substance above it:

1. Cat-skin.	7. Silk.
2. Polished glass.	8. Sealing-wax.
3. Woollen cloth.	9. Amber.
4. Feathers.	10. Roughened glass. .
5. Wood.	11. Sulphur.
6. Paper.	

As a result of experiments mentioned in the latter part of § 241, and of the facts just mentioned, the following law has been established: substances charged with like electricities repel each other, while those charged with unlike electricities attract each other. By *like electricities* is meant those having similar names, whether the terms used are positive, negative, vitreous, or resinous. Thus, a pith ball negatively electrified repels another ball charged with the same kind of electricity, and attracts one positively electrified.

244. **Electricity Everywhere Active.**—Electricity exists in all substances, each having its own capacity for it, but in the usual condition of substances the electricity is in a state of equilibrium, and therefore of quiet. We see this quiet disturbed during a thunder-storm, when we rub glass or silk, or a cat's back, or when we work an electrical machine. But the active state of electricity is not limited to such palpable demonstrations as these. Electricity is undoubtedly in action everywhere and always, although we can seldom appreciate and measure its action. Wherever there is motion the equilibrium of electricity is disturbed, and there is a consequent return

to this equilibrium. And this change from the one state to the other must be the constant cause of important changes and operations in the world around us, and in our own bodies. Let us look at some of the indications of this universality of electrical action. Friction constantly awakens it. The friction of the belts upon the drums in cotton factories develops it quite freely. Every stroke of India-rubber upon paper as you erase a pencil-mark excites electricity. The blowing of air upon glass does the same, as well as the blowing-off of steam from an engine. Electricity has been excited even upon ice by rubbing it when cooled down to −25° Centigrade. Experiments upon the air have shown that there is usually some free electricity in it, the atmosphere being generally in a positive state, especially when the air is dry and clear. It is constantly generated from one source and another. It is generated everywhere by evaporation. Every gust of wind, causing friction of the particles of the air upon various substances, generates it. Chemical action, as you will learn in another chapter, generates it everywhere. It is produced also in the operations of life, and in some animals there are special organs—electrical batteries—for the generation of this mysterious force.

245. Conductors and Non-Conductors.—Electricity passes over the surface of some substances very readily; while over others it moves with very great difficulty, and therefore very slowly and sparingly. The former are termed conductors, and the latter non-conductors. As in the case of heat, there are no perfect non-conductors. The best of all the conductors are the metals, those least liable to oxidation being the most perfect. Next come charcoal, water, living substances, flame, moist earth, ice. The best non-conductors are dry air and gases. Then come gumlac and gutta-percha, amber, resins, sulphur, glass, silk, wool, hair, feathers, cotton, paper, leather, porcelain, marble, and oils, very nearly in the order named. Non-conductors are sometimes called *insulators* (from the Latin word *insula*, an island), as they serve to confine electricity within certain bounds and prevent its escaping. Thus in the experiments with pith balls already cited, the silk threads by which they are suspended prevent the escape of electricity. The glass

knobs on which the wires of the telegraph rest are insulators, preventing the electric fluid from escaping down the
poles into the ground.

Electrics and Non-Electrics.—It will be observed, on looking over the
list of conductors and non-conductors, that among the non-conductors are
those substances in which electricity is easily excited by friction, such as
glass, amber, silk, etc. These were therefore called electrics. The conductors, on the other hand, were called non-electrics, it being supposed that
electricity could not be excited with them. But this has proved to be
incorrect. For example, if a metal be insulated by being placed on a pillar of glass or of gumlac, so that the electricity, when excited, cannot pass
off readily, its generation can be made manifest. It is probably true that
every substance is more or less an electric, it being difficult to make this
manifest in the case of conductors, because the electricity passes off as fast
as generated.

246. **Electricity Always on the Surface.**—There is a marked difference between heat and electricity in the manner in
which they are distributed. Heat pervades all the particles
of substances, and by conduction spreads through them,
while electricity in its ordinary movements operates altogether on the surface. A hollow ball, therefore, can contain
as much electricity as a solid, and a hollow conductor of
electricity is just as ef-
fectual as a solid one.
The following experi-
ment exhibits in a very
striking manner this
disposition of electric-
ity to occupy the sur-
face alone. Fig. 299
represents a hollow
metallic ball support-
ed by a glass stand,
and two metallic caps
which will just cover

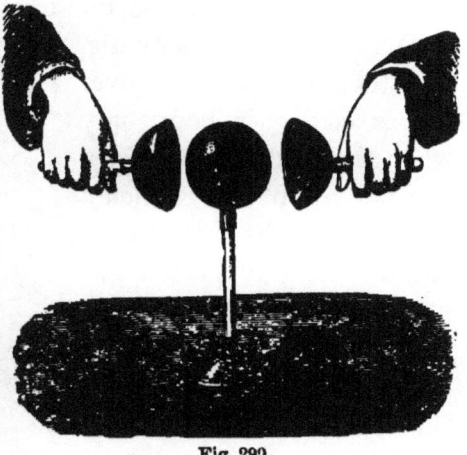

Fig. 299.

the ball, having non-conducting handles, of either glass or gumlac. Now, after having charged the ball with electricity, let the caps held by the insulating handles be carefully placed over the ball. On withdrawing them, it will be found that the electricity of the ball has all passed to the outer surface of these caps.

The illustrious Faraday devised another simple and ingenious experiment to show that electricity does not penetrate into the mass of a body. Fig. 300 represents a conical sack of woollen gauze attached to a metallic ring, which is supported on an insulating stand. A silk thread is fastened to the point of the cone by means of which the bag may be turned inside out. When the conical sack is charged with electricity, the outer surface only is found to be electrified; and if the bag be turned inside out, the exterior will again be the only portion containing electricity.

Fig. 300.

Electricity spreads or distributes itself uniformly over the surface of a body only when that body has the form of a ball or sphere. When a body having the oblong form (shown in Fig. 301) is charged with electricity, the greatest density of the fluid (borrowing a phrase from the fluid theory) is at either end. This is said to result from the re-

Fig. 301.

pulsion of the fluid in the central part of the body. The signs + and − placed at either end signify that one end is charged with positive and the other with negative electricity. This method of arrangement is sometimes called polarity. When a body smaller at one end than at the other is charged with electricity, the latter accumulates at the smaller end; and if this be a point, the density becomes so great that the electricity is forced into the air and gradually escapes. Hence, apparatus intended to confine electricity to its surface is generally provided with knobs at terminal points, and apparatus for collecting or discharging electricity— such as lightning-rods—is furnished with sharp points. To this we shall again refer in § 250.

247. Induction. — The experiments described in § 241 teach us that a body may be electrified by friction and by contact with an electrified body. Besides these two ways, however, bodies in their usual state may be electrified by influence without contact: this is known as *induction*.

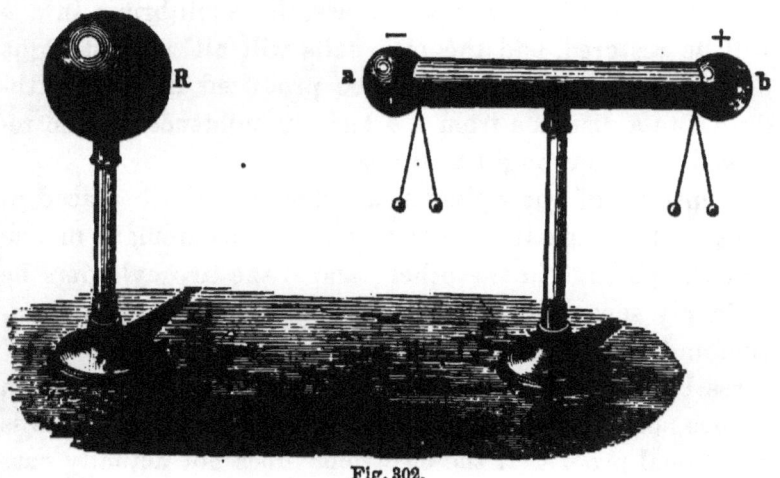

Fig. 302.

Fig. 302 will aid in illustrating this subject. R represents a metallic sphere insulated by a glass support, and charged

with positive (+) electricity; $a\,b$ represents a metallic cylinder insulated in the same manner, to which pith balls are
attached by silken threads. Now, if R be placed near $a\,b$,
but not near enough for the electric spark to pass from one
to the other, it will destroy the equilibrium of the two
electricities in $a\,b$—the negative electricity being accumulated at the end near the sphere R, and the positive at the
remote end. This is because the positive electricity in R
repels its like in $a\,b$, and attracts the unlike fluid. The two
pith balls at the positive end repel each other because they
are charged with the same electricity, and the balls at the
negative end are mutually repellent for the same reason.
But balls hung from the middle would not be affected, because they are on middle ground between the two electricities. There is no communication of electricity from R to
$a\,b$, but only an influence upon the quiescent balanced electricities of $a\,b$. Accordingly, if the surplus electricity of
R be discharged by putting the hand or any good conduct-
or upon it, the influence will cease, the equilibrium in $a\,b$
will be restored, and the pith balls will all hang straight
down. The same effect will be produced if R be withdrawn to a distance from $a\,b$, and the influence will be re-
newed if R be brought near again.

If instead of one cylinder we should use two placed in
contact, the negative electricity would accumulate in one
and the positive in the other; and if the two cylinders be
suddenly separated, one will be found charged with negative and the other with positive electricity. If a pane of
glass be held between R and $a\,b$ (Fig. 302), the induction
ensues notwithstanding the intervening insulator. This is
additional proof that the electricity does not actually pass
from one body to the other.

Electroscope.—The pith balls serve as tests of the presence of electricity;
but a far more delicate means is supplied by the simple instrument known

as the *gold-leaf electroscope*. As shown in Fig. 303, it consists of an insulated glass vessel into which a metallic rod is fastened, terminating at the exterior portion in a knob, and having two pieces of thin gold-leaf attached to the end within. A band of tinfoil is glued to the outside of the glass sphere. When an electrified body is brought near the knob of the electroscope, the gold leaves fly apart, owing to induced electricity. The amount of divergence is, to a certain extent, a measure of the force of the electric charge. When used for this purpose the instrument is called an *electrometer* (§ 248). Moreover, when a body having like electricity is brought near the electrified leaves, the divergence increases; and if a body

Fig. 303.

charged with electricity of a contrary character, the leaves fall together. Hence the electroscope may be used to determine the kind of electricity excited in any substance.

248. **The Electrical Machine.**—You are now prepared to understand the operation of the common frictional electrical machine. There are two kinds—the plate and the cylindrical; but they are both constructed on essentially the same principle. Fig. 304 represents a plate machine. P is a plate of glass which can be revolved by means of the crank, M; K and K are rubbers of silk, the pressure of which is regulated by screws. These rubbers are connected by

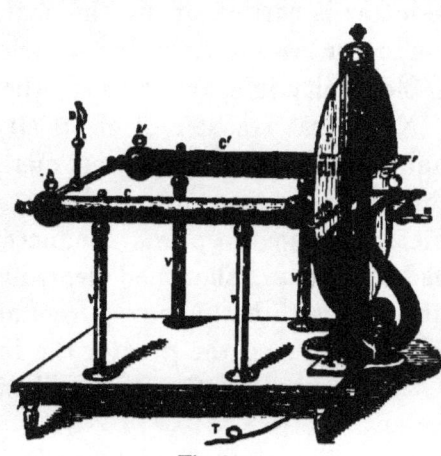

Fig. 304.

the support, m, and the chain, T, with the floor; or, in other words, with the earth. C and C' are the so-called prime conductors—hollow cylinders—of brass, insulated by the pillars of glass, V V V V. These conductors terminate at the ends next to the glass plate in rods bearing sharp points, F and F', which serve to collect the electricity excited by friction on the glass plate. Certain portions of this plate are covered with envelopes of silk, which, being non-conductors, prevent the electricity on the glass from being lost in the air, and also serve to keep the plate free from dust. The rubbers are cushions of horse-hair covered with leather, and coated with a mixture of 1 part of tin, 1 part of zinc, 2 parts of mercury, and sufficient lard—this amalgam being found very effectual in exciting electricity.

The operation of the machine depends, first, on the development of electricity by friction, and, secondly, on induction. When the glass plate is revolved, the friction of the rubbers causes positive electricity to collect upon the glass and negative upon the rubbers. The former acts by induction on the prime conductor, and the latter is carried off by the chain to the earth. Thus the conductor loses its negative electricity, and gives up positive electricity in sparks to any other conductor held near it. No actual transfer of electricity from the plate to the conductor takes place; there is merely a change in the equilibrium of the two electricities.

The intensity of the electricity on the prime conductor may be roughly determined by means of the small electrometer B (Fig. 304), consisting of a pith ball hanging from an upright rod. As with the electroscope, the greater the intensity of the electricity, the farther will the pith ball diverge from the perpendicular.

The cylinder machine is much older than the plate machine just described. The first electrical machine was constructed about 1650, by Otto von Guericke, the inventor of the

air-pump. A common form of
the cylinder machine is shown
in Fig. 305. The glass cylinder, *a a*, can be turned rapidly by the multiplying-wheel,
b b. At *c* is a piece of silk,
and on the rear part of the
cylinder is the rubber. At *d*
is the prime conductor, insu-

Fig. 305.

lated by the glass support, *e*. Its operation is precisely
similar to that of the plate machine.

249. Experiments with the Electrical Machine.—The first
thing noticeable about an electrical machine, when in full
operation, is the succession of vivid sparks which leap
with a crackling sound from the rubber to the prime conductor. If any conductor be brought near the end of the
prime conductor, the electricity darts to it at intervals with
a loud snap. These sparks passing to the hand of a person
standing by produce a keen pricking sensation, or perhaps
a painful numbness along the arm.

The Insulating-Stool. — If a person stand on a wooden
stool supported by glass legs
(Fig. 306), and
hold in his hand
a chain connect-
ed with the prime
conductor, he will
become highly

Fig. 306.

charged with electricity. His hair will
stand up on his head, and he will be able
to give electric shocks to other persons
from any part of his body. Small comic
figures charged with electricity present the
appearance shown in Fig. 307.

Fig. 307.

Q

Other Experiments.—Let a metallic plate, *a,* Fig. 308, be suspended by a chain to the prime conductor, and another plate, *b,* be supported upon a conducting stand. If figures of paper or pith be placed between these plates, as the machine is worked, they will move about briskly between the plates, being alternately attracted and repelled by the communication of the electricity.

The experiment represented in Fig. 309

Fig. 308.

is a very beautiful one. Let *a b* be a brass rod with an arch, *g,* by which it can be suspended from the end of the prime conductor. To this rod are suspended three bells, the two outer ones by chains, and the middle one by a silk thread, *c f;* also two clappers, *d* and *e,* by silk threads. The

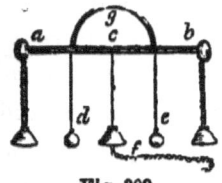

Fig. 309.

middle bell has a chain, *f,* connecting it with the table—that is, with the earth. The operation of the apparatus is as follows: As soon as the outer bells become electrified, they attract the clappers; these, on touching the bells, receive a portion of their electricity, and are repelled. They therefore strike against the middle bell, to which they impart the electricity received from the outer bells. They then swing back again in the same state that they were in at first, and are attracted again by the outer bells. This goes on so long as the electricity is communicated.

Paste upon a slip of glass a continuous line of tinfoil, going back and

Fig. 310.

forth as represented in Fig. 310, and connect a ball, G, with one end of the foil. Then make the word LIGHT upon it by cutting out with a sharp knife little portions of the foil. Placing your finger on one end of the line of foil at *a,* present the ball G to the prime conductor, and the electric fluid will run along the whole length of the line from G to *a.* In doing this the letters are beautifully illuminated, a spark being produced at each interruption of the line. So rapid is the passage of the electricity that the whole appears to the eye simultaneously illuminated.

250. Electricity Discharged from Points. — We have already (§ 246) spoken of the readiness with which electricity is received by points. It is discharged from them with equal readiness; so that if a metallic point be attached to the prime conductor, the electricity will be carried off into the air nearly as fast as received; and in passing off it creates a current. The reaction of the air upon the electrical currents can be very prettily exhibited with the apparatus represented in Fig. 311, which consists of a cap, resting upon the point of a rod, and having pointed wires branching out from it in a wheel-like arrangement. You observe that the points are all bent one way. If this apparatus be set upright upon the prime conductor, the wheel can be made to revolve rapidly by working

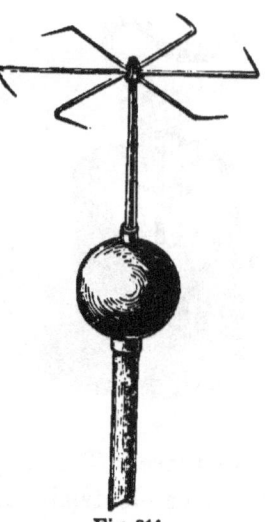

Fig. 311.

the electrical machine. In the same way that the reaction of the air against gases issuing from a rocket makes it rise, reaction against the electricity issuing from these points

Fig. 312.

causes the circular motion. If electricity be discharged from a point in a darkened room, it appears like a brush of light, as represented in Fig. 312.

251. **Leyden-Jar.**—The Leyden-jar is a contrivance for storing up or accumulating electricity. It is so named because the principle upon which it was constructed was discovered in Leyden, Holland. An experimenter endeavored to charge with electricity a vial of water: after passing many sparks into it through a brass rod placed in the vial, he took hold of the knob and received a very severe shock. Subsequent experiments showed that the charge was not in the water, and that a metallic coating within and without the glass produced better effects. A common form of the Leyden-jar is

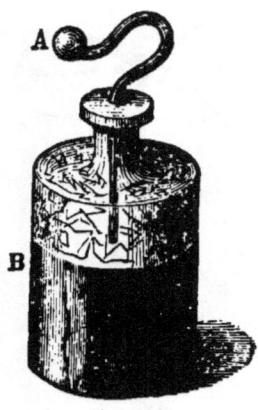

Fig. 313.

shown in Fig. 313. It consists of a glass jar, B, coated within and without to near the top with tinfoil, and having a metallic rod passing through the cork, with one end touching the inner coating, and the other surmounted by a brass ball or knob, A. The jar is charged by holding the knob near the prime conductor of an electrical machine in operation. The electricity passes by the metallic rod to the inside coating of the jar, and accumulates there. This is positive electricity. In the meantime negative electricity accumulates on the outside coating, induced by the positive electricity within. The positive and negative electricities are prevented from uniting by the non-conducting glass between the two metallic surfaces. If a slip of tinfoil were made to connect the inside foil with the outer, there would be no accumulation of electricity on the inside, for as fast as it passed from the prime conductor to the inside it would pass out over the bridge of foil to the outside, and down your arm and body to the earth.

If there were no communication of the outside with the earth, the jar would not be charged. No electricity would pass to it, because the positive electricity which is on the outside cannot be driven off, and no negative electricity can be received. To make this plain, suppose that the jar a, Fig. 314, be suspended to the prime conductor, b. By this arrangement the inside tinfoil is connected with the source of positive electricity and the outside is insulated. No electricity can pass from it or to it. It has both positive and negative electricity, but they are in equilibrium. If there were a preponderance of negative electricity there, it would attract positive electricity to it as near as possible, and the latter would enter the jar from

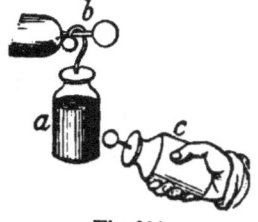

Fig. 314.

the conductor. But there is no such preponderance, and although a little may enter—a spark or two—there will not be enough to charge the jar sensibly. But bring now another jar, c, near to the outside coating of a, or connect the outer coating with the ground by means of a chain, and there is a movement at once in the electricities. The positive electricity has a chance now to pass off from the outside of a to the inside of c, leaving therefore a preponderance of negative electricity on the outside of the jar. (See § 248.)

252. **Discharge of the Leyden-Jar.**—The jar may be discharged by making a communication between the inside and outside by means of any conductor. This may be done with the discharging-rod, Fig. 315. This consists of two slender metallic rods, with brass knobs at their ends, and jointed at a, so that the knobs can be separated to different distances. The handle is made of glass, so that none of the electricity passing through the rods may be communicated to the hand. In discharging the jar one knob is placed upon the outside foil, and the other is brought near to the knob of the jar. Discharge follows, electrical equilibrium is restored, and a bright flash is produced, going from the knob of the jar to that of the discharging-rod, and this is accompanied with a report. You can yourself be the conductor to discharge the jar. If you place one hand upon the outside of the jar, and bring the other near its knob, the fluids will meet in you as they do in the discharging-rod, and a shock will be experienced in proportion to the amount of electricity in the jar. Any number of persons can simultaneously receive the same shock. To do this they must join hands, and the person at one end of the row must touch the knob of the jar while the person at the other end has his hand upon the outside.

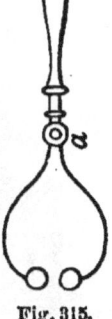

Fig. 315.

You may touch either the knob of the jar or the outside coating *separately*, and the power in it remains quiet; but the moment you touch both it bursts forth, because a bridge is made over which the electricity can pass.

In a dry air the charge in the jar can be retained for some time, the communication between the two electric fluids being very slow through the medium of air. It is otherwise when there is much moisture in the air, for water is a good conductor. For this reason, if you let the moisture from your breath come upon the jar between the outside coating and the rod, the jar will be discharged soon, though imperceptibly, the moisture making a medium of communication between the inner and outer electricities.

Electrical Battery. — By combining together a number of jars, having the insides all connected together, as

Fig. 316.

seen in Fig. 316, with metallic rods, and the outsides connected together in a similar manner, we form what is termed an electrical battery. With such an arrangement we can accumulate a large amount of electricity, which can be discharged in essentially the same way as in the case of the single jar. Persons experimenting with a large battery are obliged to exercise care in avoiding an accidental discharge of the whole electricity through the body, the effects of which are very uncomfortable, if not positively injurious.

253. **Light of Electricity.** — The light produced by electricity is not occasioned by anything like combustion. It depends obviously upon the resistance offered to its passage. Thus, when the electric fluid passes through air from the prime conductor to the knob of the Leyden-jar,

it causes a flash of light; but when it arrives at the knob, the flash ceases. What is the reason of the difference? In both cases it meets the resistance of the air, for when it reaches the knob it passes over the *surface* of the knob and rod; but in the latter case it is so diffused over the metallic surface that it meets with much less resistance from the air. By experiments with the air-pump it is found that the denser the air, the more vivid the spark; and if electricity be passed through a glass vessel exhausted of air, it forms streams of light resembling the aurora borealis, which are so strikingly in contrast with the vivid flashes of the lightning. In the experiment, § 249, in which the word LIGHT is made by the passing electricity, we have a striking illustration of the production of the spark by the resistance of the air. If the foil were one continuous surface, the electricity would be diffused over it without giving any light. It is only where the electric fluid leaps through the air from one portion of foil to another that light is seen.

254. **Other Phenomena of Electricity.**—The report of electricity is a sort of crack or snap from the sudden condensation of the air by the rapid passage of the fluid. The rolling of thunder is occasioned by the reverberation of the first sound among the clouds. The nearer the flash, the more like a crack is the first sound which reaches our ears.

Mechanical Effects.—When any great amount of electricity meets in its passage with any imperfect conductor, it does much violence to it. Thus it rends wood, scatters water, breaks glass, etc. Various experiments illustrate the manner in which mechanical effects result from electricity. Thus, if it be passed through a card or several leaves closely pressed together, a burr forms on each side of such a character as to show that two forces moving in opposite directions have forced their passage.

Production of Heat.—Electricity always produces in its

passage a certain amount of heat, probably by its me. chanical effect. When diffused over a large conducting surface, the heat is not noticeable ; but if confined to the surface of a small wire, the heat may be sufficient to melt or even burn it. Various effects can be produced by the heat thus caused by the passage of electricity. Gunpowder may be exploded by it. Alcohol and ether may be readily ignited by it, especially the latter. Gas can sometimes be lighted by touching with the finger an opened burner after walking across the room two or three times briskly, rubbing the feet upon a thick carpet.

The convertibility of the three forces—heat, light, and electricity — is illustrated by the phenomena above described ; they are all modes of motion. (See § 165.)

255. **Franklin's Discovery.** — It had very early been conjectured that the electricity produced by the electrical machine is identical with lightning; but it was reserved for our countryman Benjamin Franklin to prove the fact. He thought of making use in his investigations of a tall spire which was erecting in Philadelphia (in 1752), but before it was completed the sight of a boy's kite in the air suggested to him another plan. He made a kite by stretching a silk handkerchief over a frame, and raised it during an approaching thunder-shower, his only companion being his son. Having raised the kite, he attached to the end of the hempen string a key, and also a silk ribbon, by which he insulated his apparatus, as seen in Fig. 317. He then watched with much anxiety the result. A cloud arose, which he supposed, from its appearance, was well charged with electricity, and yet no effect was seen. Franklin began to despair; but at length he saw some loose fibres of the hempen string bristling up, and, applying his knuckle to the key, received just such a spark as he had often received from the conductor of an electrical machine. The discovery was accomplished, and Franklin was at once overcome with emotion at the thought of the immortality which it would give his name. The fame of the discovery, made in a manner so simple and yet so original, spread everywhere, and prompted to many experiments by other philosophers. One, Professor Richman, of St. Petersburg, fell a victim to his investigations. While attending a meeting of the Academy of Sciences he heard the sound of

Fig. 317.

distant thunder, and hastened home to make some observations with an apparatus which he had erected. While doing this a charge of electricity flashed from the conducting-rod, and, piercing his head, killed him instantly. His assistant, who stood near, was struck down, and remained senseless for some time, and the door of the room was torn from its hinges.

256. **Lightning-Rods.**—The discovery of Franklin led to the custom of attaching lightning-rods to buildings. The object of a lightning-rod is to conduct any electricity in a cloud that may come over the building down into the ground. For this purpose the rod should terminate in the air in points, as these readily receive the electric fluid (§ 250). The rod should be separated from the house by glass supports, and it should pass so far into the ground as to terminate in moist earth. The points should be gilded in order to be preserved from corrosion; or they may be made of silver or platinum. Lightning-rods are undoubtedly often of

Q 2

service when there is no obvious passage of the lightning down them, by quietly and continuously receiving electricity upon their points, and passing it down into the earth.

QUESTIONS.

241. What is the origin of the term electricity? What is nature? Name the simple apparatus necessary to illustrate experimentally the fundamental facts of electricity. Describe the experiments with the glass rod and silk handkerchief. How is attraction shown? How repulsion? What is meant by an electrified body? Describe the experiment with the poker, wine-glasses, etc. Describe the remaining experiments.—242. What theories have been proposed to account for electrical phenomena? Give Franklin's theory. State the two-fluid theory. What is meant by resinous and vitreous electricity? What by negative and positive? What is the modern theory?—243. What substances generate positive electricity? What ones negative? State the law as to like and unlike electricities.—244. Give in full what is said of the active presence of electricity.—245. What is meant by conductors and non-conductors? Name the best. What are insulators, and why so called? What is said of electrics and non-electrics? Is this a proper means of classification?—246. Describe an experiment showing the manner in which electricity is disposed in a body. Describe Faraday's experiment. How does electricity distribute itself over a sphere? How over an oblong body? What is the advantage of knobs to apparatus? What of points?—247. What is meant by induction? Explain in full the manner in which an electrified body induces electricity in another. What is an electroscope? Describe one form, and show how it may be used.—248. Describe the plate electrical machine, stating the use of each part. How is electricity generated by it? Describe the cylinder machine.—249. Describe some experiments with the electrical machine. What is an insulating stool? Explain the action of the electric bells. Describe the experiment showing the light of electricity.—250. What is said of the escape of electricity from points?—251. Describe the Leyden-jar. How was it discovered? What is its use? Show why an insulated Leyden-jar cannot be charged with electricity.—252. Explain the use of the discharging-rod. How can a large number of persons take a shock simultaneously? Explain the effect of moisture upon the charged jar. What is the electrical battery?—253. What is said of the light produced by electricity?—254. To what is the report of electricity owing? What is said

of mechanical injuries caused by electricity? What of the heat caused by it? What effects may be produced by this heat?—255. What was the discovery of Franklin, and how did he make it? Relate the accident which occurred at St. Petersburg.—256. What is said of lightning-rods? How. should they be constructed?

CHAPTER XIX.

GALVANISM.

257. Galvanic or Voltaic Electricity.—In the preceding chapter you learned something of the effects of electricity generated by friction. There is another form of electricity, commonly called Galvanic or Voltaic Electricity, produced by chemical action. Frictional electricity, as we shall now call that form previously studied, was recognized as a peculiar force more than six hundred years before Christ; but galvanism was not discovered until the end of the last century. The history of its discovery is interesting. The dawn of galvanism is found in an observation made by Sulzer, a citizen of Berlin, in 1767. He states that if a piece of zinc be placed under the tongue, and a piece of silver upon it, when the two metals are brought in contact a metallic taste is perceived, and a shock felt by the tongue. Sulzer attributed the effect to some vibratory motion occasioned by the contact of the metals, and, satisfied with this fanciful explanation, pursued the inquiry no further. The statement excited but little notice until other facts of a similar character were brought out in 1790 by Galvani, professor of anatomy at Bologna. He observed that the legs of some frogs, which had been obtained for his invalid wife, were convulsed by touching the nerves with a knife when near an excited electrical machine. In contrast with the example of Sulzer he was led to pursue the matter further. He found that similar muscular contractions took place even when no electricity was communicated from the machine. Having suspended some partially dissected frogs by copper hooks from the iron bar of the balcony of his window, he observed that whenever, from accidental causes, the muscles of the frogs' legs came into contact with the iron bar, the limbs were convulsed much in the same way as with frictional electricity. In consequence of this observation Galvani made the following experiment, which has since become famous. He cut a frog directly in two, skinned its hind-legs, and then arranged pieces of zinc and copper, as shown in Fig. 318. The zinc is in contact with

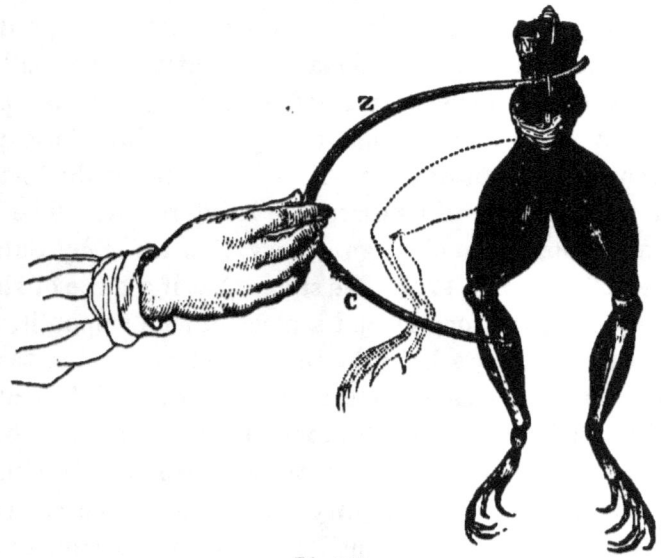

Fig. 318.

the nerves which connect the legs and the spine, and the copper is in contact with the muscles of the legs. Now, whenever the two ends of the metal strips are made to touch each other, the muscles are strongly convulsed, and

each leg is thrown up into the position shown by the
dotted lines. Galvani erroneously supposed this to be an
exhibition of animal electricity, regarding the muscles as
a kind of Leyden-jar, the nerve being the medium of communication with the inside.

258. **Volta's Pile.**—The observations of Galvani awakened
much interest in all scientific minds, and, of course, there
was much inquiry, observation, and experiment. Professor
Volta, of Pavia, went a step further than Galvani towards
the true explanation, in referring the effects to the contact of dissimilar metals; and he was led by this view of
the subject to construct his *pile*, or battery—called after
him the voltaic pile—the object of which is to produce a
much greater amount of electricity than can be obtained
by the contact of only two pieces of metal. The pile is
made of circular pieces of copper, zinc,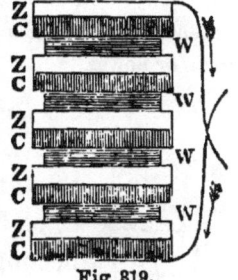
and cloth, the cloth being moistened
with salt-water or a weak acid. They
are arranged as represented in Fig. 319.
First a disk of copper is laid down, then
upon this one of zinc, then one of cloth,
and so on in the same order, the top of
the pile ending in a plate of zinc. If you
touch one end of the pile with a moist-

Fig. 319.

ened finger, and the other end with a finger of the other
hand, you will feel a shock somewhat like that from a Leyden-jar. The communication between the two ends of the
pile may be made by wires, as in the figure. That connected with the zinc disk furnishes positive electricity,
and that attached to the copper disk, negative. When the
ends of these wires are joined, sparks of galvanic electricity pass between them, the intensity being in proportion to
the number of *couples*, as the pairs of zinc and copper are
called. This arrangement is essentially the same as in the

experiments of Sulzer and of Galvani, the cloth taking the place of the tongue in the one, and of the frog's flesh in the other—moisture being present in each case.

Volta's explanation of this development of electricity—the so-called "contact theory"—was long received as the true one; but it has gradually been abandoned, and the production of electricity is now believed to depend upon chemical action.

To distinguish this form of electrical force from frictional electricity, it is often called *Galvanism*, after its illustrious discoverer. It has also been called *Voltaism*, after Volta, who added greatly to the discoveries concerning this wonderful agency. Another term often applied to it is *dynamical* electricity as opposed to *statical* electricity, a name given to frictional electricity. Statical signifies *stationary*, because the electricity accumulated by friction or other means remains stationary or at rest until some way be provided for its escape. It exerts no force until this is done. Dynamical electricity, on the other hand, is in motion or action at the instant of its production, the word dynamical being derived from a Greek word meaning force or power.

259. Voltaic Circle.—If a bar or slip of zinc, Z, Fig. 320,

Fig. 320.

and a bar of copper, C, be placed in a mixture of water and sulphuric acid, and a metallic rod, *a*, be laid across the top of them, the liquid will become turbid, and a current of electricity will flow through the liquid and across the metallic rod in the direction indicated by the arrows. If the rod be cautiously lifted from either the zinc or the copper, a spark will pass at the moment that the contact is broken, which will be visible if the room be darkened. If the rod be not placed upon the bars, the electrical action excited in the first moment soon ceases, and there will be no current, for the circuit must be perfect to obtain this result. If a rod of glass be used instead of the metallic rod, the circuit will be incomplete—glass being a non-conductor of the electric fluid—and therefore no effect

will be produced. The simple apparatus represented in Fig. 320 contains all the essentials
of what is called the voltaic or galvanic circle.
It may be made even more simple than this,
as shown in Fig. 321, the ends of the slips of
zinc and copper which are out of the liquid

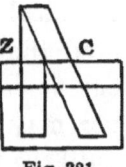

Fig. 321.

being made to touch. Commonly in galvanic experiments
the connection between the two metals is
made by metallic wires, as seen in Fig. 322.
Copper wire is ordinarily used. Platinum
is often employed, because in experiments
it is sometimes necessary to introduce the
ends of the wires into corrosive liquids.
The arrangement indicated is called the
simple voltaic circle. The *compound* circle will be described shortly.

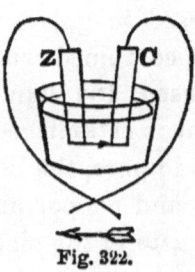

Fig. 322.

The electricity produced as described above is the result of a chemical action between the liquid and the
zinc. Through the agency of the acid, water is decomposed into its constituents—oxygen and hydrogen,* the
oxygen uniting with the zinc to form an oxide of zinc, and
the hydrogen going to the copper and appearing there
in bubbles, which escape into the air. Just as fast as the
oxide of zinc is formed the sulphuric acid combines with
it, forming sulphate of zinc, which remains dissolved in the
water. This is quite an essential part of the operation, for
if the oxide were not removed, it would very soon coat
over the zinc, and so protect it from the liquid as to stop
the chemical action. The copper acts simply as a conduct-
or of the electricity developed by the oxidation of the zinc.

Although the electricity cannot be produced unless this

---

* For explanation of these and other chemical terms, see Part II.,
Chemistry.

chemical action take place, and therefore may be considered the result of it, yet the chemical action will not occur unless the circuit be complete. The zinc and the copper may stand in the liquid, but everything will be quiet until the two metals are connected. The electrical passage or way must be opened, or there will be no chemical action to produce the electricity.

260. **Amalgamation of Zinc.**—The zinc used in galvanic apparatus is usually amalgamated—that is, combined with mercury. When perfectly pure zinc is used, the liquid does not attack it until the electrical current is established. But since this metal, as usually obtained, is impure, the liquid does act upon it even before the zinc and copper are connected, and therefore there is a useless waste of the zinc. There is waste in another way. The particles of other metals mingled with the zinc occasion chemical and electrical action when the acid liquid reaches them. There is, therefore, a local production of electricity at these points, which is disposed of just where it is produced, and does not pass to the conducting wires. It is to prevent this *local* chemical and electrical action that the zinc is amalgamated, the acid liquid having little or no effect upon the amalgamated zinc until the circuit is completed.

Sheathing of Vessels.—In the copper-zinc battery the copper is not attacked by the acid liquid, the presence of the oxidizable zinc acting as a protection. Knowing this, Sir Humphry Davy suggested that pieces of zinc connected here and there with the copper sheathing of vessels would prevent it being corroded by the sea-water. The experiment was tried and was successful; the copper remained bright and clean. At the same time, however, the plan was abandoned, because the uncorroded copper became covered with shell-fish and marine plants, the accumulation of which seriously impeded the progress of the vessel through the water.

261. **Polarity in Galvanism.**—Polarity is manifested in

galvanism, as indicated by Fig. 323. Positive electricity "flows" from the immersed
end of the zinc marked + (*plus*), while the
immersed end of the copper is negative,
marked — (*minus*). The ends in the air exhibit polarity the reverse of those in the liquid.
The terminal wires of a battery are called the
poles, being positive or negative, according

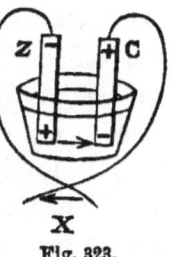

Fig. 323.

to the metals with which they are connected. The terms
"fluid" and "current" are very generally used, but their
use must not be considered as implying that there is any
real fluid flowing in a current. These and other kindred
terms are convenient in indicating processes and results,
and are used solely on this account, and not as denoting a
belief in the existence of a fluid. Electricity is a *force* of
matter, but the nature of this force is not known. That it
moves in circuits and manifests polarity is obvious to every
one; and hence the terms fluid and current, and positive
and negative electricities, may be used in a somewhat fig-
urative manner without the violation of truth.

The terms electro-positive and electro-negative may be
here defined. We explained in § 259 that in the decomposition of the acid liquid the oxygen goes to the zinc and
the hydrogen to the copper. Now, as shown in Fig. 323,
the immersed end of the zinc is positive, and that of the
copper is negative; and, since attraction is between *opposite* electricities (§ 242), the oxygen, which is attracted by
the positive zinc, is called electro-negative; while the hydrogen, which is attracted by the negative copper, is called
electro-positive. And all substances are put into one class
or the other, according to their galvanic affinities. (See
§ 264.)

The polarity of galvanism will be further illustrated
when we speak of the effects of this agent.

262. Galvanism Produced in Various Ways.—We have thus far mentioned only one way of producing galvanism, but the means of producing it may be very greatly varied. Other metals can be substituted for copper in the voltaic circle, provided they are less oxidizable than zinc. Thus zinc and lead will answer, but by no means so well as zinc and copper. Zinc and platinum are more active in producing galvanism than zinc and copper, though the latter combination is more generally used because copper is so much cheaper than platinum. Other acids besides sulphuric can be used, as hydrochloric, etc. Other substances besides metals are competent to produce galvanic electricity. Even vegetable and animal substances have been used with success, showing how universal galvanism is as an agent in nature. In all cases the positive electricity comes from the substance most readily acted upon by the fluid used, and the negative from the other. Though galvanic electricity is ordinarily produced by an arrangement of two solids in a fluid, it may be produced by a solid with two fluids. Thus, when a plate of zinc is brought into contact with salt-water on one side, and hydrochloric acid on the other, a current of electricity will flow from the former to the latter liquid.

263. Volta's "Crown of Cups."—Soon after Volta invented his pile he made the first voltaic battery—known as the "crown of cups." It is essentially the same as the pile, being the combination of a number of simple voltaic circles, and hence sometimes called a *compound voltaic circle*. It is represented in Fig. 324. The cups contain a weak acid or a saline liquid, and in each is placed a plate of zinc and one of copper, the arrangement being such that the copper plate of the first cup is connected by a wire with the zinc plate of the second, and the same with the second and third, and so through the series, however long it may

Fig. 324.

be. The positive electricity flows in each cup from the zinc to the copper, then onward by the connecting slip of metal to the zinc of the next cup, and so on until it passes out by the wire connected with the copper in the last cup to meet the electricity coming from the opposite direction. The course of the positive electricity is indicated by the arrows.

By thus uniting a number of single cups a far greater quantity of electricity is generated, and very remarkable effects can be produced. A modification of this battery is shown in Figs. 325 and 326. It consists of a long wooden

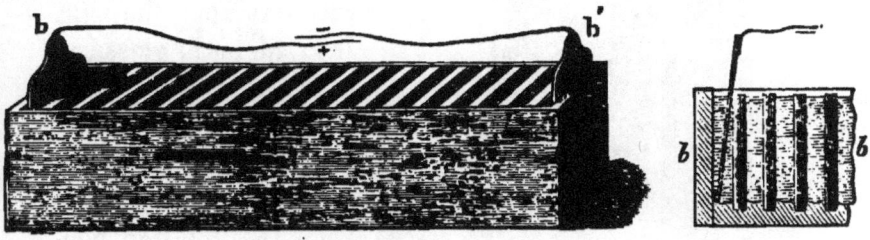

Fig. 325.　　　　　　　　　　　　　　　　Fig. 326.

trough divided into narrow cells by pairs of square plates of zinc and copper soldered together. Into these cells a saline solution is poured. Wires connected with either end serve to conduct the voltaic current. This form of battery was found more convenient than the "crown of cups," on account of its portability.

264. Other Batteries.—Batteries have been varied much, both as regards the material and the arrangement. Some of these forms we shall briefly describe. You learned in § 262 that electricity may be generated by contact of a solid with two fluids; it is also commonly produced by a peculiar arrangement of two solids and two fluids. A battery in which only one fluid is used soon loses its power, or runs down—as it may be expressed; but those in which two fluids are used are more reliable and constant, as well as far more powerful. One of these forms, known as Daniell's battery, is represented in Fig. 327. The external glass

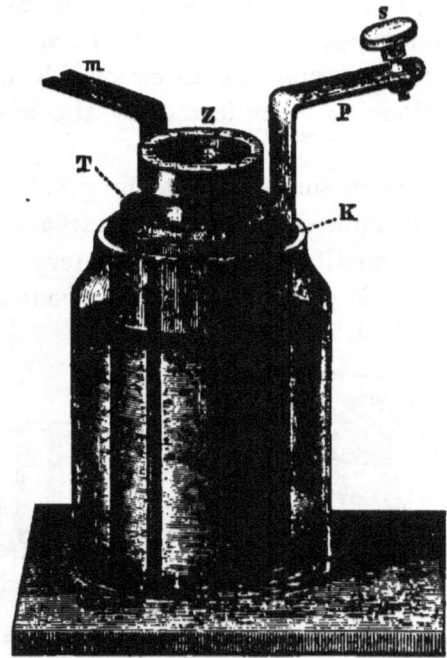

Fig. 327.

cup contains a cylinder of copper, K, within which is another cylinder of unglazed porcelain, T, which latter contains a rod or cylinder of cast zinc, Z. Dilute sulphuric acid poured into the porous cell acts upon the zinc. Into the outer glass cup is poured a strong solution of sulphate of copper, some of the solid crystals being added to keep the solution concentrated. For the convenience of connection with wires or with other similar batteries, the metallic strips, m and p, are fastened to the zinc and the copper cylinders respectively; the screw, s, assists in joining two such cups. By this compound arrangement the porous cell prevents the passage

of metallic particles between the copper and the zinc; and
the deposition of copper, resulting from the decomposition
of the sulphate, keeps the copper cylinder clean and always ready for action. Daniell's battery generates galvanism with great uniformity and constancy, and in its
varied forms has been much used for telegraphic purposes.
By combining a number of these cups by the aid of the metallic strips and screws, a very powerful battery is obtained. The copper in each cup is connected with the zinc
of the next, as shown in Fig. 328. The negative and positive poles are marked *n* and *p* respectively.

Fig. 328.

The batteries of Grove and Bunsen are very important
on account of their superior power. Bunsen's cup or cell
differs from Daniell's, both in the material and arrangement. The copper is replaced by a cylinder of coke or carbon, and the sulphate of copper solution by strong nitric
acid. The outer cylinder is of zinc; within is a porous
cell containing the carbon and filled with nitric acid. The
latter serves to remove the hydrogen liberated at the car-

bon, or positive, end, and prevents the battery from losing its power so rapidly as it otherwise would.

Grove's battery is similar in all respects to Bunsen's, except that platinum is substituted for carbon; and, although this metal is quite expensive at the outset, so great is the efficiency that it is economical in the end.

The metals used in these batteries, as well as the carbon, are not selected for the purpose at hazard, but in accordance with a well-defined law—viz., the greater the difference in the oxidability of the metals taken, the stronger the electric current obtained. The order of oxidability of the principal metals is as follows:

1. Zinc.	7. Antimony.
2. Tin.	8. Copper.
3. Lead.	9. Silver.
4. Iron.	10. Gold.
5. Nickel.	11. Platinum.
6. Bismuth.	12. Carbon (a non-metal).

The most readily oxidized by immersion in dilute acid is the zinc; the least, carbon. Any two of these metals placed in an acid liquid will produce a current of electricity. From the law just given, a zinc-copper couple will be much stronger than a zinc-iron one. The cause, too, of the efficiency of the zinc-carbon (Bunsen's) and zinc-platinum (Grove's) batteries is evident. The order of the metals depends upon the nature of the exciting liquid.

Batteries are arranged differently for different purposes; the decomposing power or intensity of galvanism depends upon the number of the plates, while its quantity and power of generating heat and magnetism (Chapter XX) depend upon their extent of surface. When great heat is wanted, all the zinc plates are connected together, so as in effect to form one great plate, and the same is done with the copper plates. This is quite different from the arrangement represented in Fig. 328, which is calculated for the production of chemical decomposition and for giving shocks.

265. **Characteristics of Voltaic Electricity.**—Several pecu-

liarities distinguish voltaic electricity from that developed
by friction. Frictional electricity is much more intense,
and that produced by chemical action is greater in quantity; the latter flows in a current as fast as produced,
while the former is discharged with a kind of explosion.
The difference has been compared to that between a fire
running through a long train of gunpowder and the firing
of a mass of it at once. Another peculiarity of voltaic electricity is that, unlike frictional electricity, it will not pass
through non-conductors. Thus you can handle the wires
used for communication without receiving a shock; but
this is impossible with frictional electricity. This property
makes voltaic electricity eminently suitable for telegraphing. As it flows with a steady current, its action can be controlled at the receiving station; and it pertinaciously adheres to the wires until it reaches its destination. The rain
may patter upon the wire without sensibly diminishing the
electricity; birds may rest upon it and receive no shock.
The principal effects of galvanism are the production of
heat and light, as well as of magnetism (Chapter XX), and
the decomposition of chemical compounds. Besides these,
it has a peculiar effect on the nervous system of animals.
These effects we will now describe in the order named.

266. **Heating Effects of Galvanism.**—Galvanism is capable
of producing the most intense heat known. By means of
a battery comprising a number of large cups, many of the
most refractory substances known may be fused and even
volatilized. Galvanic electricity, unlike frictional, does not
run over the surface of conductors; and, consequently, if
the terminal wires of a large battery be connected by a
wire of very small diameter, the resistance offered by the
latter to the passage of the electrical current causes great
heat; the wire becomes red-hot, and may even melt. If
two leaves of gold are suspended at the ends of two rods

Fig. 329.

in a jar, Fig. 329, and the rods connected with the terminal wires, or *poles*, of
a large battery, they will move towards
each other until they touch, and then
melt, or possibly burn up with a brilliant flash. This experiment, contrived
by Sir Humphry Davy, also shows that
galvanism as well as frictional electricity causes attraction (§ 241).

That this attraction is really the effect of galvanism may be shown by
suddenly cutting off the connection
between the gold leaves and the battery before they come
in contact; they will at once fall back to their original
position.

Leaves of various metals may be burned by the same force, each metal
giving a different color to the flame. If you place a piece of silver on a
lump of charcoal connected with one pole of a battery, and bring the other
pole armed with a charcoal pencil in contact with it, the silver will burn
most brilliantly, as seen in Fig. 330. The burning of iron by galvanic electricity is very splendid. Take a
glass partly filled with mercury,
and, immersing one pole of a battery, bring a thin piece of steel or
iron connected with the other pole
in contact with the mercury, and
the iron will take fire, throwing out
bright sparks, the mercury being
rapidly volatilized by the great

Fig. 330.

heat produced. Almost every substance can be fused, burned, or volatilized
between the poles of a powerful battery. Even platinum, which can withstand the heat of the hottest furnace, can be fused. The heat thus produced has been applied in firing gunpowder in blasts, and this is sometimes
done even under water. The powder, enclosed in a vessel, has a fine platinum wire running through it, which is connected at its two ends with two
wires that extend to the galvanic apparatus. The battery being put in
operation, the moment the connection is established the platinum becomes

red-hot and fires the powder. The battery may be at a great distance from the powder, and yet the requisite effect is produced.

267. Production of Light.—Whenever the heat produced by galvanism becomes very intense, it is accompanied by light, as shown in the preceding section.

Generally speaking, the production of light is the result of chemical action, as will be shown in Part II., Chemistry. But galvanism may produce light without any chemical change taking place at the point where the light is seen. For example, if points of charcoal be fastened to the ends of the two wires of a powerful galvanic battery, a splendid light will appear when the charcoal points are brought in contact. The charcoal does not burn up, but is merely mechanically transferred from one pole to the other in an incandescent state. That this light is not the result of a combustion of the charcoal may be proved in two ways. First, the moment that the connection is broken the light goes out, and there is no glowing of the charcoal, as there would be if it had been set on fire. Then, again, the light can be produced in a vacuum, as represented in Fig. 331. Here the light is in a vessel which has been exhausted by the air-pump, and there can be, of course, no combustion without air. The same splendid light can be obtained with the charcoal points under water. Although there is no combustion of the charcoal, nevertheless the electric light is indirectly the result of combustion. As will be explained in Part II., Chemistry, the oxidation of the zinc in the battery is really a combustion. These facts illustrate

Fig. 331.

R

again the close connection between the forces of heat, light, electricity, and chemical attraction, to which we have already referred (§§ 165 and 212). The chemical action in the battery cup generates electricity, which itself produces both heat and light. The transmission of force is very wonderful.

268. **Chemical Effects.**—Galvanic electricity not only originates in chemical action, but it also produces chemical action between its poles. A great variety of compound substances can be decomposed, and simple substances can be made to unite and form compounds. In the production of the electrical light there is no chemical effect, for there is neither decomposition nor composition. In fusing metals or volatilizing them, there is no chemical action, for they are not made to enter into combination with anything, but remain simple elements, having their form alone changed. But when any metal is actually burned, there is chemical action, for the metal unites with the oxygen of the air. In the fusing of platinum there is no chemical action, for it is not oxidized. Mercury is so volatile that it merely flies off in vapor, and is condensed again in liquid form in the cool air.

Most interesting results follow when water is submitted to the action of voltaic electricity. This liquid, usually regarded as a simple substance, is actually a compound of two gases, oxygen and hydrogen, in definite proportions, both as regards weight and volume. It is readily decomposed into these gases by a battery of two cups constructed on Grove's or Bunsen's plan. Fig. 332 shows a form of apparatus which permits the gases to be collected separately. Through the bottom of a glass vessel are introduced, water-tight, two platinum wires, $a\,b$ and $a\,c$. Over each of these wires a tube, with its upper end closed, is placed. The tubes and the dish are filled with water, which is slightly acidulated in order to make it a better conductor.

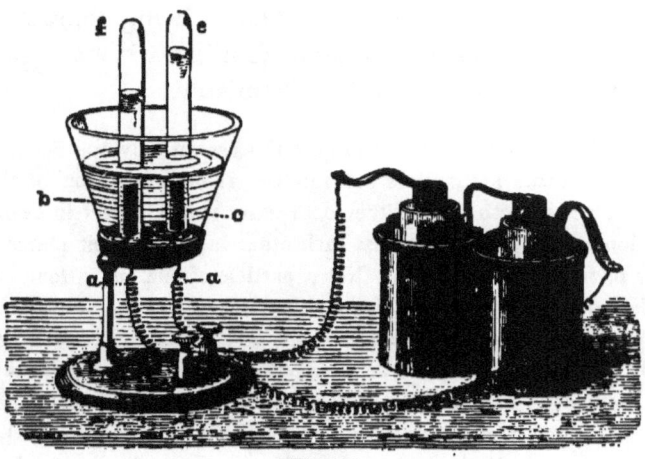

Fig. 332.

If the wire $a\,c$ be connected with the positive pole of a battery, and the wire $a\,b$ with its negative pole, some of the water will be decomposed, and the resulting gases, oxygen and hydrogen, will collect in the tubes, e and f respectively, driving the water down in them. Twice as much gas accumulates in f as in e, water being composed of two volumes of hydrogen and one of oxygen. As the decomposition of the water proceeds, the hydrogen, being electropositive, goes to the negative pole, while the oxygen, being electro-negative, goes to the positive pole. This is in accordance with the law that substances charged with opposite electricities attract each other (§ 243). After a sufficient amount of the gases has collected, cautiously remove the tube e, closing its mouth with your finger, and, turning it upside down, introduce into it a slip of wood with a spark on its end—the wood will burst into a flame, showing that the gas is oxygen. If you then remove f, and apply a light to its mouth, the gas will ignite and burn with a pale flame. Or if you mingle with it an equal quantity of atmospheric air, and then apply the light, an

explosion will ensue—these phenomena being characteristic
of hydrogen. You will become familiar with the properties of these gases in Part II., Chemistry.

Explanation. — Let us now examine the reason why the gases in the
above experiments accumulate separately. If decomposition of the water took place at both poles, either both gases would collect in each tube,
or portions of the gases must pass each other in taking their places. But
neither of these results occurs. Not a particle of one gas is found mixed
with the other—one tube contains pure hydrogen and the other pure oxygen. And there is no appearance of any bubbles of gas passing between
the poles. Though the gases collect briskly, the water is perfectly quiet,
which would not be the case if gases were passing through it. These
things being so, we must conclude that all the oxygen is produced at the
positive pole, and all the hydrogen at the negative. Let us see how this
can be. Suppose the decomposition begins in a particle of water at the
positive pole, an atom of oxygen uniting with the zinc. This will, of
course, leave two atoms of hydrogen free. Now, what becomes of the
atoms of hydrogen? Do they travel over to the other pole? No. It is
supposed that they seize an atom of oxygen from the adjoining molecule of
water, forming with it a particle of water. This sets two more atoms of
hydrogen free, which take an atom of oxygen from the next molecule of
water, and so on through a chain of particles extending to the negative
pole. You see what the result will be at this pole. The last atoms of hydrogen set free have no particle of water to take oxygen from, and hence
the gas collects in the tube.

269. **Electrolysis.** — A great many other substances besides water can be decomposed by galvanism, each of the
constituents appearing at the pole opposite to it in electrical character. Thus oxygen, chlorine, iodine, sulphur,
and the various acids, being electro-negative, are attracted
by the positive pole; while hydrogen, potassium, sodium,
copper, silver, lead, and the various bases are attracted by
the negative pole. It was by galvanism that Sir Humphry
Davy made his grand discovery that the alkalies and earths
are oxides of metals. Great must have been the joy of the
discoverer when, as he subjected the potash to the action

of the battery, he saw the metallic globules of potassium appear, turning the conjecture which had so long burdened his mind into reality in an instant. The process of decomposing chemical substances by voltaic electricity has received the name *electrolysis*.

270. **Electrotyping.** — Solutions of chemical compounds are also decomposed by galvanism; sulphate of copper, for example, submitted to the action of an electric current is separated into its constituents, and metallic copper is deposited at one pole (negative) of the battery. This has given rise to an important branch of applied science called *electro-metallurgy*, whereby one metal may be coated with a uniform and brilliant film of another (plating) or a fac-simile of a metallic mould may be obtained. The latter process is termed *electrotyping*. The arrangement of the battery and bath is represented in Fig. 333. The glass trough contains a solution of sulphate of copper, and the galvanic battery is of the form known as Smee's, in which platinized silver is used in place of copper. Suppose we wish to obtain a copy in copper of one of the faces of a medal. Covering with wax all the medal except that part which is to be copied, we

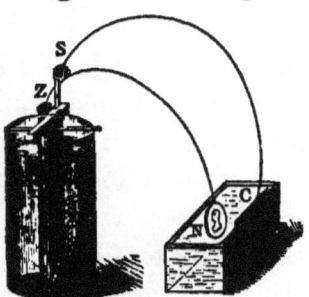

Fig. 333.

immerse it in the liquid attached to the negative wire, Z, and attach a piece of copper, C, to the positive wire, S. Observe now what takes place. The sulphate of copper is decomposed by the voltaic current, the copper going to the negative pole (§ 269), and therefore being deposited on the medal, N. The sulphuric acid, on the other hand, goes to the positive pole, and, finding copper there, unites with it, forming sulphate of copper. This explains the use of the lump of copper. It is for the purpose of keeping a good supply of sulphate of copper in the liquid. We obtain in this way, however, only a *mould* of the medal; but by pursuing the same course with the mould we get the fac-simile. So in obtaining a fac-simile of a page of printer's type or of an engraving a mould is first made. This is commonly done with plaster of Paris or wax, and then the mould is electrotyped. We get in this way a thin coat of copper showing distinctly every line of the types or the engraving. Then to fit this for use melted lead is poured into the back of it, so as to make the thin coat a firm plate. Since the wax and the plaster are not good conductors, we are obliged to coat the surface of the mould with finely powdered graphite in order to secure the deposit of the copper.

271. **Gilding and Silver-Plating.** — Common metals may be plated with silver, or silver and other metals may be gilded, by this process of electrotyping. Suppose, for example, that a silver spoon is to be gilded. It is attached to the negative pole of a battery, and immersed in a solution of chloride of gold, while a plate of gold is attached to the positive pole. The chloride is decomposed, and as fast as the gold is deposited upon the silver the chlorine set free unites with some of the gold plate, thus keeping the solution of the chloride of uniform strength. The chlorine is the carrier of the gold to the pole connected with the silver in the same way that the sulphuric acid is the carrier of the copper in the process of copying medals described in § 270.

The arrangement of a bath for silver-plating a number of articles at one operation is shown in Fig. 334. The trough C contains a solution of cyanide of silver, in which are suspended the silver plates o′ o and o′ o, connected with the positive pole of the battery, E E, and the pitchers and other articles connected with the negative pole. The operation proceeds exactly as with the gold-plating above described. Electroplating, as illustrated in the processes men-

Fig. 334.

tioned and in others, is one of the most beautiful and valuable presents which science has made to the arts.

272. **Physiological Effects of Galvanism.**—The muscular
contractions produced by galvanism have been already alluded to in speaking of Sulzer's and Galvani's experiments.
The force of the contraction depends upon the number of
the pairs of plates, and not upon their size. The shock
received from a galvanic battery is not like the sharp and
instantaneous one given by frictional electricity, but it is
more of a continued sensation. It is felt only at the moment when contact is made or broken, a steady, continuous
current being maintained so long as the connection exists.
With a battery of some hundred couples, as the pairs are
termed, the shock is painfully severe, and may be even fatal. As muscular contractions were produced in Galvani's
dead frogs, so can they be produced in the human subject
after death. Because, although the body may be dead as a
whole, as a *system* of organs, some of the properties of life
still remain in some of its parts. The irritability of the
muscles is such a property, and it is through this that the
culprit who has been hanged can be galvanized into apparent life—the countenance exhibiting frightful contortions,
and the limbs being thrown violently about. Galvanism

has also been used successfully in cases where animation was suspended, but not destroyed; the muscles of respiration and the heart, which had ceased to act, being awakened again into action by the stimulus of the electric current. Physicians sometimes employ galvanism in certain diseases with beneficial effects.

QUESTIONS.

257. What is said of the history of galvanism? What was Sulzer's observation? Describe Galvani's experiment. What was his explanation? —258. Describe the construction of Volta's pile. How does it operate? What was Volta's explanation? What is said of the different names given to galvanism?—259. Describe the action of the voltaic circle. Explain the chemistry of this action.—260. What is the advantage of amalgamating the zinc? What is said of Davy's plan of protecting the sheathing of vessels? —261. What is said of polarity in galvanism? What is meant by the terms electro-negative and electro-positive?—262. In what other ways may galvanism be produced?—263. Describe Volta's "Crown of Cups." How does it operate in generating galvanism? Describe the modification known as the trough battery. — 264. What is said of other forms of batteries? Describe Daniell's. Also Grove's and Bunsen's. What determines the choice of metals used in a battery? Upon what does the intensity and the quantity of galvanism depend?—265. What are the chief characteristics of galvanism? What are its principal effects?—266. Give examples of the heating effects of galvanism. Describe Sir Humphry Davy's experiment. Show how silver may be burned. How may steel be burned by the heat of galvanism? What is said of firing gunpowder by electricity?— 267. What is said of the production of light? What proof is given that the light does not result from combustion? Where, however, does the combustion really take place?—268. Mention some of the chemical effects of voltaic electricity. Describe the apparatus used for the decomposition of water. How does the decomposition proceed? How are the gases examined? Explain why the gases accumulate at separate points. What substances are attracted to the positive pole? What to the negative pole? —269. What remarkable discoveries were made by Sir Humphry Davy by means of voltaic electricity? What is meant by electrolysis?—270. What is meant by electro-metallurgy? Explain the manner of producing an electrotype.—271. How is electro-plating conducted? Describe the method

of silver-plating.—272. What is said of the physiological effects of galvanism? How is the human body affected by galvanism? How are dead bodies affected?

CHAPTER XX.

MAGNETISM.

273. Natural Magnets.—Many centuries ago, a certain ore of iron was discovered possessing the property of attracting pieces of common iron or steel. This iron ore, sometimes called loadstone, has also the power of communicating its peculiar properties to other pieces of iron. These are called magnets; and the force residing in them, magnetism. These facts were probably considered at first as mere curiosities, and the world was slow to find out their great value. It has been recently discovered that in magnetism we have one of the great forces of the earth; and even now we know probably but little of the real extent and variety of its action. New and important discoveries are frequently made in regard to the agency and the laws of this mysterious power, and its connections with the other grand forces of nature.

The terms magnet and magnetism come from the fact that the loadstone was first found near Magnesia, an ancient village in Asia Minor. This ore is an oxide of iron, differing in constitution from ordinary iron rust, and commonly called magnetite. It occurs in abundance in the iron-mines of Sweden, England, and America. Large beds of it are worked in New York, New Jersey, and elsewhere in the United States and Canada.

The property of attracting iron, possessed by magnets, may be exhibited in many different ways. A magnet brought near a heap of iron filings, needles, tacks, etc.,

R 2

will attract them and cause them to adhere to it when removed. Many toys for children depend upon magnetism for their attractiveness. Thus the toy fishes and swans follow the small iron-bar magnet because a small piece of iron is concealed in their heads. Simple experiments with a magnet soon convince us that the nearer the magnet and the iron are to each other, the stronger the attraction. Indeed, the attractive influence is governed by the same law in regard to distance as the common attraction of matter—viz., it is inversely as the square of the distance. The attraction also is mutual—the iron attracting the magnet as much as the magnet does the iron.

274. **Polarity of the Magnet.**—Every magnet has two poles, one at each extremity, about which the chief power resides. For this reason, if a magnet be rolled in iron filings, these collect about the ends, as represented in Fig. 335. There is a diminution of attraction from the ends to

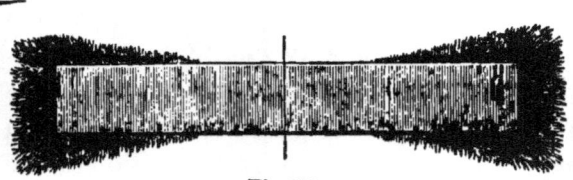

Fig. 335.

the middle line, which is called the *neutral line.* These poles are called north and south poles, because if a magnet be suspended, or be supported upon a pivot, so that it can revolve, it will take a north and south direction, one of its ends invariably pointing towards the ·north. Fig. 336 represents a magnet supported upon a pivot. If a magnet be broken, each piece becomes immediately a perfect magnet, having poles of its own.

275. **Magnetism by Induction.**—The magnet in exerting its attraction temporarily makes a magnet of the body attracted. Actual contact is not necessary to produce

this result. Thus if a large iron key be brought only very near to a powerful magnet, it will support small keys, as represented in Fig. 337. When the key is moved away from the magnet, the keys attached to it fall. You see the analogy to the induction of electricity noticed in § 247. The two ends of the body in which the influence is induced are in opposite states. If the end of the magnet to which the first key is near or at-

Fig. 336.

tached be the north pole, the end of the key next to the magnet will be the south pole, and its farther end the north pole. The same is the case with the small key attached to the end of the large one. And if a nail should hang from the small key, and a needle from the nail, both of these would have the same polarities. But all this would be reversed if the large key were attached to the south pole of the magnet. In this case the upper end of each of these articles would be the north pole, and its lower end the south pole.

Fig. 337.

As a result of these experiments we learn that *like poles repel* and *unlike attract*. But this law can be more strikingly illustrated. If a magnet be placed on a pivot, as in Fig. 336, and another magnet be brought near it, attraction or repulsion will be manifested according to the mode of presentation. If a north pole be presented to a north pole, or a south to a south, repulsion will be the result. But if a north pole be presented to a south, or a south to a north, then attraction will be manifested.

Magnetic Curves.—The polarity of magnetism causes a very singular arrangement of iron filings when gently agitated upon a sheet of paper over a magnet, as represented in Fig. 337. The production of these curves is owing entirely to the fact that each bit of filing is polarized by the bit next preceding it in the r o w, reckoning from the m a g n e t outward, the nearest one in each row deriving its magnetic state from the magnet itself. Since the chief power resides in the ends of the magnet, it is easy to see how such a disposition of the lines of magnetic filings is effected. These curves may be beautifully and curiously varied by arranging several magnets under the paper.

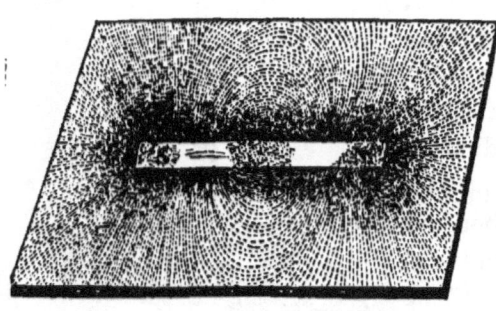

Fig. 337.

276. **Artificial Magnets.**—As already mentioned, the power residing in the loadstone can be communicated readily to iron and steel. Though soft iron becomes magnetic more readily than steel, it does not retain the power so well, and the latter is therefore used in making artificial magnets. When a magnet imparts its magnetic influence, it loses none of its own power, whether it be an original loadstone or an artificial magnet. There are many ways of imparting magnetism permanently to steel, of which we will notice only two. If you wish to magnetize a bar or needle, pass one pole of a magnet from one end of it to the other a number of times, always in the same direction. A more effectual way is to take two magnets, and, placing the south pole of one and the north pole of the other in contact over the middle of the bar or needle, draw them slowly and steadily

apart towards the opposite ends. This process must be repeated several times.

Horseshoe Magnets.—One of the most common forms of the magnet is the horseshoe magnet, Fig. 338. A piece of soft iron, called the *keeper*, is attached to the end of this, held there by attraction. So long as it is suffered to remain there it is itself a magnet, having its north pole (+) attached to the south pole (−) of the magnet which holds it, while the reverse is the case with its south pole. The object of the keeper is to preserve the power of the instrument.  Indeed, it is found that the exertion of the

Fig. 338.

magnet's power not only preserves but actually increases it. If, therefore, you attach to a magnet a keeper having a hook, as shown in Fig. 339, you can add to the weight gradually from day to day, and thus considerably augment the power of the magnet.

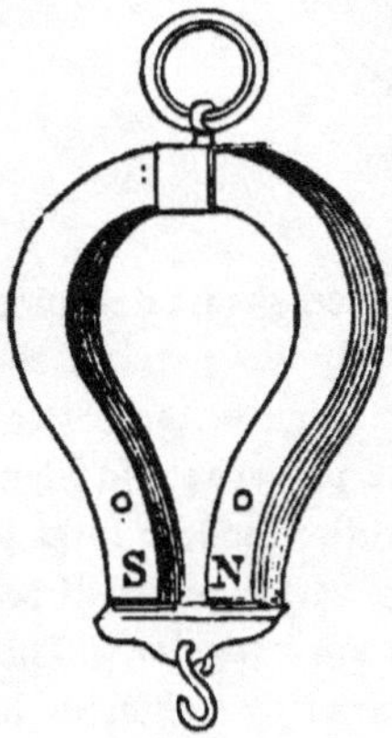

277. **Magnetic Needle.**—The magnetic needle is a very small magnet fixed upon a pivot. On account of its property of invariably pointing north and south, it is

Fig. 339.

of great use to sailors. The mariner's compass is a round box having a magnetic needle balanced in it, and provided with a disk of cardboard on which is drawn a circle divided into thirty-two parts, as shown in Fig. 340. These divisions, called the "points of the compass," have received names with reference to the four directions—north, east, south, and west. The original compass was a rude affair, consisting of a small piece of loadstone laid upon a cork floating in water. The date and place of its first use are unknown, but it is commonly ascribed to the Chinese.

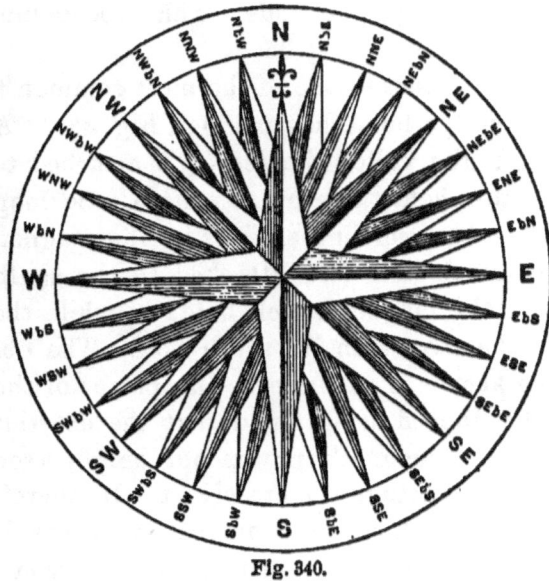

Fig. 340.

Declination of the Needle.—The declination of the needle is its deviation from a north and south line. There are comparatively few parts of the earth's surface in which there is no deviation from this line to the east or the west. "True as needle to the pole" has become a proverb; and when first uttered, it was supposed to be founded in strict truth; modern investigation, however, has shown that the needle not only varies in its direction in different localities, but that even its variations are irregular. This irregularity in the declination of the needle was observed by Columbus in his first voyage of discovery, and it occasioned great alarm among the sailors, who, as Irving states, "thought the laws of nature were changing, and that the compass was about to lose its mysterious power." Notwithstanding these and other observations of a similar character, little attention was paid to the declination of the needle till the middle of the seventeenth century. But, since that time, extensive records of its declinations at different localities have been made, and tables and charts have been constructed exhibiting them. These declinations are not constant, but vary somewhat every day, from the influence, it is supposed, of the sun upon the earth.

Dip of the Needle.—It is found in most parts of the earth that if a needle be balanced before it is magnetized, and then suspended in the same manner after magnetization, one end will dip downward, as shown in

Fig. 341. This fact was discovered by Norman, a London optician, in 1576. He found that the dip at London was towards the north at an angle of 72°. In pursuing the investigation of this phenomenon it was found that going from the north towards the equator the dip constantly lessened, until a point was reached where the needle was horizontal. Then, on going south of this, a reverse dip occurred towards the south pole; and the farther south the needle was carried, the greater the dip. In the north, Captain Ross, in 1832, came to a locality north of Hudson's Bay, in latitude 70° 5′ N., longitude 96° 45′ W., where the magnetic needle, freely suspended, was in a vertical line. No such locality has yet been discovered towards the south pole.

Fig. 341.

278. **The Earth a Great Magnet.** — You can readily see, from all that has been stated in regard to the magnetic needle, that the earth acts as a magnet. The dip of the needle shows that the two poles of this magnet are somewhere near the north and south poles of the earth. The locality which Captain Ross found must be near the north pole of the magnet in that quarter of the world. The vertical position of the needle is analogous to the straight lines of iron filings, shown in Fig. 337, near the poles of the magnet. And it is easy, also, to trace the analogy between the dip of the needle at different distances from what is called the magnetic equator of the earth, where the needle is horizontal, and the curves extending from pole to pole. The different declinations of the needle and the different intensities of the magnetic force in different localities corresponding in latitude show that the magnetic force in the earth is irregular in distribution, or in some way its influence varies much in different parts of the earth's crust.

The Earth a Magnetizer.—Since the earth is really a mag-net, it might be expected to impart magnetism by induction as other magnets do. And this is found to be the fact. If you hold a bar of soft iron in the direction of the dip of the needle, it becomes a magnet—its lower end being the north pole and its upper the south. That this is so can be ascer-tained by bringing a small magnetic needle near each end. No effect of this kind is produced when the bar is held hori-zontally east and west. Lightning-rods, pokers, upright iron bars in fences, etc., are often found to be magnetized because they have continued so long nearly in the required position for magnetization. When a bar of iron has been magnetized in the manner indicated, its magnetism may sometimes be rendered permanent by giving it a stroke with a hammer. It is a curious and inexplicable fact that this vibration of the particles of iron should have this effect. But though such vibration helps to impart magnetism, it is not at all favorable to its retention, for magnets are always injured by blows or falls, or, indeed, any rude treatment. For this reason care is requisite in removing a keeper from a magnet. If pulled off abruptly, the power of the magnet is lessened. If a magnet be heated, its power is impaired; but, on cooling, it returns. A red heat, however, destroys it completely.

279. **Theory of Magnetism.**—Magnetism is now generally conceded to be a molecular affection. "For, in the first place, if a steel-bar magnet be broken in two, each part is as complete a magnet as the original; and, however often we break it, the minutest fragment is a perfect magnet—showing that the polarity or duality of character which the original bar possessed is equally a property of every mole-cule." This may be roughly illustrated by reference to Figs. 342, 343, and 344, in which the north and south poles of each molecule in a magnet are represented by black and

white parts of the circular particles composing it. In Fig. 342, some of the poles are turned one way and some another; and if the number be

Fig. 342.

equal, their effect is neutralized and the steel has no magnetic properties. In the next figure you see some of the molecules turned half-way around, as in an imperfectly magnetized bar;

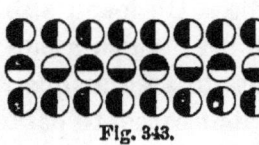

Fig. 343.

and, finally, in Fig. 343, we have represented a complete magnet, in which each molecule is turned in the same direction.

Fig. 344.

A second reason for the theory of molecular affection is that "when a bar of iron is magnetized it becomes slightly elongated, while its width is correspondingly reduced, just as if there were a re-arrangement of the molecules among themselves; as if each turned around and set with its greatest length in the axis of the bar." In the third place, as mentioned in the previous section, heating and striking a magnet destroy its power —in other words, disturb its molecular arrangement.

Diamagnetism.—It was formerly thought that only iron and its companion metals cobalt and nickel were affected by magnetism; but the researches of Faraday have shown that all bodies are affected by it to a greater or less degree. Experiments with a very powerful magnet proved that all bodies may be divided into two classes—the *paramagnetic* which are attracted by a magnet, and the *diamagnetic*, which are repelled. To the former class belong iron, nickel, cobalt, platinum, oxygen gas, air, etc.; to the latter, bismuth, mercury, silver, gold, water, alcohol, wax, sugar, wood, leather, bread, and a great variety of substances.

280. **Electro-Magnetism.**—In 1819, Professor Oersted, of

Copenhagen, discovered a remarkable relation between electricity and magnetism, which has led to inventions of immense benefit to mankind. His first observation was that a current of electricity passing over a wire near a magnetic needle affected the position of the needle. He found also that iron filings would adhere to a wire over which a current of electricity is passing, just as they do to a magnet, dropping off, however, as soon as the current ceases to pass. Such facts led to a great variety of investigations and arrangements of apparatus by Oersted and others. Some of these we will indicate. Let a wire be wound in a spiral form, leaving the two ends of the wire free, one at each end of the coil. Such a coil, made with a great length of wire, called a *helix*, is represented in Fig. 345. Since the electricity is to pass through all the length of the

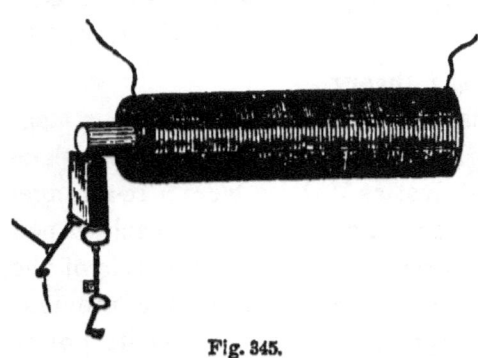

Fig. 345.

wire uninterruptedly, it must be covered with some non-conducting substance, as silk or cotton. If a bar of iron be introduced into the helix, and the wires be connected with the poles of a battery, the iron becomes at once a magnet, and will attract various iron articles, as shown in the figure. But the moment you cut off the connection with the battery, these articles fall, showing that the magnetism of the iron depends wholly upon the electric current. A magnet thus temporarily made is called an *electro-magnet*, and the power thus developed is called electro-magnetism.

Experiments with an Electro-Magnet.—The power of an electro-magnet is very amusingly exhibited in the experiment represented in Fig. 346. A bar of iron having the common horseshoe form has a wire coiled round it. If the connection be made with the two poles of a battery, on bringing it near to a heap of nails they will become attached to the two poles of the temporary magnet thus formed, and a bridge of nails will be constructed between them, the nails being grouped together in all

Fig. 346.

kinds of positions, as the magnetic influence extends among them. You can, however, at once demolish this bridge by cutting off the connection with the battery.

A Bar Suspended in Air.—The electro-magnetic power can be still more strikingly exhibited. Let the helix, A, Fig. 347, be placed vertically, and the bar of iron, B, be held upright in the circular space in the helix. On making the connection with the battery you can let go of the bar, and it will remain suspended, being held there by the magnetic power created by the electric current. If the wire in the coil be very long, and the battery powerful, a considerable weight may be attached to the bar without making it fall. "Science," says Professor Porter, in relation to this experiment, "has thus realized the fable of Mohammed's coffin, which was said to have been miraculously suspended in air."

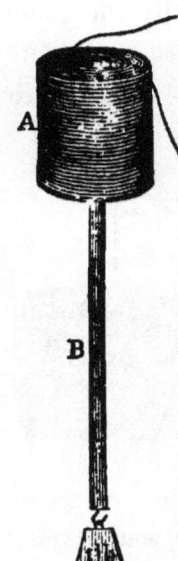

Fig. 347.

Fig. 348 represents an apparatus which exhibits electro-magnetism very prettily. Two pieces of soft iron, when put together, form a ring, *d*, and each piece has a handle. If the pieces be put together with the coil, *c*, in the position rep-

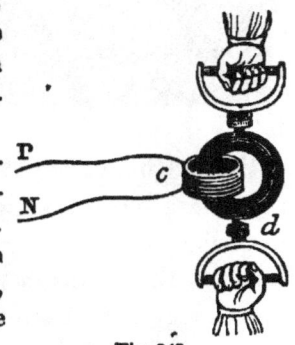

Fig. 348.

resented, on connecting the wires, P and N, with a battery in action, magnetism is brought into play and the adhesion is so strong as to resist a great
force; but as soon as the connection is broken, the pieces come apart at once.

The attractive power of electro-magnets is determined
by the number of coils of insulated wire wound around the
bar, and the strength of the current of electricity passing
through these wires. Since these may be indefinitely increased, the magnetic power is almost unlimited. Profess-
or Henry constructed a horseshoe electro - magnet which
sustained a weight of over 2000 pounds, and other experimenters have much exceeded this.

281. **The Electric Telegraph.** — In the electric telegraph
both voltaic electricity and electro-magnetism are brought
into our service, the former to transmit the message and
the latter to record it. To effect the transmission of the
galvanic current the arrangement is essentially the same
as in the simple voltaic circle (§ 259). Suppose a message
is to be sent from New York to New Haven. Let a, Fig.
349, be the battery at New York, and b the wire communi-

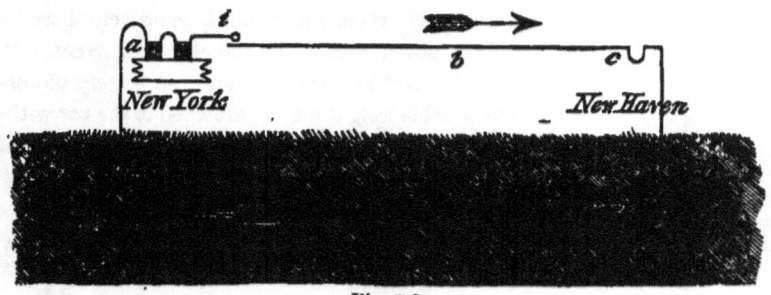

Fig. 349.

cating with the register, c, at New Haven. Two plates of
metal are buried in the moist earth at d and e, each having
a surface of several square feet. These are connected by
wires with the insulated wire between the two places, the
one through the battery at New York and the other through
the registering electro-magnet at New Haven. The cur-

rent of positive electricity from the battery runs, as the arrows indicate, first along the wire, *b*, then from the electro-magnet at *c* down to the plate *e*, then through the earth to the plate *d*, and up to the place of beginning, *a*. It was supposed when the telegraph was first projected that it would be necessary to have two wires to complete the circuit; but it was soon found that the earth answered perfectly well for the returning current with the arrangement of plates described. The control which the telegrapher has over the current with his key, as described below, is indicated at *i*. The wire is there represented as broken, and no effect will be produced at *c* until the operator at *i* presses one end of the wire down upon the other end.

The manner of recording the message is as follows: As before stated, voltaic electricity is used. This is generated at the place from which the message is sent, and passes over the wire to the place where the message is received. There it acts upon soft iron by passing through coiled wire, producing the modified power called electro-magnetism. We will make all this plain to you by describing the machine used in Morse's telegraph, Fig. 350. W W are the

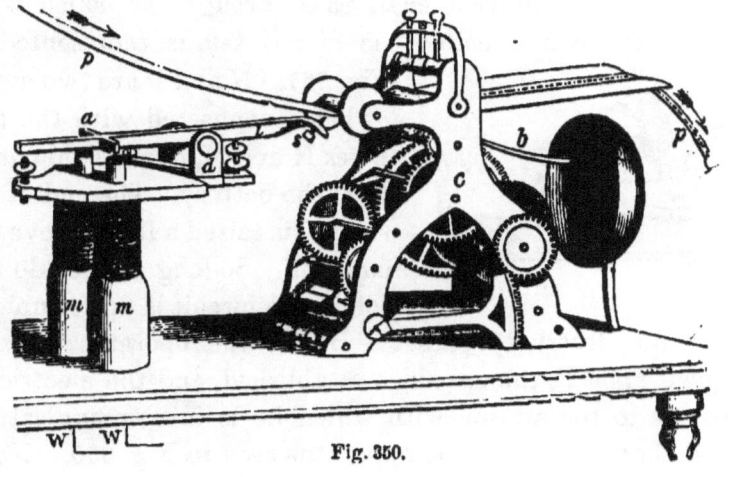

Fig. 350.

wires which connect with the station from which the message is to be received, and these connect with the copper wire coiled round the horseshoe of soft iron, *m m.* Above the magnet is a lever, *a l*, which works on a fulcrum at *d.* One end of this lever has a steel point, *s*, attached to it. At *c* is an arrangement of wheel-work, the object of which is to pass along regularly a narrow band of paper, *p*, in the direction of the arrows. Observe now how the apparatus works. When the electric current passes through the coiled copper wire it makes a magnet of the iron, *m m.* The lever, *a l*, is therefore attracted at the end, *a*, and moves downward. Of course the end, *l*, moves upward, bringing the steel point, *s*, against the paper, where it makes a mark. The length of this mark depends upon the length of time the electricity is allowed to pass along the coiled wire, for the moment that it is shut off *m m* ceases to be magnetic, the " keeper," *a*, being no longer attracted, moves upward, and the other end, *l*, of the lever moves downward, taking the point, *s*, from the paper.

In order to make the marks on the paper of different lengths, there is a contrivance for regulating the length of time that the current shall pass through the coiled wire. This contrivance, called the *signal key*, is represented in

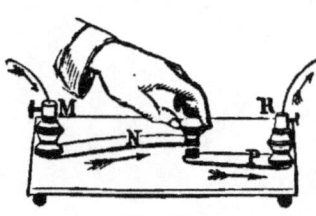

Fig. 351. N and P are two strips of brass connected with the two wires R and M, of which M comes from the battery. The end of the strip N is raised a little above the end of P. So long as they do not touch, the circuit is not complete,

Fig. 351.

and no electricity passes. But if the operator press N down upon P, the circuit is established, and the electricity passes to the station with which he is in communication, and there acts upon the apparatus seen in Fig. 350. Now, the longer the finger presses down N upon P, the longer will be the mark on the paper at the distant station. An operator then at New York, for example, controls by this key the length of the marks made on the paper in New Haven or any other place with which he is communicating.

You can readily see how a telegraphic alphabet can be constructed by combinations of marks of different lengths representing different letters and numerals. Below is the alphabet used in connection with Morse's telegraph:

		Numerals.
A · —	N — ·	
B — · · ·	O · ·	
C · · ·	P · · · · ·	1 · — — ·
D — · ·	Q · · — ·	2 · · — · ·
E ·	R · · ·	3 · · · — ·
F · — ·	S · · ·	4 · · · · —
G — — ·	T —	5 — — —
H · · · ·	U · · —	6 · · · · · ·
I · ·	V · · · —	7 — — · ·
J — · — ·	W · — —	8 — · · · ·
K — · —	X · — · ·	9 — · · —
L — —	Y · · · ·	0 — —
M — —	Z · · · ·	

282. Invention of the Telegraph. — Great as is the credit which should be awarded to our countryman Morse, he was by no means the first inventor of a telegraph. There was one invented and put in successful operation more than a century ago (in 1747) in London by Dr. J. Watson. It was over two miles long, and the earth was used for the return current. Frictional electricity was, of course, employed, as galvanism was not discovered till about half a century after. How Dr. Watson contrived to make the electricity communicate any information is not stated by Professor Silliman, from whose work on Natural Philosophy the facts here given are taken. The year following the construction of Watson's telegraph Franklin set fire to spirits of wine by electricity sent across the River Schuylkill on a wire, using, like Dr. Watson, the earth to complete the circuit. He made, however, no attempt to apply electricity to telegraphing. In 1774, Le Sage, a Frenchman, constructed a telegraph at Geneva, in which he used twenty-four wires enclosed in glass tubes, which were buried in the earth, each wire answering to some letter in the alphabet. From this time there were various trials made at telegraphing, but only partial success attended them, until the year 1837, when, as Professor Silliman says, "almost at the same time appeared Morse in the United States, Steinheil at Munich, and Wheatstone and Cooke in England, as distinct and independent claimants for the honor of the discovery." But, while credit is due to all these, Morse stands unmistakably pre-eminent, because his mode of communication and record at the receiving station is entirely original. For nearly one hundred years there were scientific men earnestly feeling after the result which Morse had the privilege of fully consummating, and how much influence their investigations had upon his mind we do not know. Neither does he know himself; but, with the truthfulness and modesty characteristic of the honest votary of science, no one would be more ready than the noble inventor to acknowledge this influence. Nay, more: he was directly indebted to some of his immediate predecessors for discoveries which were essential to the consummation which he achieved. For example, Oersted's discovery of electro-magnetism supplied Morse with the means of contriving his beautiful mode of receiving and recording messages. And here observe the distinction between discovery and invention. Oersted was a discoverer, Morse an inventor. Oersted discovered a great fact or principle, and Morse found or invented a way of applying this principle in telegraphing.

Other Applications of Electricity. — Another volume equal in size to this one might be written describing the marvellous phenomena and applications of electricity. For an account of the Atlantic telegraph cable and its method of working, of the various machines contrived for converting electricity into motive power (known as electro-magnetic engines), and of many other recent inventions, such as the telephone, we must refer the student to larger treatises.

QUESTIONS.

273. What are loadstones? Where do they abound? What is said of discoveries in magnetism? Whence come the terms magnetism and magnet? How may the property of magnetism be exhibited? What is said of the attraction of magnetism? What law is there in regard to it?—274. What is said of the poles of a magnet?—275. What of magnetism by induction? What is said of attraction and repulsion in magnets? Explain

the formation of the curves of iron filings in the experiment described.—
276. How may artificial magnets be made? What is said of the horseshoe magnet and its armature?—277. What is said of the magnetic needle
and the mariner's compass? What is the declination of the needle? When
was it first observed? What is said of observations after this? What is
said of the dip of the needle?—278. What is said of the earth as a magnet? What of it as a magnetizer? What is said of fixing magnetism?
What of impairing it?—279. What is the present theory of magnetism?
How is this theory supported? What is meant by diamagnetism? In
what other substances besides iron does magnetism exist?—280. What observations did Oersted make in 1819? What is said of the relation between
electricity and magnetism? Describe the experiments with electro-magnets. What determines the power of electro-magnets?—281. What is said
of the forces used in the electric telegraph? Explain the principle by
which the galvanic current is transmitted. Explain the manner of recording the message. What is the use of the signal key? How is the alphabet of Morse's telegraph constructed?—282. Give some points in the history of the electric telegraph. What is the distinction between invention
and discovery?

S

APPENDIX.

METRIC SYSTEM OF WEIGHTS AND MEASURES.

THE metric system of weights and measures is based upon an arbitrary unit called the *metre*. The simplicity of the system, its uniformity, decimal notation, and expressive nomenclature, besides the fact that its units of length, volume, and weight are mutually related upon scientific principles, render it peculiarly serviceable in all experiments, calculations, and writings of a scientific character.

The immense number of arbitrary weights and measures, totally devoid of uniformity, which existed in the various countries of Europe had long been felt as exceedingly inconvenient, laborious, injurious to international commerce, and as "inexhaustible fountains of diversity, confusion, and fraud." France, feeling the great evil, took advantage of the revolutionary spirit which prevailed at the close of the last century, and set herself vigorously to work to overcome it.

The proposition for the creation of a metric system originated in 1790 with Prince Talleyrand, who introduced into the National Assembly of France a decree providing for a commission to select a unit of measure and to build up a complete system. Talleyrand had favored the adoption for a linear unit of the length of a pendulum beating seconds in latitude 45°; but, for reasons upon which we cannot dwell, the committee of the Academy of Sciences decided in favor of deriving the unit of length from some one of the natural dimensions of the earth, and recommended as the standard unit of linear measure one ten millionth of the quadrant of a meridian.

Delambre and Méchain, two distinguished astronomers, were charged with the measurement of an arc of the meridian passing through Paris and extending from Dunkirk to Barcelona—an operation of immense labor, which occupied seven years.

From the length of this base line, which was determined 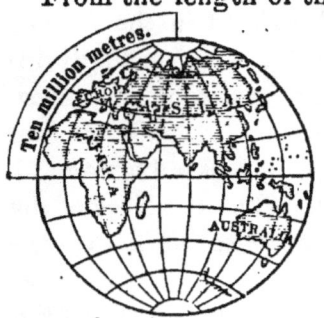 with the greatest possible accuracy, that of the distance from the equator to the pole was calculated, and the ten millionth part of this was taken as the standard unit of length: this length, equal to three feet three inches and three eighths *nearly*, was called a *metre* from the Greek *metron*—"a measure." A bar of platinum representing this length was constructed as a standard and deposited in the Palace of the Archives in Paris.

From this standard the measures of surface, capacity, and weight are derived on principles explained below. The metric system was declared legal in France in 1799, and was made obligatory in 1840. Since then it has been adopted by most of the countries of Europe, and is in partial use by every nation of Christendom.

Units of the Metric System.—The measures of surface, capacity, and weight are connected with the unit of length, the metre, by very simple relations.

The unit of *length,* called METRE, is one ten-millionth of the quadrant of a terrestrial meridian.

The unit of *surface,* called ARE, is equal to a square whose side is ten metres, and is consequently one hundred square metres.

The unit of *volume,* called LITRE, is the volume of a cube whose edge is one tenth of a metre, and is consequently one thousand cubic centimetres.

The unit of *capacity,* called STERE, is a cubic metre.

The unit of *weight,* called GRAMME, is the weight of a cubic centimetre (one thousandth part of a litre) of distilled water at 4° C.

The subdivisions and multiples of these measures are decimal, and are indicated by Latin prefixes for the former and Greek prefixes for the latter. Thus twelve words suffice to express the units, multiples, and fractions of the whole metric system. These are:

Units.
- (1) Metre, from the Greek *metron,* signifying a measure.
- (2) Litre, " " *litra,* " pound.
- (3) Gramme, " " *gramma,* " small weight.
- (4) Are, " Latin *area,* " surface.
- (5) Stere, " Greek *stereos,* " solid.

Sub-divisions.
- (6) Milli, " Latin *mille,* " one thousand.
- (7) Centi, " " *centum,* " one hundred.
- (8) Deci, " " *decem,* " ten.

Multiples.
- (9) Deka, " Greek *deka,* " ten.
- (10) Hecto, " " *hekaton,* " one hundred.
- (11) Kilo, " " *chilios,* " one thousand.
- (12) Myria, " " *myrias,* " ten thousand.

Most of these words are already in use in the English language: thus *metre* occurs in thermometer, metrology, etc.; *litre* in litrameter; *gramme* in telegram, etc.; *are* in area; *stere* in stereoscope; *milli* in millennium and mill; *centi* in century and cent; *deci* in decimal; *deka* in decade; *hecto* in hecatomb; *myria* in myriad.

The manner in which these terms are applied is shown in the following table, which comprises the whole metric system:

Metric System of Weights and Measures.

MONEY.

10 mills make one cent.
10 cents   "   "   dime.
10 dimes   "   "   dollar.
10 dollars "   "   eagle.

LENGTH.

10 millimetres make one centimetre.
10 centimetres "   "   decimetre.
10 decimetres   "   "   metre.
10 metres   "   "   dekametre.
10 dekametres   "   "   hectometre.
10 hectometres "   "   kilometre.
10 kilometres   "   "   myriametre.

WEIGHT.

10 milligrammes make one centigramme.
10 centigrammes "   "   decigramme.
10 decigrammes   "   "   gramme.
10 grammes   "   "   dekagramme.
10 dekagrammes   "   "   hectogramme.
10 hectogrammes "   "   kilogramme.

CAPACITY.

10 millilitres make one centilitre.
10 centilitres "   "   decilitre.
10 decilitres   "   "   litre.
10 litres   "   "   dekalitre.
10 dekalitres   "   "   hectolitre.

The SQUARE and CUBIC MEASURES are simply the squares and cubes of the measures of length. The square deka-metre having received the name *are*, and the cubic metre the name *stere*, as before stated; the stere, however, is not in common use, its place being taken by the litre.

The following diagram* explains itself:

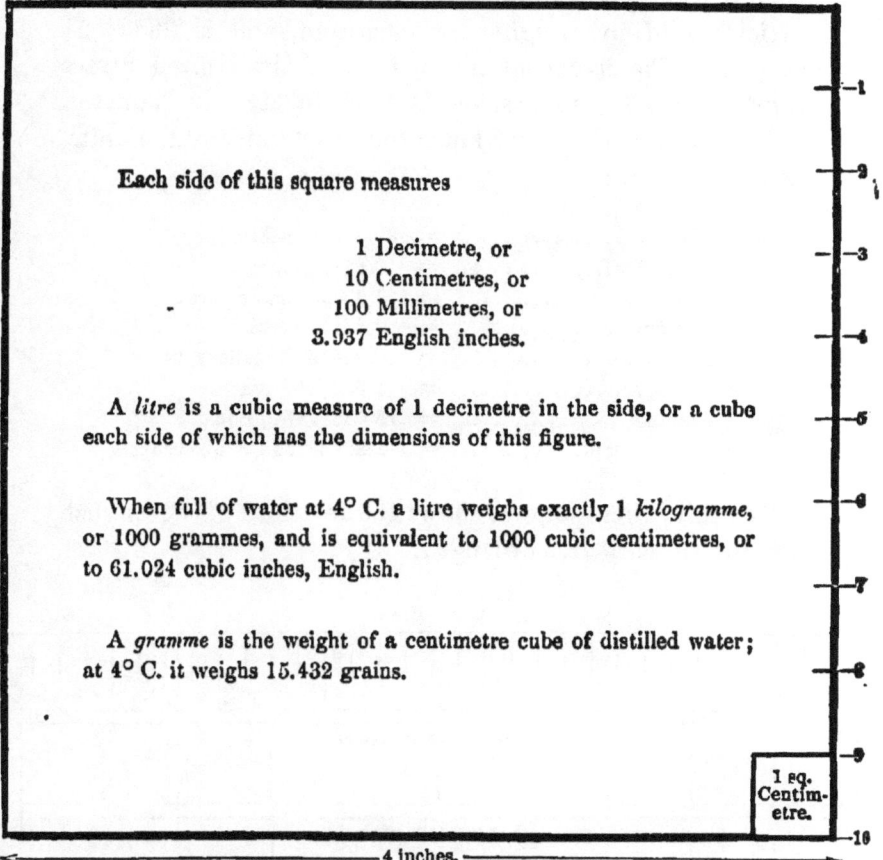

Each side of this square measures

1 Decimetre, or
10 Centimetres, or
100 Millimetres, or
3.937 English inches.

A *litre* is a cubic measure of 1 decimetre in the side, or a cube each side of which has the dimensions of this figure.

When full of water at 4° C. a litre weighs exactly 1 *kilogramme*, or 1000 grammes, and is equivalent to 1000 cubic centimetres, or to 61.024 cubic inches, English.

A *gramme* is the weight of a centimetre cube of distilled water; at 4° C. it weighs 15.432 grains.

1 sq. Centimetre.

4 inches.

Comparison of English and Metric Measures.—The metre corresponds to the English yard, and approximates to *three* feet *three* inches and *three* eighths of an inch. Five metres are nearly one rod. The kilometre, which is used for measuring distances as we use the English mile, is a little less

---

* From "Introduction to the Study of Inorganic Chemistry," by William Allen Miller, M.D.

than 200 rods, or ⅝ of a mile. The litre is a little more than a wine quart. The kilogramme is the unit by which articles sold by weight are measured, and is about 2¼ pounds. The five-cent nickel coin of the United States Mint weighs 5 grammes, and is 20 millimetres in diameter.

It is often desirable to know the exact value of the units, as given below:

1 Metre.................... = 39.37079 inches.
1 Kilometre................ = 0.62138 miles.
1 Are...................... = 119.60332 square yards.
1 Hectare.................. = 2.47114 acres.
1 Litre.................... = 0.26418635 gallons, or
1 " .... = 1.0567454 quarts.
1 Gramme................... = 15.43234874 grains.
1 Kilogramme.............. = 2.20462125 lbs. avoirdupois.

The following diagram shows the relations of the English and metric measures of length:

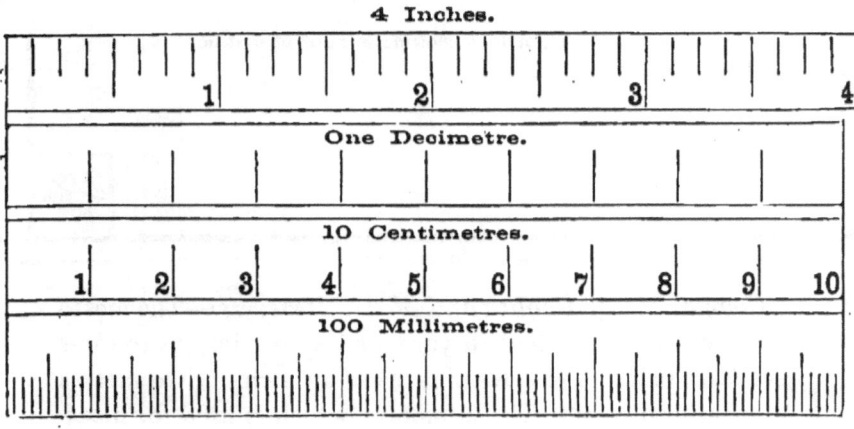

The following tables give the metric measures legalized in the United States, with the abbreviations and their equivalents now in use:

MEASURES OF LENGTH.

Metric Denominations.	Abbreviations.	Values.		Equivalents Legalized by Congress in Denominations now in Use.
Myriametre.........	Mm.	10,000	m.	6.2137 miles.
Kilometre..........	Km.	1,000	m.	0.62137 " or, 3280 ft. 10 in.
Hectometre.........	Hm.	100	m.	328 ft. 1 in.
Dekametre..........	Dm.	10.	m.	393.7 "
Metre...............	m.	'1	m.	39.37 "
Decimetre..........	dm.	.1	m.	3.937 "
Centimetre.........	cm.	.01	m.	0.3937 "
Millimetre.........	mm.	.001	m.	0.03937 "

MEASURES OF SURFACE.

Metric Denominations.	Abbreviations.	Values.	Equivalents Legalized by Congress in Denominations now in Use.
Hectare.............	Ha.	10,000 sq. m.	2.471 acres.
Are.................	a.	100 sq. m.	119.6 sq. yards.
Centare.............	ca.	1 sq. m.	1550 sq. inches.

MEASURES OF CAPACITY.

Metric Names.	Abbreviations.	No. of Litres.	Dry Measure.	Liquid or Wine Measure.
Kilolitre, or Stere.	Kl., st.	1000	1.308 cu. yds.	264.17 gals.
Hectolitre.........	Hl.	100	2 bu. 3.35 pks.	26.417 "
Dekalitre.........	Dl.	10	9.08 qts.	2.6417 "
Litre	l.	1	0.908 qt.	1.0567 qts.
Decilitre..........	dl.	.1	6.1022 cu. in.	0.845 gill.
Centilitre.........	cl.	.01	0.6102 " "	0.338 fld. oz.
Millilitre.........	ml.	.001	0.061 " "	0.27 fld. dr.

WEIGHTS.

Metric Denominations and Values.			Weight of what Quantity of Water at Maximum.	Equivalent in Denominations now in Use. Avoirdupois W'ght.
Names.	Abbreviations.	No. of Grammes.		
Millier, or Tonneau.	M., or T.	1,000,000	1 cu. metre.	2204.6 lbs.
Quintal.............	Q.	100,000	1 hectolitre.	220.46 "
Myriagramme	Mg.	10,000	10 litres.	22.046 "
Kilogramme, or Kilo.	Kg.	1,000	1 litre.	2.2046 "
Hectogramme......	Hg.	100	1 decilitre.	3.5274 oz.
Dekagramme......	Dg.	10	10 cu. centim.	0.3527 "
Gramme...........	g.	1	1 " "	15.432 grs.
Decigramme.......	dg.	.1	.1 " "	1.5432 "
Centigramme......	cg.	.01	10 cu. millim.	0.1543 "
Milligramme......	mg.	.001	1 " "	0.0154 "

INDEX.

THE END.

www.ingramcontent.com/pod-product-compliance
Lightning Source LLC
Chambersburg PA
CBHW021324110726
47900CB00005B/1343